Prediction, Analysis and Design of Environmental Geomechanics

Prediction, Analysis and Design of Environmental Geomechanics

Editor:

Navodita Bhatnagar

Delve Publishing

www.delvepublishing.com

Prediction, Analysis and Design of Environmental Geomechanics

Editor: **Navodita Bhatnagar**

708 3rd Avenue, 6th Floor
New York, NY 10017
United States of America
www.delvepublishing.com

ISBN: 978-1-68095-838-6

Library of Congress Control Number: 2016951592

About the Editor

Navodita Bhatnagar

Navodita Bhatnagar finished her masters in technology and currently pursuing her PhD in environmental sciences from Institute of Technology, Carlow, Ireland. She's a keen enthusiast of environment protection and management. An alumunus of Indo-German centre for sustainibility which is essentially a collaborative interface to gather scientists in the field of environment sciences to come together and work for the common cause which is environment protection.

Contents

List of Contributors xi
Preface xv

Chapter 1 Geomechanical Rock Properties of a Basaltic Volcano 1
1. Introduction 1
2. Geologic Setting 2
3. Materials and Experimental Methodology 3
4. Results 6
5. Discussion 13
Conclusions 14
Acknowledgments 15
References 15

Chapter 2 Geomechanical Aspects and Suitability of the Limestone (Sulaiy Limestone Formation) For Foundation Bedrock, Sulaiy Region, Saudi Arabia 19
Abstract 19
1. Introduction 19
2. Geological Setting 20
3. Purpose of the Study 21
4. Site Investigation 21
5. Geotechnical Field Work 21
6. Physical and Geomechanical Properties 22
7. Bearing Capacity and Settlement Calculations 24
Conclusion and Recommendations 25
References 25

Chapter 3 Geomechanical Properties of Unconventional Shale Reservoirs 27
Abstract 27
1. Introduction 27
2. Methodology 28
3. Results and Discussions 31
Conclusions 35
Nomenclature 36
Highlights 36
References 36

Chapter 4 Factors Influencing Collection Performance of Near Surface Interseasonal Ground Energy Collection and Storage Systems 37
Abstract 37
1. Introduction 37
2. Modelling Framework 38
3. Case Study: Interseasonal Heat Storage System 41
4. Validation 42
5. Numerical Investigation 44
Conclusions 50
Acknowledgements 50
References 50

Chapter 5 **Rockslides on Limestone Cliffs With Subhorizontal Bedding in the Southwestern Calcareous Area of China** **53**

Abstract 53

1. Introduction 53
2. Failure Mode 54
3. Underground Mining 63

Conclusions 64

Acknowledgements 64

References 64

Chapter 6 **Estimation of the Effect of Anisotropy on Young's Moduli and Poisson's Ratios of Sedimentary Rocks Using Core Samples in Western and Central Part of Tripura, India** **67**

Abstract 67

1. Introduction 68
2. Methodology 68
3. The Relation Between Stress (S_{ij}) And Strain (E_{kl}) In Vertical Transverse Isotropy (Vti) Is Given by The Following Relation As: 70
4. Result And Discussion 70

Conclusions 73

Acknowledgements 77

References 77

Chapter 7 **Stress Rise Precursor To Earthquakes In The Tibetan Plateau** **79**

Abstract 79

1. Introduction 79
2. Stress Changes Before and After Large Earthquakes 81
3. Conclusions and Discussion 85

Acknowledgements 86

References 86

Chapter 8 **Test Study on Airtight Capability of Filter Cakes For Slurry Shield and its Application in a Case** **89**

Abstract 89

1. Introduction 89
2. Experimental Methods and Materials 90
3. Experimental Results 92
4. A Case Study on Opening a Slurry Shield Excavation Chamber in the Nanjing Yangtze River Tunnel 94

Discussion 96

Conclusions 97

Acknowledgments 98

References 98

Chapter 9 **Influence of Fracture Roughness on Aperture Fracture Surface and in Fluid Flow on Coarse-Grained Marble, Experimental Results** **99**

Abstract 99

1. Introduction 99
2. Materials and Method 100
3. Results 102
4. Conclusions and Discussion 104

Acknowledgements 105

References 105

Chapter 10 **An Energy-Based Safety Evaluation Index of Blast Vibration** **107**
Abstract 107
1. Introduction 107
2. Hysteretic Energy of Structures 108
3. Maximum Instantaneous Input Energy of Structures 111
4. Dynamic Parameters of Structures 112
5. Application of the Energy-Based Safety Evaluation Method 112
Conclusions 115
Acknowledgments 116
References 116

Chapter 11 **Analysis of the Rheological Behavior of Aging Bitumen and Predicting the Risk of Permanent Deformation of Asphalt** **119**
Abstract 119
1. Introduction 119
2. Materials 120
3. Experimental Methods 120
4. Results and Discussions 121
Conclusions 124
References 125

Chapter 12 **New Approach of Geomechanical Properties by Scale Effect and Fractal Analysis in the Kedougou-Kenieba Inlier (Senegal-West Africa)** **127**
Abstract 127
1. Introduction 127
2. Geological Framework 128
3. Fractal Theory 128
4. Theory of Scale Effect of the Rocks 129
5. Structural Analysis of Discontinuities on Studied Rocks 129
6. Geometric and Spatial Studies 130
Conclusions 136
References 137

Chapter 13 **Model Test of Anchoring Effect on Zonal Disintegration in Deep Surrounding Rock Masses** **139**
Abstract 139
1. Introduction 139
2. Similarity Theory and Analogical Material 140
3. Development of the Model Test System 143
4. Construction of the Model 145
5. Burying of the Measuring Components 145
6. Results of Model Test and Discussion 149
Conclusion 152
Acknowledgments 152
References 152

Chapter 14 **Structural Characteristics and Physical Properties of Tectonically Deformed Coals** **155**
Abstract 155
1. Introduction 155
2. Geologic Background And Coal Structure In Huainan And Huaibei Mine Areas 156

3. The Sample and Methods ... 156
4. Results ... 157
5. Discussion ... 160
Conclusions ... 163
Acknowledgments ... 164
References ... 164

Chapter 15 Multiscale Erosion Surfaces of the Organic-Rich Barnett Shale, Fort Worth Basin, USA ... 167
Abstract ... 167
1. Introduction ... 167
2. Geologic Setting ... 168
3. Data and Methodology ... 169
4. Results ... 169
5. Origin of Erosion Surfaces ... 174
Acknowledgments ... 179
References ... 179

Index ... 183

List of Contributors

Lauren N. Schaefer
Department of Geological and Mining Engineering and Sciences, Michigan Technological University, Houghton, MI, USA

Jackie E. Kendrick
Department of Earth, Ocean and Ecological Sciences, University of Liverpool, Liverpool, UK

Thomas Oommen
Department of Geological and Mining Engineering and Sciences, Michigan Technological University, Houghton, MI, USA

Yan Lavallée
Department of Earth, Ocean and Ecological Sciences, University of Liverpool, Liverpool, UK

Gustavo Chigna
Departamento de Vulcanología, Instituto Nacional de Sismología, Vulcanología, Meteorología e Hidrología, Guatemala City, Guatemala

Abd El-Aal AK
Civil Engineering Department, Faculty of Engineering, Najran University, Najran, Saudi Arabia
Geology Department, Faculty of Science, Al Azhar University, Assuit Branch, Cairo, Egypt

Mohammad O. Eshkalak
Department of Petroleum and Geosystems Engineering, The University of Texas at Austin, Austin, TX 78712, USA

Shahab D. Mohaghegh
Department of Petroleum and Natural Gas Engineering, West Virginia University, Morgantown, WV 26506, USA

Soodabeh Esmaili
Occidental Petroleum Corporation, Bakersfield, CA 93311, USA

José Javier Muñoz-Criollo
Cardiff School of Engineering, Cardiff University, Cardiff, CF24 3AA, Wales, UK

Peter John Cleall
Cardiff School of Engineering, Cardiff University, Cardiff, CF24 3AA, Wales, UK

Stephen William Rees
Cardiff School of Engineering, Cardiff University, Cardiff, CF24 3AA, Wales, UK

Z. Feng
Key Laboratory of Neotectonic Movement and Geohazard of MLR at Institute of Geomechanics, Chinese Academy of Geological Science, Beijing, 100081, China

B. Li
Key Laboratory of Neotectonic Movement and Geohazard of MLR at Institute of Geomechanics, Chinese Academy of Geological Science, Beijing, 100081, China

Y. P. Yin
China Institute of Geo-Environment Monitoring, Beijing, 100081, China

K. He
Chang'an University, Xi'an, 710054, China

Jwngsar Brahma
Pandit Deendayal Petroleum University, Gandhinagar Raisan Village, Raisan, India

Anirbid Sircar
Pandit Deendayal Petroleum University, Gandhinagar Raisan Village, Raisan, India

Zhenhan Wu
Chinese Academy of Geological Sciences, Beijing, China

Qunce Chen
Institute of Geomechanics, Chinese Academy of Geological Sciences, Beijing, China

Patrick J. Barosh
P.J. Barosh and Associates, Bristol, USA

Hua Peng
Institute of Geomechanics, Chinese Academy of Geological Sciences, Beijing, China

Daogong Hu
Institute of Geomechanics, Chinese Academy of Geological Sciences, Beijing, China

Fanlu Min
Key Laboratory of Geomechanics and Embankment Engineering of the Ministry of Education, Hohai University, Nanjing 210098, China
State Key Laboratory for Geomechanics and Deep Underground Engineering, China University of Mining and Technology, Xuzhou 221116, China

Wei Zhu
College of Environment, Hohai University, Nanjing 210098, China

Shengquan Xia
Key Laboratory of Geomechanics and Embankment Engineering of the Ministry of Education, Hohai University, Nanjing 210098, China

Rui Wang
Key Laboratory of Geomechanics and Embankment Engineering of the Ministry of Education, Hohai University, Nanjing 210098, China

Daiwei Wei
Key Laboratory of Geomechanics and Embankment Engineering of the Ministry of Education, Hohai University, Nanjing 210098, China

Teng Jiang
Key Laboratory of Geomechanics and Embankment Engineering of the Ministry of Education, Hohai University, Nanjing 210098, China

G. Dimadis
Department of Civil Engineering, Laboratory of Technical Geology, Aristotle University, Thessaloniki, Greece

A. Dimadi
Department of Civil Engineering, Laboratory of Technical Geology, Aristotle University, Thessaloniki, Greece

I. Bacasis
Department of Civil Engineering, Laboratory of Technical Geology, Aristotle University, Thessaloniki, Greece

Mingsheng Zhao
Guizhou Xinlian Blasting Engineering Group Co. Ltd., 78 Xinhua Road, Fuzhong Building 17, Guiyang, Guizhou 550002, China

Dong Huang
Key Laboratory of Mountain Hazards and Surface Process and Institute of Mountain and Environment, Chinese Academy of Science and Ministry of Water Conservancy, Chengdu, Sichuang 610041, China

Maosen Cao
College of Mechanics and Materials, Hohai University, 1 Xikang Road, Nanjing, Jiangsu 210098, China

En-an Chi
Guizhou Xinlian Blasting Engineering Group Co. Ltd. and Mining College, Guizhou University, 78 Xinhua Road, Fuzhong Building 17, Guiyang, Guizhou 550002, China

Jun Liu
Key Laboratory of Ministry of Education for Geomechanics and Embankment Engineering and Institute of Engineering Safety and Disaster Prevention, Hohai University, 1 Xikang Road, Nanjing, Jiangsu 210098, China

Qiang Kang
Institute of Geotechnical Engineering, Hohai University, 1 Xikang Road, Nanjing, Jiangsu 210098, China

Samia Saoula
Laboratory of Environment, Water Geomechanics and Structures, Faculty of Civil Engineering, University of Sciences and Technology Houari Boumediene, Algiers, Algeria

Khedoudja Soudani
Laboratory of Environment, Water Geomechanics and Structures, Faculty of Civil Engineering, University of Sciences and Technology Houari Boumediene, Algiers, Algeria

Smail Haddadi
Laboratory of Environment, Water Geomechanics and Structures, Faculty of Civil Engineering, University of Sciences and Technology Houari Boumediene, Algiers, Algeria

Maria Eugenia Munoz
Department of Science and Technology of Polymers, Faculty of Chemistry, University of the Basque Country, San Sebastian, Spain.

Antxon Santamaria
Department of Science and Technology of Polymers, Faculty of Chemistry, University of the Basque Country, San Sebastian, Spain.

Déthié Sarr
Laboratoire de Mécanique et Modélisation, UFR Sciences de l' Ingénieur, Université de Thiès, Thiès, Sénégal

Meissa Fall
Laboratoire de Mécanique et Modélisation, UFR Sciences de l' Ingénieur, Université de Thiès, Thiès, Sénégal

Papa Malick Ngom
Département de Géologie, FST-UCAD, Université Cheikh Anta Diop de Dakar, Dakar, Sénégal

Mamadou Gueye
Institut des Science de la Terre, FST-UCAD, Université Cheikh Anta Diop de Dakar, Dakar, Sénégal

Xu-Guang Chen
Institute of Tunnel and Urban Railway Engineering, Hohai University, Nanjing 210098, Key Laboratory of Ministry of Education for Geomechanics and Embankment Engineering, Hohai Univ., Nanjing 210098, China
State Key Laboratory for GeoMechanics and Deep Underground Engineering, China University of Mining & Technology, Xuzhou 221000, China
Research Center of Geotechnical and Structural Engineering, Shandong University, Jinan 250061, China

Qiang-Yong Zhang
Research Center of Geotechnical and Structural Engineering, Shandong University, Jinan 250061, China

Yuan Wang
Institute of Tunnel and Urban Railway Engineering, Hohai University, Nanjing 210098, Key Laboratory of Ministry of Education for Geomechanics and Embankment Engineering, Hohai Univ., Nanjing 210098, China

De-Jun Liu
Research Center of Geotechnical and Structural Engineering, Shandong University

Ning Zhang
Research Center of Geotechnical and Structural Engineering, Shandong University, Jinan 250061, China

Yiwen Ju
Key Laboratory of Computational Geodynamics, College of Earth Science, Graduate University of Chinese Academy of Sciences, Beijing 100049, China

Zhifeng Yan
Key Laboratory of Computational Geodynamics, College of Earth Science, Graduate University of Chinese Academy of Sciences, Beijing 100049, China

Xiaoshi Li
Key Laboratory of Computational Geodynamics, College of Earth Science, Graduate University of Chinese Academy of Sciences, Beijing 100049, China

Quanlin Hou
Key Laboratory of Computational Geodynamics, College of Earth Science, Graduate University of Chinese Academy of Sciences, Beijing 100049, China

Wenjing Zhang
Key Laboratory of Computational Geodynamics, College of Earth Science, Graduate University of Chinese Academy of Sciences, Beijing 100049, China

Lizhi Fang
Key Laboratory of Computational Geodynamics, College of Earth Science, Graduate University of Chinese Academy of Sciences, Beijing 100049, China

Liye Yu
Key Laboratory of Computational Geodynamics, College of Earth Science, Graduate University of Chinese Academy of Sciences, Beijing 100049, China

Mingming Wei
Key Laboratory of Computational Geodynamics, College of Earth Science, Graduate University of Chinese Academy of Sciences, Beijing 100049, China

Mohamed O. Abouelresh
Faculty of Petroleum and Mining Engineering, Suez University, Salah Naesem St., Etaka, Suez 43721, Egypt

Preface

Environmental geomechanics integrates the principles of solid and fluid mechanics with problems pertaining to environment and geosystems. More specifically, it would involve issues related to deforming geomaterials, mechanical failure, fluid flow or transport of contaminants and heat transfer. The fluids here could be water, vapor or air and in association with the waste disposal these could be oil or gas as well. Concepts of engineering are increasingly being applied to solve problems of pollution, sustainable environment and natural disasters. These issues, for e.g. rockslides, debris flow, corrosion of buildings due to pollution, instability of soil and waste disposal, can be natural as well as anthropological in nature. Recently, environmental geomechanics has been relying more and more on information technology which provides quicker solutions in a more accurate manner. The book contains a comprehensive collection that deals with aforementioned issues including the geomechanical properties of geological structures, principles and techniques involved. I hope it provides the information readers are looking for.

Editor
Navodita Bhatnagar

CHAPTER 1

GEOMECHANICAL ROCK PROPERTIES OF A BASALTIC VOLCANO

Lauren N. Schaefer[1], Jackie E. Kendrick[2], Thomas Oommen[1], Yan Lavallée[2] and Gustavo Chigna[3]

[1]Department of Geological and Mining Engineering and Sciences, Michigan Technological University, Houghton, MI, USA

[2]Department of Earth, Ocean and Ecological Sciences, University of Liverpool, Liverpool, UK

[3]Departamento de Vulcanología, Instituto Nacional de Sismología, Vulcanología, Meteorología e Hidrología, Guatemala City, Guatemala

In volcanic regions, reliable estimates of mechanical properties for specific volcanic events such as cyclic inflation-deflation cycles by magmatic intrusions, thermal stressing, and high temperatures are crucial for building accurate models of volcanic phenomena. This study focuses on the challenge of characterizing volcanic materials for the numerical analyses of such events. To do this, we evaluated the physical (porosity, permeability) and mechanical (strength) properties of basaltic rocks at Pacaya Volcano (Guatemala) through a variety of laboratory experiments, including: room temperature, high temperature (935°C), and cyclically-loaded uniaxial compressive strength tests on as-collected and thermally-treated rock samples. Knowledge of the material response to such varied stressing conditions is necessary to analyze potential hazards at Pacaya, whose persistent activity has led to 13 evacuations of towns near the volcano since 1987. The rocks show a non-linear relationship between permeability and porosity, which relates to the importance of the crack network connecting the vesicles in these rocks. Here we show that strength not only decreases with porosity and permeability, but also with prolonged stressing (i.e., at lower strain rates) and upon cooling. Complimentary tests in which cyclic episodes of thermal or load stressing showed no systematic weakening of the material on the scale of our experiments. Most importantly, we show the extremely heterogeneous nature of volcanic edifices that arise from differences in porosity and permeability of the local lithologies, the limited lateral extent of lava flows, and the scars of previous collapse events. Input of these process-specific rock behaviors into slope stability and deformation models can change the resultant hazard analysis. We anticipate that an increased parameterization of rock properties will improve mitigation power.

Keywords

Volcanic materials, mechanical properties, high temperature, strength, Pacaya volcano

Citation: Schaefer LN, Kendrick JE, Oommen T, Lavallée Y and Chigna G (2015) Geomechanical rock properties of a basaltic volcano. Front. Earth Sci. 3:29. doi: 10.3389/feart.2015.00029

INTRODUCTION

Several studies of non-volcanic slopes using numerical models and geomechanical material properties have been adapted for the investigation of volcanic collapses, slope stability, and ground deformation. These include applying Limit Equilibrium Methods (LEM) (Voight et al., 1983; Iverson, 1995; Elsworth and Voight, 1996;Hürlimann et al., 2000; Reid et al., 2000; Donnadieu et al., 2001; Okubo, 2004; Apuani and Corazzato, 2010) and numerical modeling by Distinct Element Methods

(DEM) (Zimbelman et al., 2004), Finite Difference Methods (FDM) (Hürlimann et al., 2000; Apuani et al., 2005a,b, 2007; Apuani and Corazzato, 2010) and Finite Element Methods (FEM) (Sousa and Voight, 1995; Apuani and Corazzato, 2010; del Potro et al., 2013;Schaefer et al., 2013). However, numerical data on the relevant mechanical properties for volcanic rocks and rock masses used in these types of studies remain a major source of uncertainty. Quantifying these properties is challenging due to the diversity of materials that compose volcanoes, the unknown interior structures and geometry, and the often inaccessible locations of volcanic rock masses (Thomas et al., 2004; del Potro and Hürlimann, 2008).

Experimental rock mechanics has traditionally been executed in the fields of civil engineering and mining as well as structural geology, with a focus on sedimentary and intrusive igneous rock (see Paterson and Wong, 2005 and references therein). Early efforts were made to understand rock behavior and strength under high pressure (Scholz, 1968; Tapponnier and Brace, 1976; Batzle et al., 1980), high temperature (Griggs et al., 1960;Simmons and Cooper, 1978; Handin and Carter, 1979; Fredrich and Wong, 1986), and in water saturated conditions (Bauer et al., 1981). Additional work was done to connect mineralogy and microstructures with rock strength (Merriam et al., 1970; Přikryl, 2001) and to model the mechanical properties of porous material; that is, the pore-emanated crack model (Sammis and Ashby, 1986) and the wing-crack model (Ashby and Sammis, 1990). Only recently have laboratory studies begun to systematically investigate the microstructural, physical, and mechanical properties of volcanic rocks, a material infamously known for its formation under disequilibrium conditions, thereby rich in heterogeneities at all scales. These include experiments on a range of material under relevant volcanic conditions such as thermal stressing (Vinciguerra et al., 2005; Kendrick et al., 2013a; Heap et al., 2014a), cyclic inflation-deflation cycles by intrusions (Heap et al., 2009, 2010; Kendrick et al., 2013a), fragmentation (Spieler et al., 2004; Kueppers et al., 2006; Scheu et al., 2008), and flow or fracture at high temperatures and/or pressures (Balme et al., 2004; Rocchi et al., 2004; Smith et al., 2005; Lavallée et al., 2007, 2008; Benson et al., 2008; Cordonnier et al., 2009; Loaiza et al., 2012; Kendrick et al., 2013b).

These investigations have shown that porosity is a fundamental control on rock properties. At volcanoes, the porous network comprises of vesicles and fractures that build the permeable network and which controls degassing (Ashwell et al., 2015). Porosity also controls the mechanical response of materials whereby their strength decreases and elasticity increases with porosity (e.g., Al-Harthi et al., 1999; Spieler et al., 2004;Paterson and Wong, 2005; Heap et al., 2014a,b). Thermal stressing of rock has also been argued to affect its strength, although damage imparted by thermal cracking has been shown to lead to unsystematic changes in mechanical response (Jones et al., 1997; Rocchi et al., 2003; Balme et al., 2004; Kendrick et al., 2013a; Heap et al., 2014a); the ambiguity of this claim has been linked to the extent of pre-existing microcrack damage in a rock (Vinciguerra et al., 2005; Heap et al., 2014a). Other rocks can debilitate upon thermal stressing, as is the case during decarbonation (Heap et al., 2013b) and dehydroxylization (Heap et al., 2012) of certain rock-forming minerals. Cyclic loading can lead to mechanical weakening (Kendrick et al., 2013a) and an increase in heterogeneity of elastic modulus values, likely due to an increase in anisotropic crack damage within the samples (Heap and Faulkner, 2008; Heap et al., 2010).

This paper focuses on the challenge of characterizing volcanic materials for numerical analyses and addressing the scarcity of published geomechanical data for volcanic materials. To do this, we evaluated the mechanical properties of rocks at Pacaya Volcano (Guatemala) using a variety of laboratory experiments, including room temperature, high temperature, and cyclically-loaded uniaxial compressive strength tests (UCS) on as-collected and thermally treated rock samples. Pacaya is an ideal case study for material characterization for use in deformation analysis due to its persistent activity, significant morphological changes, and history of slope instability (Vallance et al., 1995; Matías Gómez et al., 2012; Schaefer et al., 2013). By considering the non-unique geomechanical behaviors when the materials are subjected to this wide range of stressing conditions, we may allow the determination of more encompassing and realistic models that will help better predict volcanic processes and imminent catastrophes.

GEOLOGIC SETTING

Pacaya is a basaltic stratovolcano complex in the Central American Volcanic Arc (Figure 1A) that has been persistently active since 1961 with over 250 lava flows, intermittent strombolian activity, and ash and fumarolic plumes (Matías Gómez et al., 2012; Rose et al., 2013). Between 0.6 and 1.5 ka B.P., the SW sector of the initial cone failed in a major edifice collapse, forming a debris avalanche that traveled 25 km SW, containing approximately 0.65 km^3 of debris (Kitamura and Matías Gomez, 1995; Vallance et al., 1995). The edifice rebuilt within the collapse amphitheater until the nineteenth century, when it went into a period of repose. Activity renewed once again in 1961, loading the SW flank of the edifice with 100–150 m thick of primarily effusive lava flows. Although the volcano has grown substantially, the collapse scarp is still exposed today on the north and east sides of the cone on account of the asymmetric loading of material on the SW flank.

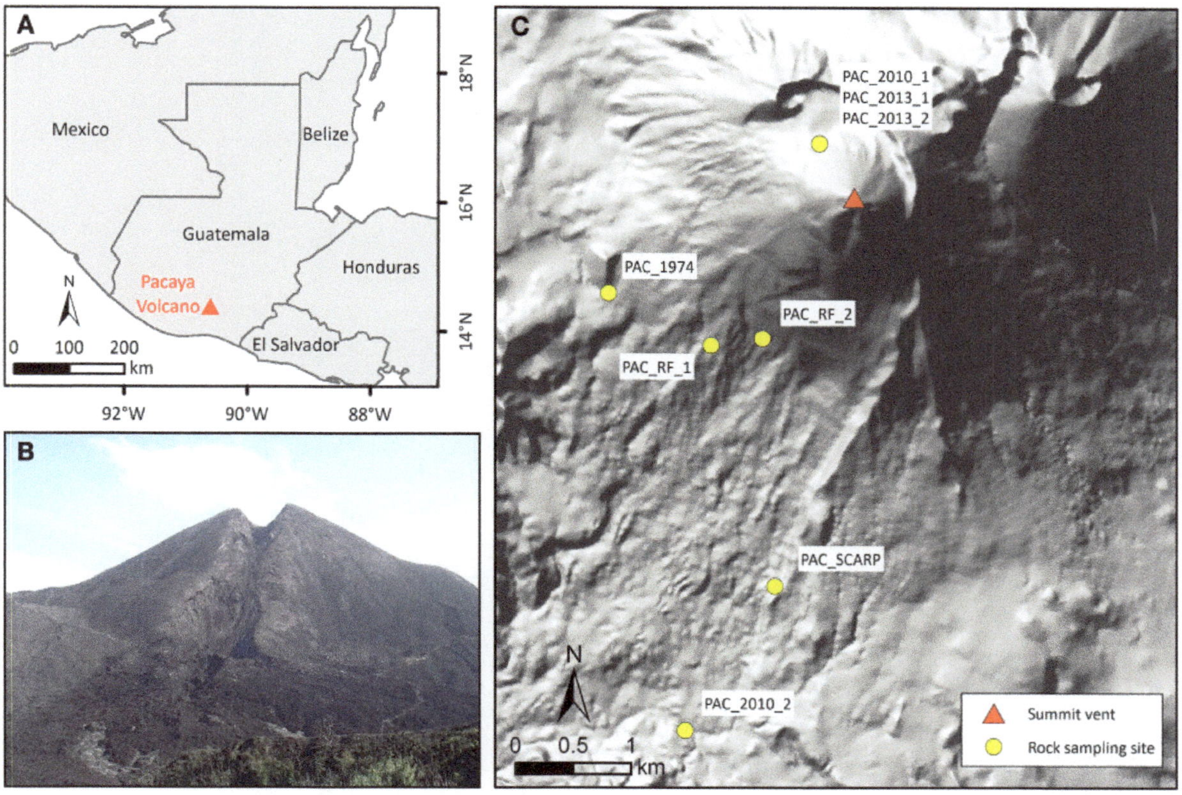

Figure 1. (A) Map showing the location of Pacaya in Guatemala. **(B)** Photograph of Pacaya taken in January 2012, showing the 2010 collapse scar and **(C)** map showing the location of samples chosen for this study. See the Supplementary Image for a detailed volcanological map from Matías Gómez et al. (2012) with field site locations.

Explosive and effusive eruptions have caused constant morphometric changes to the edifice: in 1962, an oval-shaped area, approximately 300 m by 200 m, subsided near the summit. This subsidence has since been in-filled by erupted material. In 2010, a large collapse trough 50–80 m deep and 100 m wide, extending 600 m from the summit, developed concurrently with explosive activity on May 27, 2010 (Figure 1B). Although the exact causes of these collapses are unknown, shallow magma systems and high growth rates leading to over-steepening of the cone have been postulated as factors affecting edifice stability (Eggers, 1983; Vallance et al., 1995; Matías Gómez, 2009; Schaefer et al., 2013). The rock characterization determined in this study can help to better understand events such as these and assess potential future intrusions, collapses, and deformation events.

MATERIALS AND EXPERIMENTAL METHODOLOGY

Experimental Materials

Sample Characterization

Pacaya volcano is made of diverse lithologies (lavas and volcaniclastic deposits) which vary in extent and character significantly, though non-systematically (Wunderman and Rose, 1984; Matías Gómez et al., 2012). This heterogeneity at the large scales calls for a characterization of equally diverse rock types. For this experimental study, eight sample blocks were collected from different eruptive deposits (Figure 1C, see Supplementary Image for detailed volcanological map from Matías Gómez et al., 2012), which cover a range of macroscopically representative physical properties of the edifice-forming basalt at Pacaya (Table 1). These samples, which vary in porosity, crystal size, and vesicle size distribution include: two samples north of the summit from lava flows in January and August–October of 2013; two samples of blocks from rock fall collected from on top of the 2010 deposits; two samples from lava flows associated with the May-June eruption in 2010; one sample from the brecciated top of a lava flow that occurred in February of 1974; and one sample from the (>1.5 ka B.P.) southern portion of the ancestral scarp (Figures 2A–H). Typically, basalt at Pacaya is porphyritic with olivine and plagioclase phenocrysts (45 wt%), with lesser amounts of clinopyroxene and iron-titanium oxides and plagioclase dominated microlites (Bardintzeff and Deniel, 1992). No significant petrographic differences have been found between prehistoric and modern lava flows. Each of the sample blocks was drilled to prepare a series of cylindrical cores with a diameter of 26 mm and a length of 52 mm to maintain a length-to-diameter ratio of 2:1. Although other samples were collected in the field (see Supplementary Image), the eight sample blocks were chosen from which cores could be drilled to provide the widest range of representative samples while ensuring reliable and repeatable measurements of strength. Based on the abundance of effusive products, and the similarity of eruptive product composition over time, we feel as though our samples represent the majority of edifice-building material variability.

Table 1. Sample names and eruption date

Sample name	Eruption date
PAC_2013_1	August–October, 2013
PAC_RF_2	2010–2013
PAC_SCARP	>0.2–4 ka
PAC_2010_2	May–June, 2010
PAC_RF_1	2010–2013
PAC_2013_2	January, 2013
PAC_2010_1	May–June, 2010
PAC_1974	February, 1974

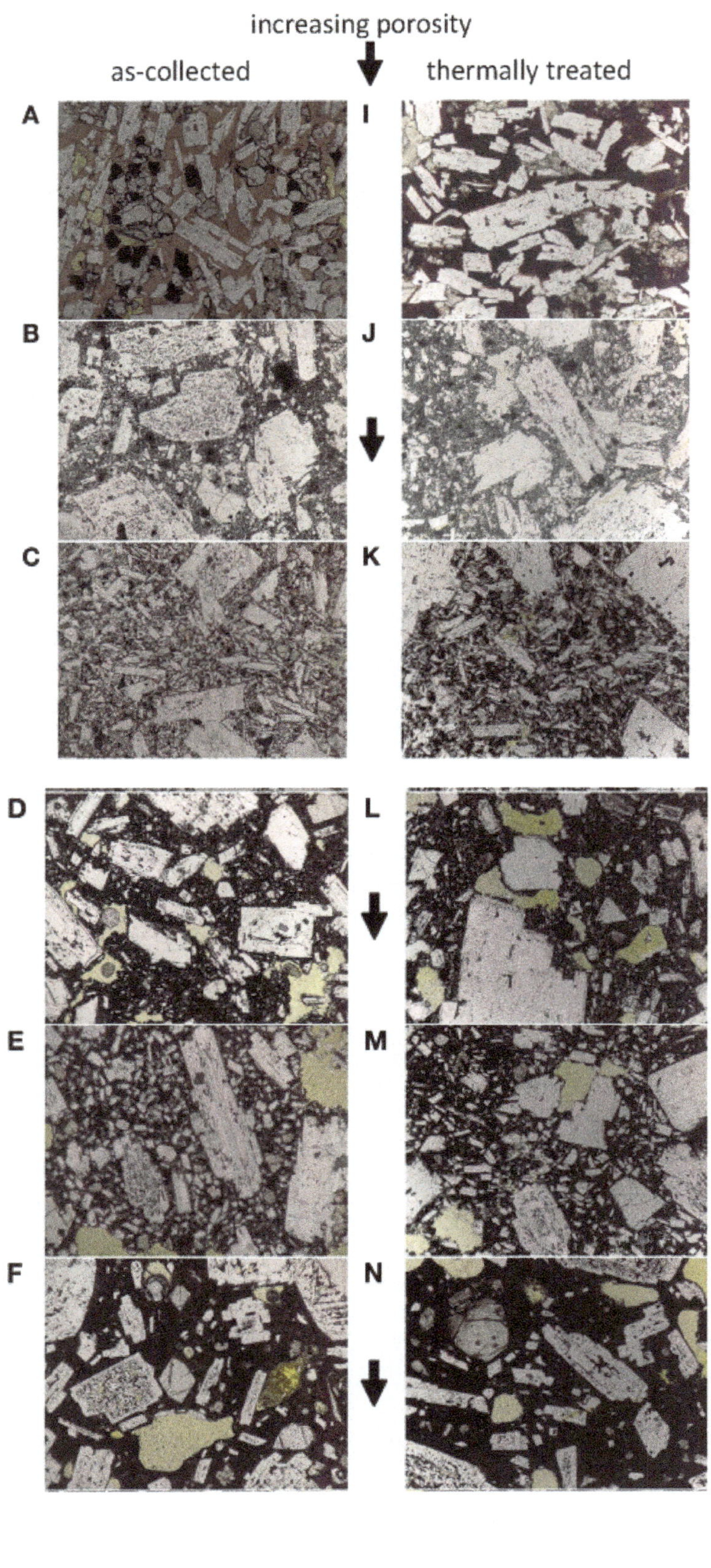

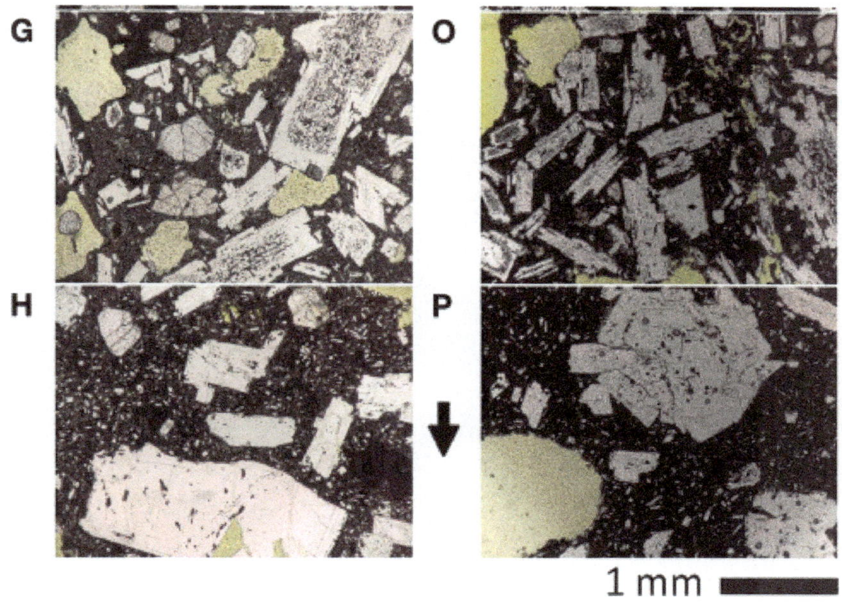

Figure 2. Photomicrographs of the samples chosen for this study (Left, As-collected samples; Right, Thermally stressed samples). (A) PAC_2013_1 shows intact phenocrysts and few microlites set in a glassy groundmass; (B) PAC_RF_2 is made of large plagioclase phenocrysts (broken, sieved, and zoned) surrounded by fine grained groundmass; (C) PAC_SCARP contains fewer and smaller plagioclase phenocrysts surrounded by a mildly developed trachytic texture in the microcrystalline groundmass; **(D)**PAC_2010_2 contains well-defined plagioclase and pyroxene phenocrysts as well as large deformed pores in a fine grained groundmass host of plagioclase and iron oxides; **(E)** PAC_RF_1 hosts large phenocrysts of plagioclase with poikilitic inclusions and equant microlites in a fine grained groundmass rich in iron oxides;**(F)** PAC_2013_2 shows large phenocrysts of plagioclase and pyroxene, with some microlites in a groundmass hosting very fine grained crystals. **(G)** PAC_2010_1 hosts large phenocrysts, rounded pores and very fine grained groundmass. **(H)** PAC_1974 hosts the largest phenocrysts (of the collected samples) and a fine groundmass containing equant grains. **(I–P)** photomicrographs of thermally stressed samples, showing a tendency for the samples groundmass to darken due to oxidation of iron oxides and possibly micro-crystallization of the groundmass, as well as the minor development of micro-fractures.

Porosity and Permeability

Before each test, porosity was determined for each core using a 100cc helium pycnometer from Micromeritics, providing measurements with an accuracy of 0.1%. First, the density of a cylindrical core (ρ_{rock}) was estimated and used to calculate the connected porosity (ϕ_c). Total porosity (ϕ_T) was also determined for each sample type by measuring the density of a powdered sample (ρ_{powder}), and computing:

$$\phi_T = (\rho_{powder} - \rho_{rock})/\rho_{rock}$$

From this, isolated porosity was calculated by subtracting connected porosity from total porosity.

Permeability was determined for each sample block using a TinyPerm II Permeameter from New England Research Inc., which was also used to measure the permeability of 97 *in-situ* lavas and breccias (unconsolidated air-fall tephras to brecciated surfaces of lava flows) at Pacaya during fieldwork in October 2013.

Thermal Stressing

Thermal stressing occurs in volcanic edifices due to heat radiating from the magma in the conduit that may be upwards of 1100°C. Rocks in direct contact with conduit magma may experience partial melting, and further afield temperatures exceeding 900°C are likely common. Thermal analysis was performed on cores of samples in a box furnace by heating them to 1050°C and then re-cooling to room temperature at a rate of 10°C/min to assess physical changes, including oxidation and the onset of melting. This data was used to constrain the maximum experimental temperature for uniaxial compressive strength tests, as it revealed partial melting between 1000 and 1050°C for some samples; therefore a temperature of 935°C was chosen for the high-temperature tests as well as for the thermal stressing tests.

For thermal stressing, three cores from each block were placed in a box furnace and heated at a rate of 4°C/min to the target temperature, left to dwell for 1 h to allow thermal equilibration and completion of any reactions which might have taken place, and then cooled to room temperature using the same rate. A rate of 4°C/min was chosen to minimize damage that could result from temperature gradients within the sample with the presumption that heating in volcanic scenarios occurs relatively steadily, except for rocks in direct contact with magma. Porosity of samples was measured before and after thermal treatment. Thin sections were made from each block with and without thermal stressing (Figures 2A–P).

Experimental Methods

Uniaxial compressive strength (UCS) experiments were conducted at constant strain rates in a 100 kN Instron uniaxial loading frame (model 1362) in the Experimental Volcanology Laboratory at the University of Liverpool. A summary of the experiment type, temperature, and strain rate is shown in Table 2. Applied load (in kN) and displacement (mm) were measured throughout at a rate of 10–100 Hz. Following calibration, to take into account the compliance of the experimental set-up at the given conditions, axial stress (σ_a; MPa) and strain (ε_a) were calculated. The Static Young's modulus (E_s) was calculated from the linear portion of the stress/strain curve (at 30% of the peak stress) via:

$$E_s = d\sigma_a/d\varepsilon_a$$

Table 2. Sample, strain rate and temperature conditions for each experiment

Experiment I.D.	Test	Strain rate (s^{-1})	Temperature (°C)
a	UCS on original cores	10^{-1}	20
b	UCS on original cores	10^{-3}	20
c	UCS on original cores	10^{-5}	20
d	UCS on thermally stressed cores	10^{-1}	20
e	UCS on thermally stressed cores	10^{-3}	20
f	UCS on thermally stressed cores	10^{-5}	20
g	Cyclic loading tests	10^{-5}	20
h	UCS at high temperature	10^{-1}	935
i	UCS at high temperature	10^{-5}	935

Room Temperature Experiments

UCS experiments were performed on dry specimens at room temperature by testing three cores (52 mm length and 26 mm diameter) from each of the eight original sample blocks as well as three thermally stressed cores from seven sample blocks. The cylindrical cores were axially loaded at strain rates of 10^{-1}, 10^{-3}, and 10^{-5} s^{-1} (5, 0.05, and 0.0005 mm/s, respectively) until failure occurred, as defined by a sharp stress drop.

Cyclic Experiments

In cyclic experiments, the uniaxial compressive stress was applied at a constant strain rate of 10^{-5} s^{-1} until 15% of the sample's value of UCS (as determined from previous room temperature UCS tests at a strain rate of $10^{-5}s^{-1}$), then unloaded to 5% of this stress. Subsequently, a further loading/unloading cycle was imposed, but this time reaching an additional 15% load increment (i.e., 30% of UCS), then again unloaded to 5%. These cycles were repeated by increasing the peak stress in 15% stress increments of the UCS until failure. All cyclic tests were conducted at room temperature on dry, as-collected specimens at a strain rate of 10^{-5} s^{-1}.

High Temperature Experiments

High temperature tests were conducted by placing a core in the same uniaxial press surrounded by a three-zone split cylinder furnace, and heated at a rate of 4°C/min until a temperature of 935 ± 3°C was reached. Before undertaking the experimental program, the compliance of the setup at high temperature was measured and considered to remove any of the mechanical artifacts of the machine on the dataset. Samples were held to thermally equilibrate at this temperature for an hour before the experiment began. Temperature was consistently monitored during heating, cooling, and while the compressive load was applied by a thermocouple in contact with the outer surface of the specimen. Samples at high temperature were tested at strain rates of 10^{-1} and 10^{-5} s^{-1} to determine end-member strain-rate behavior.

RESULTS

Porosity and Permeability

Samples range from 2 to 48% total porosity, and isolated pores makes up a varying but sometimes substantial amount of total porosity (Figure 3) as a result of the varying abundance of microcracks and vesicles (Figure 2). Although porosity varied greatly between samples, it remained relatively constant between cores from the same sample, with standard deviations of <3.3% for five or more measurements. Pre- and post-thermal stressing pycnometer measurements showed only minor changes in porosity of −0.4 to +0.2% despite visible differences in sample structure that resulted from thermal cracking and/or oxidation (Figure 2). Permeability measurements showed a bi-modal distribution (Figure 4A) that approximately split into coherent lavas which were sampled for this study (peak at 10^{-13}), and brecciated lava flow surfaces and unconsolidated air-fall deposits (peak at 10^{-10}). The sample's permeability was observed to scale to porosity via the relationship for effusive lavas constrained by Mueller et al. (2005) (Figure 4B).

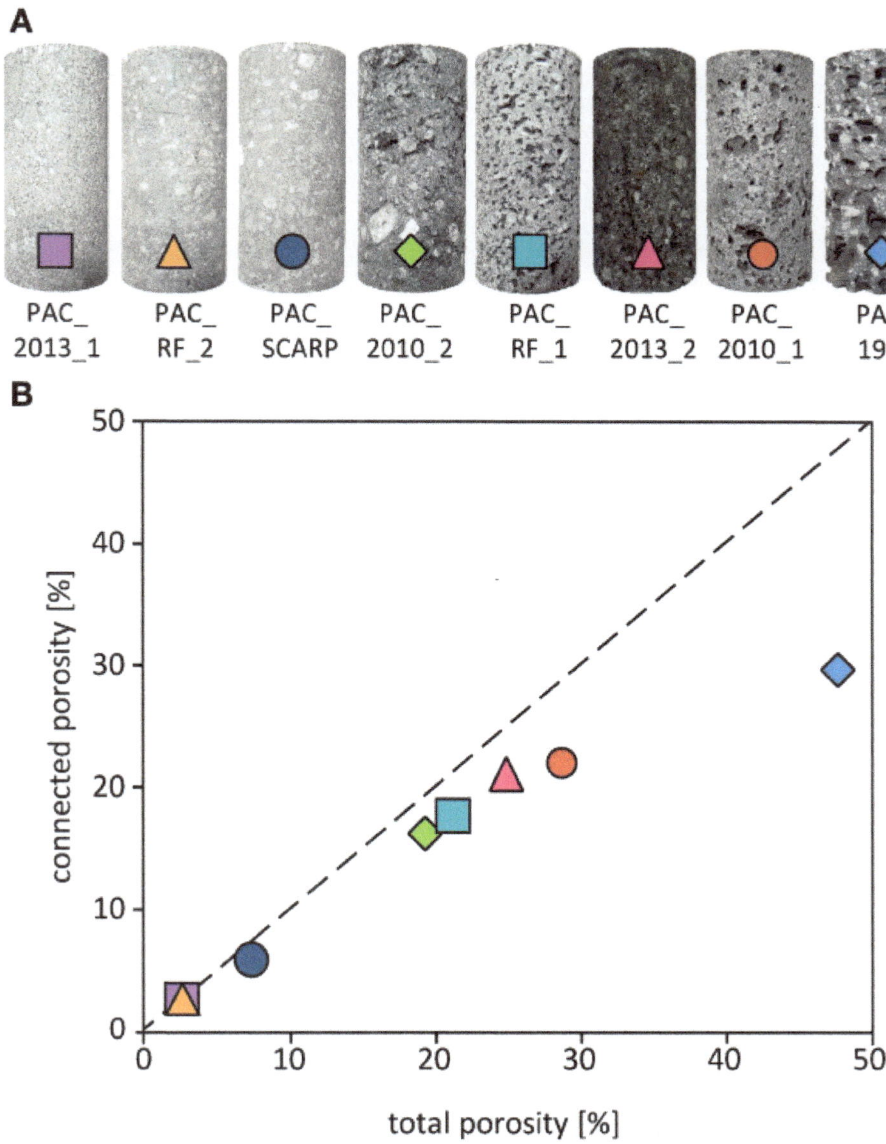

Figure 3. (A) Photographs of the samples selected for this study and **(B)** Porosity of the samples as determined by helium pycnometry. The figure displays the fraction of connected pores against the total porosity of the samples.

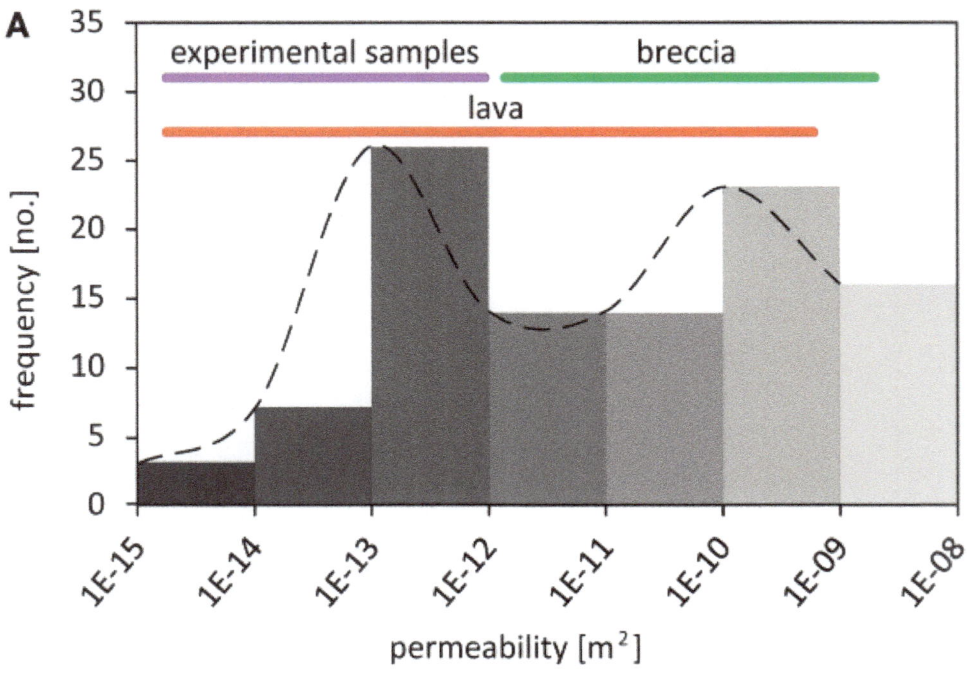

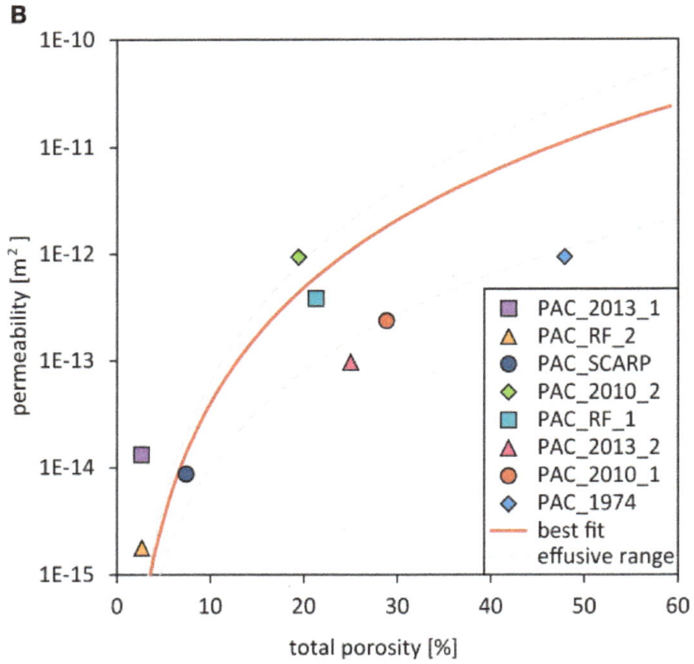

Figure 4. (A) Permeability frequency of *in-situ* rocks at Pacaya from 97 measurements made using a portable permeameter during fieldwork in October 2013. The data shows that the permeability of lavas spans a wider range than that of breccias (defined, here as unconsolidated volcanic deposits). The samples collected for the experimental study had low relative permeability so as to constrain the strength of intact material associated with failure, such as in edifice collapse. **(B)** Permeability vs. porosity for each sample selected for this study. The best fit regression of the dataset falls within the porosity-permeability relationship for effusive lavas described by Mueller et al. (2005).

Uniaxial Compressive Strength Results

Here we present the results of 64 UCS experiments and the effects of (1) strain rate, (2) porosity, (3) cyclic loading, (4) thermal stressing, and (5) temperature on the strength of the tested rocks. For all UCS stress-strain curves.

A comparison of UCS stress-strain curves for all as-collected, room temperature cores at 10^{-1} and 10^{-5} s^{-1}strain rates is shown in Figure 5. These curves display the shape of typical brittle failure under compression, outlined in Heap et al. (2014a) as follows: (1) convex shape as microcracks aligned sub-perpendicular to the direction of loading begin to close (leading to a large strain despite little stress accumulation); (2) linear shape as the rock deforms elastically; (3) concave shape as the rock deforms inelastically and experiences strain hardening; and (4) rock failure marked by a substantial stress drop. Higher strain rate tests fail in a more brittle manner (marked by sharper curve peaks at the stress drop) than slower strain rates (Figure 5) and reach both a higher axial strain and a higher peak stress before failure occurs (Figure 6A). Indeed, room temperature, thermally stressed, and high temperature tests all show a clear pattern of apparent strengthening with increasing strain rates, which is more significant at lower porosity (Figure 6B, Table 3). On average, room temperature cores tested at a strain rate of 10^{-1} s^{-1} failed at axial stresses 5% higher than strain rates of 10^{-3} s^{-1}, and strain rates of 10^{-3} s^{-1} failed 29% higher than strain rates of 10^{-5} s^{-1}. In a similar pattern, thermally stressed cores tested at a strain rate of 10^{-1} s^{-1} failed at axial stresses 4% higher than strain rates of 10^{-3} s^{-1}, and strain rates of 10^{-3} s^{-1} failed at a stress 22% higher than the stress at strain rates of 10^{-5} s^{-1}. Finally, high-temperature cores failed at an average 52% higher axial stress at 10^{-1} s^{-1} as compared to strain rates of 10^{-5} s^{-1}(Table 3).

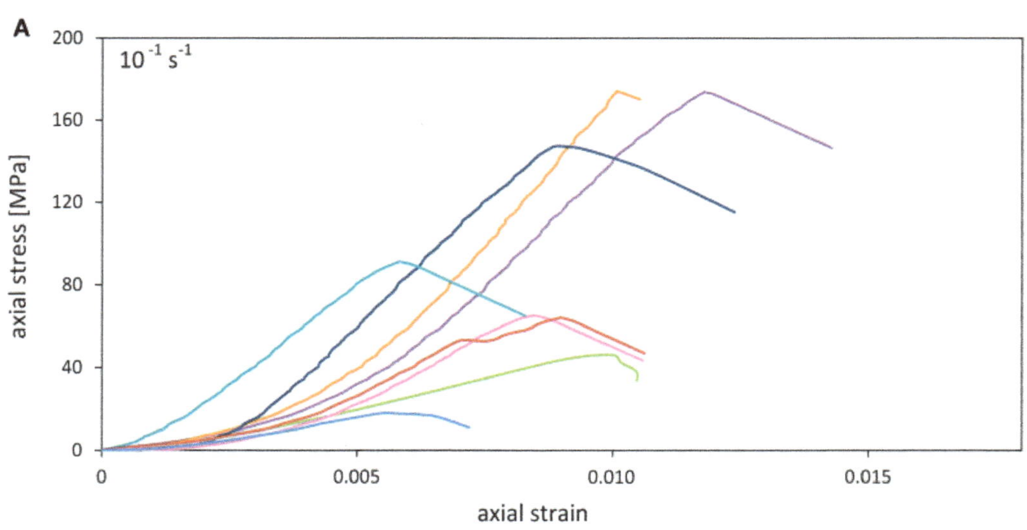

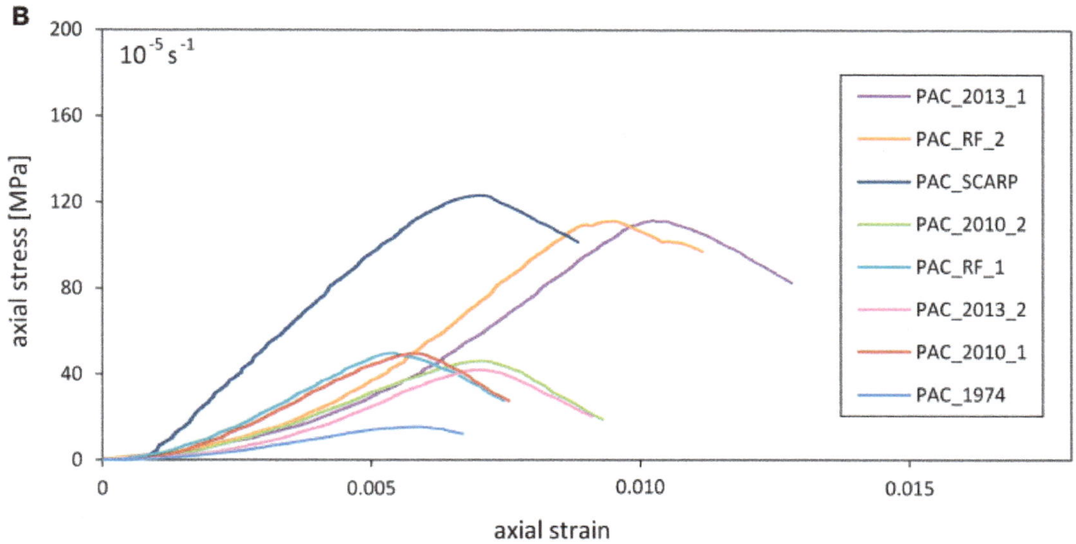

Figure 5. Stress-strain curves for uniaxial compressive strength tests conducted on original cores (i.e., without thermal stressing) at room temperature and strain rates of (A) 10^{-1} s^{-1}, and (B) 10^{-5} s^{-1}. (A) Compression at this high constant strain rate, results in an initial compaction of the rock and closure of cracks (leading to an initially small stress build up despite a relatively large strain); beyond this point stress increases linearly until values approaching the peak stress, where the onset of strain hardening due to sample dilation is triggered by fracture nucleation, propagation, and coalescence, after which the test concludes with failure and a rapid stress drop. The data suggests that the densest rocks reached a higher peak stress and accumulated the most strain before yielding to failure. **(B)** Compression at a lower strain rate of 10^{-5} s^{-1}, suggest a similar behavior although we note more strain hardening as the stress approaches the peak stress that results in sample failure and less distinct stress drops.

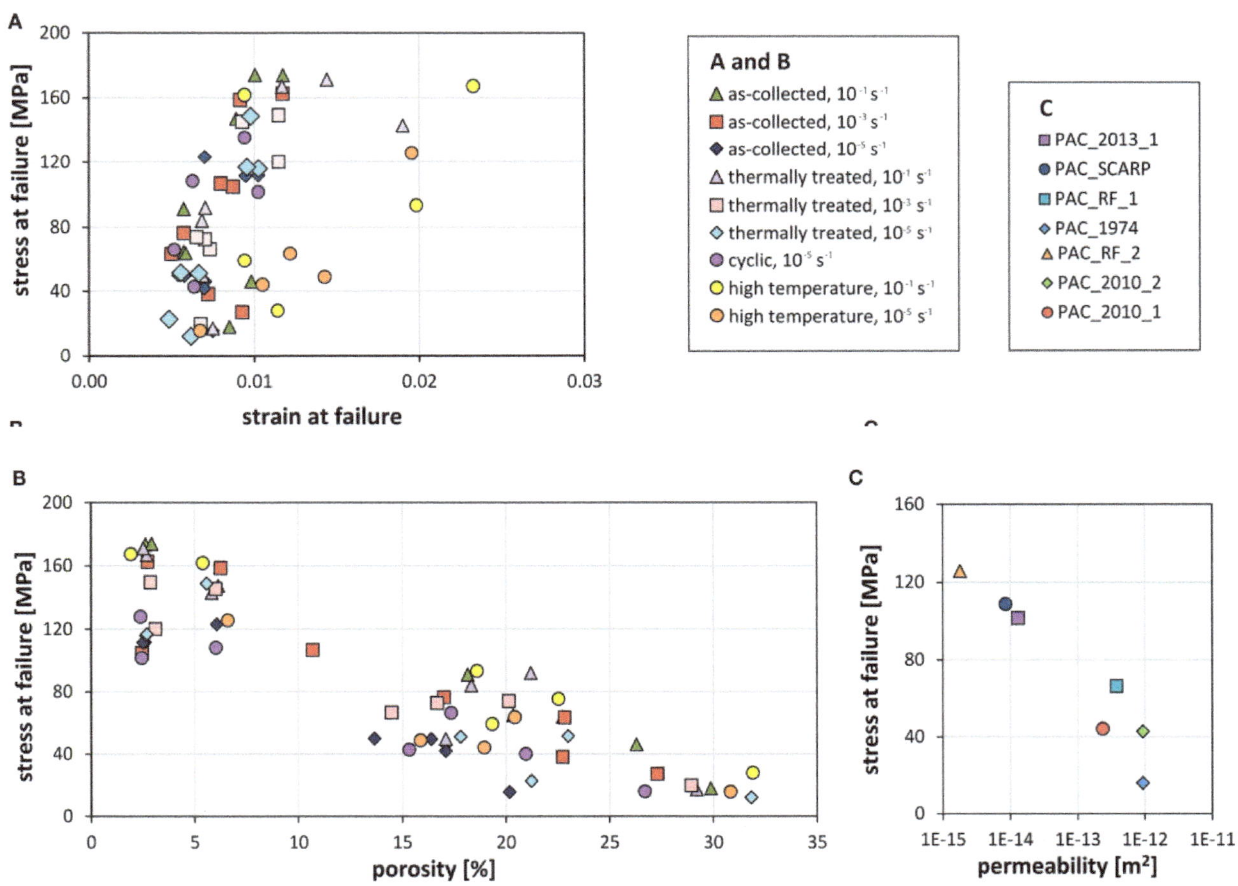

Figure 6. Uniaxial compressive strength of materials as a function of (A) strain at failure, (B) porosity and (C) permeability. (A) The data show that sample responses are generally rate-strengthening, as samples deformed at 10^{-1} s^{-1} typically have higher failure stresses. (B) The samples exhibit a non-linear decrease in strength with porosity. The data shows that rate-strengthening is more important during deformation of the densest rocks. The plot also shows that generally the strength of material at high-temperature is higher than that of the rocks (whether original or thermally stressed) deformed at room temperature. **(C)** The strength of samples shares a linear relationship with the logarithmic of permeability, suggesting that cracks as well as pores influence the strength of these materials.

Table 3. Summary of all results, including porosity, uniaxial compressive strength and Young's Modulus for each sample type by experiment I.D

Experiment I.D.	Sample	Test	Strain rate (s^{-1})	Temperature (°C)	Open porosity (%)	Failure stress (MPa)	Young's modulus at 25% UCS (GPa)
a	PAC_2013_1	Room temperature	10^{-1}	20	2.64	173.61	16.84
b	PAC_2013_1	Room temperature	10^{-3}	20	2.75	162.52	15.80
c	PAC_2013_1	Room temperature	10^{-5}	20	2.62	111.63	11.04
d	PAC_2013_1	Thermally stressed	10^{-1}	20	2.70	166.95	16.14
e	PAC_2013_1	Thermally stressed	10^{-3}	20	2.87	149.51	14.22
f	PAC_2013_1	Thermally stressed	10^{-5}	20	2.66	115.74	12.44
g	PAC_2013_1	Cyclic	10^{-5}	20	2.44	101.46	13.35–17.80
h	PAC_2013_1	High temperature	10^{-1}	935	1.93	167.33	7.87
a	PAC_RF_2	Room temperature	10^{-1}	20	2.94	173.91	17.48
b	PAC_RF_2	Room temperature	10^{-3}	20	2.46	104.55	13.33
c	PAC_RF_2	Room temperature	10^{-5}	20	2.51	111.26	12.54
d	PAC_RF_2	Thermally stressed	10^{-1}	20	2.54	171.08	17.87
e	PAC_RF_2	Thermally stressed	10^{-3}	20	3.12	120.03	13.00
f	PAC_RF_2	Thermally stressed	10^{-5}	20	2.70	116.69	14.22
g	PAC_RF_2	Cyclic	10^{-5}	20	2.39	135.63	16.22–22.40
a	PAC_SCARP	Room temperature	10^{-1}	20	6.16	147.38	22.43
b	PAC_SCARP	Room temperature	10^{-3}	20	6.26	158.68	21.64
c	PAC_SCARP	Room temperature	10^{-5}	20	6.08	123.18	24.69
d	PAC_SCARP	Thermally stressed	10^{-1}	20	5.84	143.10	7.14
e	PAC_SCARP	Thermally stressed	10^{-3}	20	6.03	145.58	16.12
f	PAC_SCARP	Thermally stressed	10^{-5}	20	5.60	148.87	16.18
g	PAC_SCARP	Cyclic	10^{-5}	20	6.03	108.12	17.26–22.66
h	PAC_SCARP	High temperature	10^{-1}	935	5.41	161.75	8.21
i	PAC_SCARP	High temperature	10^{-5}	935	6.62	125.60	7.18
a	PAC_2010_2	Room temperature	10^{-1}	20	26.30	46.01	4.99
b	PAC_2010_2	Room temperature	10^{-3}	20	10.70	106.51	15.41
c	PAC_2010_2	Room temperature	10^{-5}	20	13.67	45.99	7.66
d	PAC_2010_2	Thermally stressed	10^{-1}	20	17.11	49.66	8.53
e	PAC_2010_2	Thermally stressed	10^{-3}	20	14.50	66.49	10.22
f	PAC_2010_2	Thermally stressed	10^{-5}	20	21.23	22.75	6.00
g	PAC_2010_2	Cyclic	10^{-5}	20	15.34	42.76	8.39–10.97
h	PAC_2010_2	High temperature	10^{-1}	935	19.35	59.29	4.26
i	PAC_2010_2	High temperature	10^{-5}	935	15.88	48.77	3.87
a	PAC_RF_1	Room temperature	10^{-1}	20	18.17	90.96	17.52
b	PAC_RF_1	Room temperature	10^{-3}	20	17.03	76.48	19.02
c	PAC_RF_1	Room temperature	10^{-5}	20	16.41	49.92	11.11
d	PAC_RF_1	Thermally stressed	10^{-1}	20	18.34	84.00	13.78
e	PAC_RF_1	Thermally stressed	10^{-3}	20	16.70	72.72	11.95
f	PAC_RF_1	Thermally stressed	10^{-5}	20	17.84	51.15	10.04
g	PAC_RF_1	Cyclic	10^{-5}	20	17.37	66.27	14.51–20.07
h	PAC_RF_1	High temperature	10^{-1}	935	18.62	93.22	8.91
i	PAC_RF_1	High temperature	10^{-5}	935	18.96	44.07	4.59
a	PAC_2013_2	Room temperature	10^{-1}	20	20.35	64.73	8.78
b	PAC_2013_2	Room temperature	10^{-3}	20	22.74	38.21	5.92
c	PAC_2013_2	Room temperature	10^{-5}	20	20.17	41.92	7.23
a	PAC_2010_1	Room temperature	10^{-1}	20	22.73	63.78	7.91
b	PAC_2010_1	Room temperature	10^{-3}	20	22.84	63.45	12.51

Experiment I.D.	Sample	Test	Strain rate (s^{-1})	Temperature (°C)	Open porosity (%)	Failure stress (MPa)	Young's modulus at 25% UCS (GPa)
c	PAC_2010_1	Room temperature	10^{-5}	20	23.34	49.60	11.82
d	PAC_2010_1	Thermally stressed	10^{-1}	20	21.20	91.76	13.95
e	PAC_2010_1	Thermally stressed	10^{-3}	20	20.15	73.95	12.98
f	PAC_2010_1	Thermally stressed	10^{-5}	20	23.01	51.49	9.93
g	PAC_2010_1	Cyclic	10^{-5}	20	20.96	40.17	8.81–11.31
h	PAC_2010_1	High temperature	10^{-1}	935	22.53	75.23	8.16
i	PAC_2010_1	High temperature	10^{-5}	935	20.44	63.62	5.63
a	PAC_1974	Room temperature	10^{-1}	20	29.90	17.93	3.44
b	PAC_1974	Room temperature	10^{-3}	20	27.30	27.15	4.52
c	PAC_1974	Room temperature	10^{-5}	20	28.94	15.58	2.90
d	PAC_1974	Thermally stressed	10^{-1}	20	29.19	17.06	2.69
e	PAC_1974	Thermally stressed	10^{-3}	20	28.94	19.73	3.52
f	PAC_1974	Thermally stressed	10^{-5}	20	31.84	12.11	2.63
g	PAC_1974	Cyclic	10^{-5}	20	26.71	15.96	1.99–6.22
h	PAC_1974	High temperature	10^{-1}	935	31.91	27.96	4.99
i	PAC_1974	High temperature	10^{-5}	935	30.84	15.67	2.53

The UCS results show that failure stress is highly dependent on porosity, with maximum stress decreasing non-linearly as porosity increases (Figure 6B). Samples can be divided into three porosity groups, with samples having <10% porosity averaging a UCS of 138 MPa, samples with 10–25% porosity with a strength decreasing by 55% to an average of 62 MPa, and samples with >25% porosity with a strength decreasing another 66% to an average of 21 MPa. Overall, axial compressive strength for Pacaya basalt ranges widely from 12 to 174 MPa (Table 3), with the lower end falling into strength categories as low as chalk (Hoek, 1994). Given that the porosity of the rocks scales with permeability, we chose to assess the relationship of UCS with permeability, as this measures the strong control that cracks have on the physical properties of rocks. The data shows a remarkable linearity in UCS decrease with logarithmic increase in rock permeability (Figure 6C). Cyclic loading tests in which the peak load of each successive cycle was increased had a variable effect on strength (Table 3, Figure 6). Each cycle peak was 15% of the failure strength of the room temperature tests higher (as-collected cores at 10^{-5} s^{-1}). The highest stress was usually reached on the final cycle, exceeding all previous stress cycles (Figures 7A,B). However, some cyclically loaded samples failed at a lower stress than the peak stress experienced in the penultimate cycle (Figures 7C,D). Compared to the mechanical behavior of the rocks during standard UCS tests at strain rates of 10^{-5} s^{-1}, four out of seven rocks failed at a lower stress during the cyclic tests, and five out of seven rocks failed at lower strain (Figure 8A). Continuous stress-strain curves (Figures7B,D) show a degree of hysteresis, showing that at a given stress, strain is higher during unloading than during loading of a particular cycle. Each successive cycle led to further accumulated (non-recoverable) strain noted when the samples were unloaded to 5% of their value of their UCS (Figure 8B).

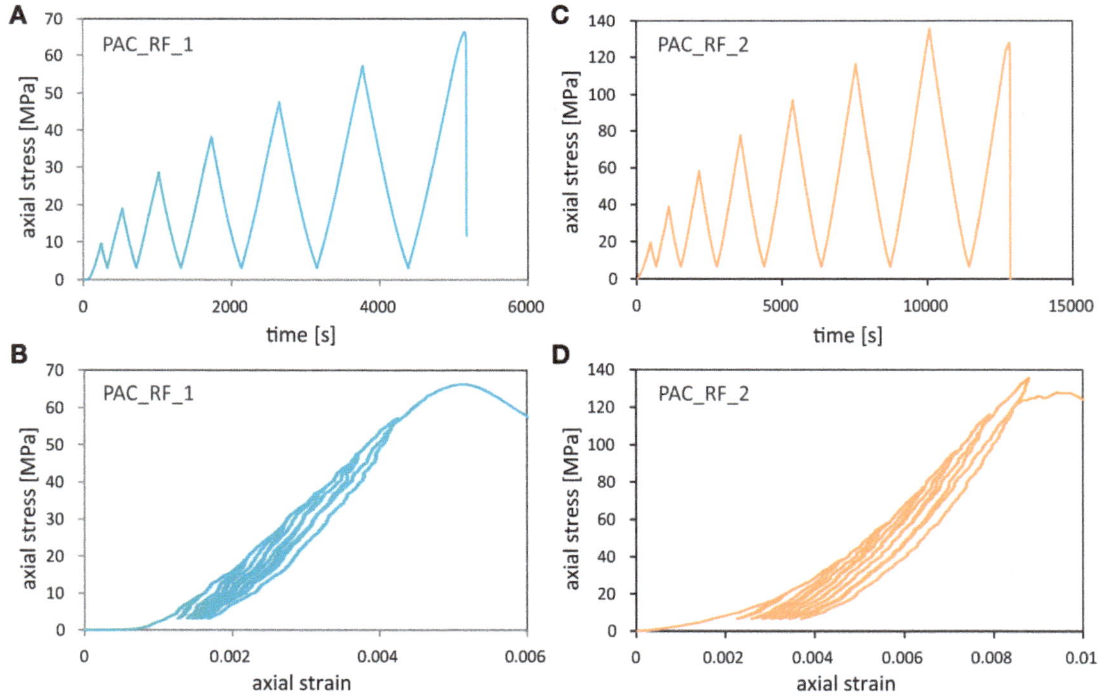

Figure 7. Results from cyclic loading experiments in which each cycle reaches higher stress increments (set to 15% of the UCS) and is unloaded to 5% of the UCS between each cycle.Evolution of axial stress through time (A, C) and strain (B, D). (A) During cyclic loading of

sample PAC_RF_1 deformed at $10^{-5}s^{-1}$ failure is achieved in the final cycle at a stress exceeding the stress value reached in the previous loading cycle. (B) Cyclic deformation and unloading shows hysteresis as strain is accumulated. (C) During cyclic loading of sample PAC_RF_2 deformed at $10^{-5}s^{-1}$ failure is achieved at a stress lower than the peak stress reached in the penultimate cycle. **(D)** The sample is subjected to a stronger damage accumulation, and hysteresis, than sample PAC_RF_1, which may explain why this sample failed at a total stress lower than previously experienced.

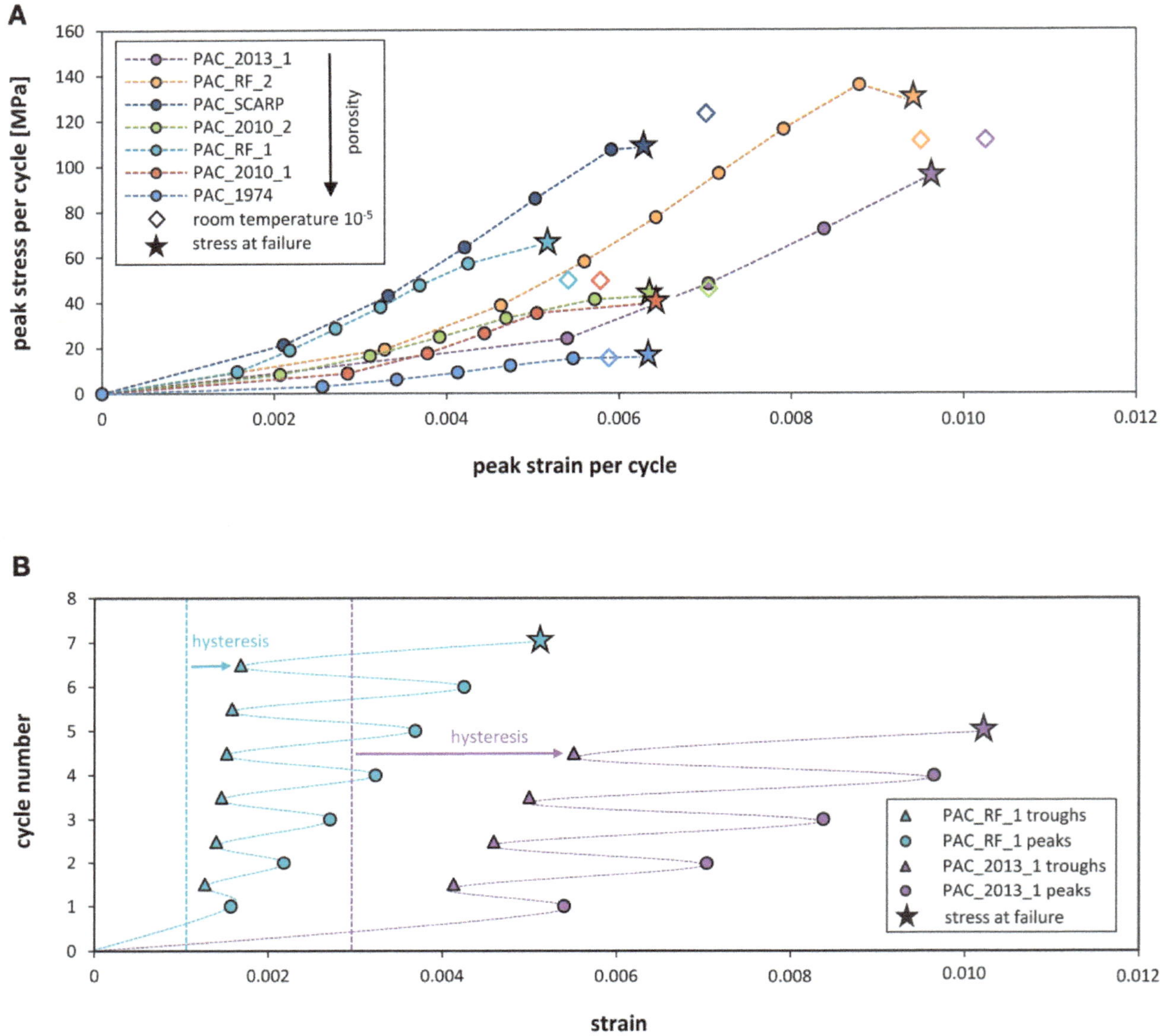

Figure 8. Cyclic loading experiments; (A) Stress-strain evolution for all samples during cyclic loading until failure (shown as a star), as compared to the strength and strain at failure of original (non-thermally stressed) samples determined at room temperature (open diamond). The data shows that generally, the highest porosity samples fail at lower stress and strain. (B)Strain accumulation in samples requiring the lowest and highest strain to failure. Strain accumulation in each stress through highlights the hysteresis induced by the damage imparted. The hysteresis seems to be highest in the stronger, denser sample PAC_2013_1.

The effect of thermal stressing of cores varied across our investigated suite of samples: 12 out of 20 cores failed at a lower stress than the original samples (Table 3). The difference in porosity before and after thermal treatment was, however, minimal (−0.4 to +0.2%) and overall, the changes in strength by thermal stressing are within the natural variability of each sample (Figures 6A,B). Additionally, there were no significant variations in the magnitude of strain to failure or style of failure upon thermal stressing (see Supplementary Information).

The deformation of rock at high-temperature exhibited an increase in strength. On average, strength at 935°C was 19 and 5% higher than the rock strength at room temperature for deformation at strain rates of 10^{-1} and 10^{-5} s^{-1}, respectively. These tests revealed that rock failure at high temperature requires a higher magnitude of strain than at room temperature; this is especially the case for the lower porosity (strongest) samples (Figure6A) The peak stress of these tests also scale inversely with porosity as do the tests at room temperature (Figure6B).

Almost all the samples had a Young's Modulus that scaled approximately linearly to peak strength (Figure 9A), the only outliers were the high-temperature experiments, which had a lower Young's Modulus for a given compressive strength (Figure 9A). This difference becomes particularly prominent at higher strength, and corresponds to a slightly higher strain to failure, especially at low porosity (Figure 9B).

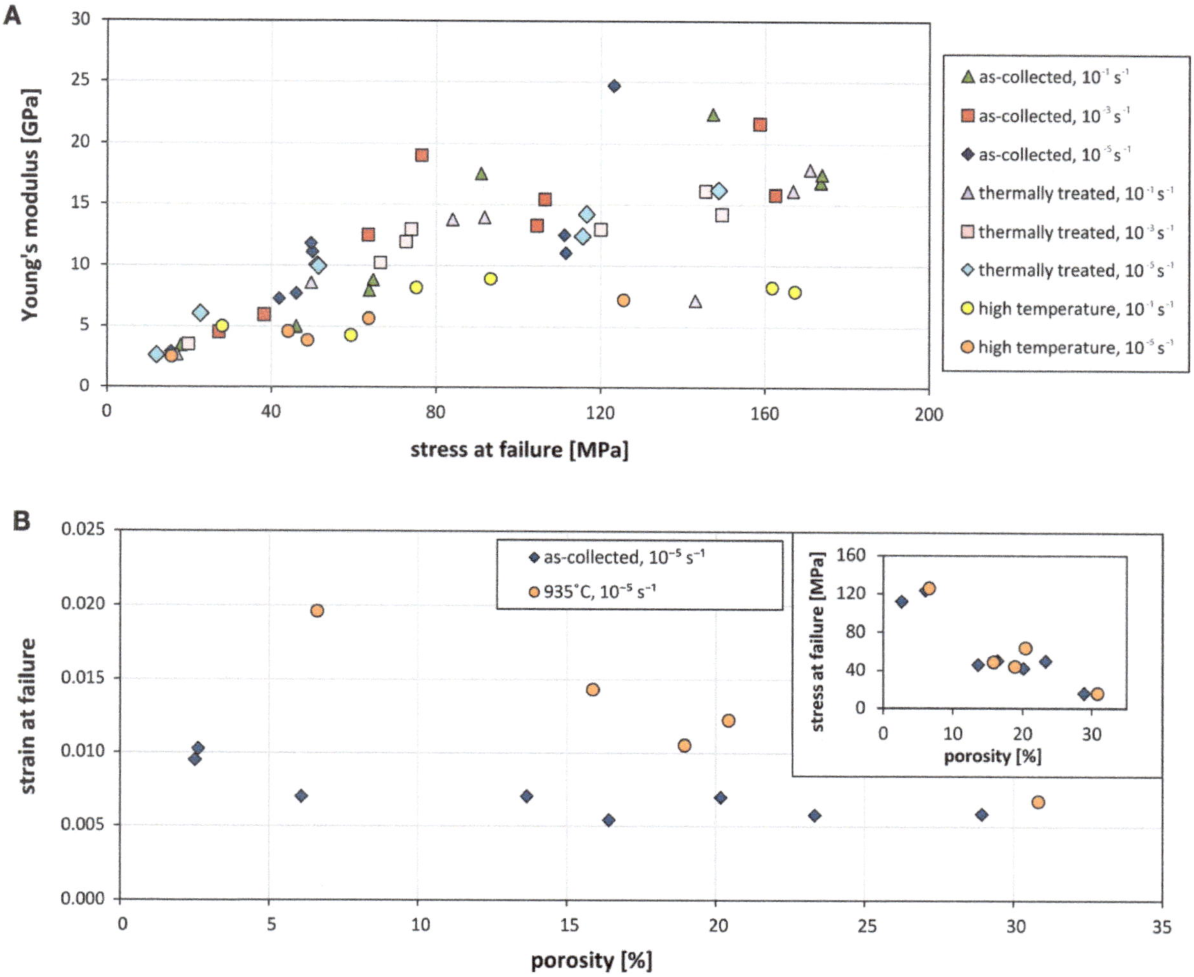

Figure 9. (A) Static Young's Modulus plotted against the sample strength for all but the cyclically loaded samples. The results show that samples with higher Young's Modulus fail at higher stress, and typically deformation at a higher rate of 10^{-1} s^{-1} results in a higher apparent Young's Modulus. Noticeably, the material response during tests at high temperature display the lowest Young's Moduli, which correlate with the observation that **(B)** at high temperature more strain is required to achieve failure than at room temperature; this, despite only a mild strengthening of the material at high temperature (insert).

DISCUSSION

Examination of the mechanical properties of eruptive products at Pacaya volcano provide us with a landscape of data from which to improve our understanding of volcano edifice properties. Of the attributes measured, porosity has the strongest control on rock strength, followed by strain rate, then temperature, which show non-negligible, second order effects. Cyclic loading and thermal stressing seem to have lesser, unsystematic influences on the mechanical behavior of the rocks tested (Figure 6B). The mechanical data show that the strength of the tested Pacaya basalts vary by slightly over one order of magnitude from 12 to 174 MPa across a range of porosities from 2 to 48%, corresponding to a permeability range of 10^{-12}–10^{-15} m^2 for the experimental samples. Given that the samples tested here only represent coherent rocks forming the volcano (Figure 4A), it is likely that incoherent lithologies such as unconsolidated breccias (not studied experimentally, but for which we measured high permeabilities in the field; Figure 4A) have negligible strength, leading to presumably very weak layers which can be destabilized and jeopardize the stability of the volcano.

The wide range of strength in volcanic rocks is commonly attributed to varying compositions and microstructures inherent in volcanic rock (e.g., Byerlee, 1978; Heap et al., 2014b). Strength of materials has long been described as a function of their porosity (see Paterson and Wong, 2005). In volcanic rock, the porosity is made up of vesicles and microcracks and as such, the relationship presented between strength and permeability, though not commonly investigated, deserves consideration. Early theoretical efforts to model the mechanical properties of porous material have proposed models describing the process of failure for pore-bearing rocks, including the pore-emanated crack model (Sammis and Ashby, 1986), as well as the process of failure of crack-bearing rocks, including the wing-crack model (Ashby and Sammis, 1990). A recent micromechanical study of volcanic rock behavior suggested that the abundance of both vesicles and cracks in volcanic rocks thwart our ability to predict failure using these models (e.g., Heap et al., 2014b). Pores and cracks are fundamental to volcanic systems; the importance of permeability and its influence on gas release vs. build-up in a volcanic systems play an important role on the explosivity of an eruption (e.g., Mueller et al., 2005) but also structural stability (Thomas et al., 2004). Here, we suggest that the strength of rock and the character of the porous and permeable network are intrinsically related (Figure 4B) and influence the mechanical properties of volcanic edifices (Figures 6B,C).

Strain-rate strengthening is observed and may be a result of increased elasticity as well as crack coalescence lagging behind axial deformation at faster rates (e.g., Jaeger and Cook, 1976). Strain-rate strengthening is observed across all conditions similar to previous studies on volcanic rocks (e.g., Rocchi et al., 2004), but is more effective at high temperature, as was suggested by Lindholm et al. (1974). The strain-rate effect implies that slow strain events, such as downslope movement, creep, or shallow magma pooling, could cause more pervasive damage and be more detrimental to the stability of an edifice than fast strain events such as dike intrusions or ground motion produced during earthquakes.

High temperature strengthening occurred across the sample set. While in some cases it is possible that chemical alteration, partial melting or viscous relaxation of interstitial glass may heal pre-existing fractures, we suggest that under the conditions investigated here, it is more plausible that increased strength relates to expansion and densification (via closure of cracks) and higher elasticity in molecular bonds (see Young's Modulus values, Table 3). For example, the experiment temperature (at 935°C) is above the order-disorder transition of plagioclase, which is known to increase seismic velocities (Kono et al., 2008) and may correlate to increased elasticity. Indeed, here we show that rock deformation in a fully brittle regime at high temperatures experience significantly more strain to failure than at room temperature. This characteristic is not unique to the basalts at high temperature, though many other rock types often become weaker with temperature, especially if they undergo alteration (e.g., decarbonation, dehydroxylization; Heap et al., 2012, 2013a,b) or significant thermal expansion (e.g., due to anisotropic expansion in crystals; Meredith and Atkinson, 1982;Zhang et al., 2009).

Experiments conducted at high temperatures cause rocks to be stronger and more elastic, suggesting that areas near magma intrusions, or areas heated over prolonged periods of time, could experience more deformation before failure than previously inferred from knowledge of rock behavior at low temperatures. According toEbmeier et al. (2013), a lack of surface deformation before eruptions at Pacaya suggests that magma rises directly from the base of the crust without a period of crustal storage. The high temperature effects may thus explain the trivial amount of pre-eruptive ground deformation associated with dike propagation and magma transport in cold rocks prior to eruption at Pacaya. When also considering the strain-rate strengthening effect, these relatively fast dike injection events could also lead to an increase in edifice elasticity, resulting in less pervasive damage and thus less detriment to the edifice forming rock. However, we must note that the results from cyclic experiments show that if dikes are continually intruded, the country rock could be mechanically weakened over time through strain accumulation, particularly in areas with high elastic modulus values.

Thermally stressing rocks has a complex effect on strength: some cores were slightly weakened, while others were slightly strengthened post-heat treatment. The variable effect of thermal stressing at similar temperatures has also been noted by changes in crack content in other rocks such as andesite (Heap et al., 2014a) and granite (Wang et al., 2013). Although heating and cooling can both induce cracking, there seems to be no systematic response as a function of starting porosity. Volcanic rocks form by cooling and as such, their microstructures and importantly here, microcracks, preserve a record of the cooling history. In our study, the cores were heated and cooled at a rate likely lower than that experienced during cooling upon eruption; it is possible that such mild thermal conditions would minimize the extent of crack damage overprint. According to Heap et al. (2009), so long as the mineralogical assemblage remains intact, microcracks (and subsequent weakening) induced by thermal stressing are likely a function of the original thermal stressing imposed by natural cooling; that is, thermal stressing must be applied at a rate that exceeds natural cooling rate to induce significant damage. Recent studies assessing the effect of fractures induced by thermal stressing concluded that its occurrence would likely not (significantly) affect volcanic edifice stability (Vinciguerra et al., 2005; Heap et al., 2009; Kendrick et al., 2013b). However, we point out that additional mechanical stresses resulting from magmatic forcing during transport would create a zone of damage, increasing the overall heterogeneity of the edifice.

Cyclic loading tests had a variable effect on failure stress and strain, however, they do suggest that repeated dike intrusion or inflation and deflation has the potential to cause edifice weakening by imparting a permanent strain on rocks. The stress-strain curves show hysteresis, and during unloading to 5% of the peak failure stress, the strain remains above zero and gradually increases from cycle to cycle (Figure 8). Such behavior has been observed in other volcanic rocks, including andesite (Kendrick et al., 2013a) and basalt (Heap et al., 2009,2010), along with sedimentary rocks (e.g., sandstone; Song et al., 2013). Our work further suggests that material with porosities above 5 vol. % can typically absorb stress with fewer permanent damage accumulation (Figure 5B). This elastic response may result from compressibility of the rocks into the porous space, attempting to shut the cracks and vesicles. Elastic response may also be due to changes in porosity isolation as pores are connected through microcracking, with the resultant higher connected porosity allowing for more fluid stress compensation between pores, simultaneously preventing stiffening. This effect would be more abundant in rocks with higher porosity. According to Gurevich (2003), approximately 10% porosity or greater provides enough space for fluid to flow into pores and escape fracture. The implications of cyclic loading/ unloading suggest that volcanic rocks may not behave entirely elastically and accumulate damage. Once imparted, damage in turn affects the resultant material response to stress (changes in strength and Young's modulus), which may favor stress rotation (e.g., Faulkner et al., 2006), with potentially more widespread consequences on the distribution of stress in the edifice.

CONCLUSIONS

An experimental investigation of the mechanical properties of basalts at Pacaya volcano has been undertaken to increase our knowledge of contributing variables on slope stability in volcanic edifices. The rocks forming Pacaya volcano are heterogeneous at a range of scales: at the edifice scale, we find the main lithologies are locally heterogeneous due to the limited spread of lava flows as well as to flow emplacement inside earlier collapse scars; at the rock scale, we find that the rocks are macroscopically and microscopically heterogeneous, showing little variation in mineralogy, but strong variation in textures, porosities (formed of pores and cracks) and permeability. The permeability is constrained to increase non-linearly with porosity. Laboratory testing of the mechanical behavior shows that the strength decreases with porosity, increases with strain rate, increases with ambient (deformation) temperature, and is unsystematically varied by both thermal stressing, and due to prolonged stressing histories where cycles of loading/ unloading can induce mechanical hysteresis. We anticipate that integration of such process-specific rock properties in modeling of volcanic events will improve hazard analysis interpretation at active volcanoes such as Pacaya.

ACKNOWLEDGMENTS

LS would like to recognize support provided by the NASA Earth and Space Science Fellowship Program (no. NNX13AO50H) and the AEG Foundation. JK and YL acknowledge support from the European Research Council Starting Grant on Strain Localization in Magma (SLiM; No. 306488). We would like to thank Oleg E. Melnik and Corrado Cimarelli for their constructive reviews of this manuscript.

REFERENCES

1. Al-Harthi, A. A., Al-Amri, R. M., and Shehata, W. M. (1999). The porosity and engineering properties of vesicular basalt in Saudi Arabia. *Eng. Geo.* 54, 313–320. doi: 10.1016/S0013-7952(99)00050-2
2. Apuani, T., and Corazzato, C. (2010). "Etna flank dynamics: a sensitivity analysis by numerical modeling" in *ISRM International Workshop on Rock Mechanics and Geoengineering in Volcanic Environments* (Puerto de la Cruz: International Society for Rock Mechanics).
3. Apuani, T., Corazzato, C., Cancelli, A., and Tibaldi, A. (2005a). Physical and mechanical properties of rock masses at Stromboli: a dataset for volcano instability evaluation. *Bull. Eng. Geol. Environ.* 64, 419–431. doi: 10.1007/s10064-005-0007-0
4. Apuani, T., Corazzato, C., Cancelli, A., and Tibaldi, A. (2005b). Stability of a collapsing volcano (Stromboli-Italy): limit equilibrium analysis and numerical modelling. *J. Volcanol. Geotherm. Res.* 144, 191–210. doi: 10.1016/j.jvolgeores.2004.11.028
5. Apuani, T., Merri, A., and Masetti, M. (2007). "Effects of volcanic seismic events on the Stromboli stability by finite difference numerical modeling" in *Proceedings of the International Workshop on Volcanic Rocks. 11th ISRM Congress* (Ponta Delgada), 101–109.
6. Ashby, M. F., and Sammis, C. G. (1990). The damage mechanics of brittle solids in compression. *Pure Appl. Geophys.* 133, 489–521. doi: 10.1007/BF00878002
7. Ashwell, P. A., Kendrick, J. E., Lavallée, Y., Kennedy, B. M., Hess, K. U., von Aulock, F. W., et al. (2015). Permeability of compacting porous lavas.*J. Geophys. Res. Solid. Earth* 120, 1605–1622. doi: 10.1002/2014jb011519
8. Balme, M. R., Rocchi, V., Jones, C., Sammonds, P. R., Meredith, P. G., and Boon, S. (2004). Fracture toughness measurements on igneous rocks using a high pressure, high temperature rock fracture mechanics cell. *J. Volcanol. Geotherm. Res.* 132, 159–172. doi: 10.1016/S0377-0273(03)00343-3
9. Bardintzeff, J., and Deniel, C. (1992). Magmatic evolution of Pacaya and Cerro Chiquito volcanological complex, Guatemala. *Bull. Volcanol.* 54, 267–283. doi: 10.1007/BF00301482
10. Batzle, M. L., Simmons, G., and Siegfried, R. W. (1980). Microcrack closure in rocks under stress: direct observation. *J. Geophys. Res. Solid. Earth* 85, 7072–7090. doi: 10.1029/JB085iB12p07072
11. Bauer, S. J., Friedman, M., and Handin, J. (1981). "Effects of water-saturated on strength and ductility of three igneous rocks at effective pressures to 50 MPa and temperatures to partial melting," in *Rock Mechanics from Research to Application*, eds H. H. Einstein and D. P. Scandriato (Rotterdam: Balkema), 79–84.
12. Benson, P. M., Vinciguerra, S., Meredith, P. G., and Young, R. P. (2008). Laboratory simulation of volcano seismicity. *Science* 322, 249–252. doi: 10.1126/science.1161927
13. Byerlee, J. (1978). Friction of rocks. *Pure Appl. Geophys.* 116, 615–626. doi: 10.1007/BF00876528
14. Cordonnier, B., Hess, K. U., Lavallée, Y., and Dingwell, D. B. (2009). Rheological properties of dome lavas: case study of Unzen volcano. *Earth Planet Sci. Lett.* 279, 263–272. doi: 10.1016/j.epsl.2009.01.014
15. del Potro, R., and Hürlimann, M. (2008). Geotechnical classification and characterization of materials for stability analyses of large volcanic slopes. *Eng. Geol.* 98, 1–17. doi: 10.1016/j.enggeo.2007.11.007
16. del Potro, R., Hürlimann, M., and Pinkerton, H. (2013). Modelling flank instabilities on stratovolcanoes: parameter sensitivity and stability analyses of Teide, Tenerife. *J. Volcanol. Geotherm. Res.* 256, 50–60. doi: 10.1016/j.jvolgeores.2013.02.003
17. Donnadieu, F., Merle, O., and Besson, J. C. (2001). Volcanic edifice stability during cryptodome intrusion. *Bull. Volcanol.* 63, 61–72. doi: 10.1007/s004450000122
18. Ebmeier, S. K., Biggs, J., Mather, T. A., and Amelung, F. (2013). Applicability of InSAR to tropical volcanoes: insights from Central America. *Geol. Soc. Lond. Spec. Publ.* 380, 15–37. doi: 10.1144/SP380.2
19. Eggers, A. (1983). *Temporal Gravity and Elevation Changes at Pacaya Volcano, Guatemala.* Ph.D. dissertation, Dartmouth College.
20. Elsworth, D., and Voight, B. (1996). Evaluation of volcano flank instability triggered by dyke intrusion. *Geol. Soc. Lond. Spec. Publ.* 110, 45–53. doi: 10.1144/GSL.SP.1996.110.01.03
21. Faulkner, D. R., Mitchell, T. M., Healy, D., and Heap, M. J. (2006). Slip on 'weak' faults by the rotation of regional stress in the fracture damage zone. *Nature* 444, 922–925. doi: 10.1038/nature05353
22. Fredrich, J. T., and Wong, T. F. (1986). Micromechanics of thermally induced cracking in three crustal rocks. *J. Geophys. Res.* 91, 12743–12764. doi: 10.1029/JB091iB12p12743
23. Griggs, D. T., Turner, F. J., and Heard, H. C. (1960). Chapter 4: deformation of Rocks at 500° to 800°C. *Geol. Soc. Am. Mem.* 79, 39–104. doi: 10.1130/mem79-p39

24. Gurevich, B. (2003). Elastic properties of saturated porous rocks with aligned fractures. *J. Appl. Geophys.* 54, 203–218. doi: 10.1016/j.jappgeo.2002.11.002

25. Handin, J., and Carter, N. (1979). Rheological properties of rock at high temperatures. *Proc. Congr. Int. Soc. Rock. Mech.* 4, 97–106.

26. Heap, M. J., and Faulkner, D. R. (2008). Quantifying the evolution of static elastic properties as crystalline rock approaches failure. *Int. J. Rock Mech. Min.* 45, 564–573. doi: 10.1016/j.ijrmms.2007.07.018

27. Heap, M. J., Faulkner, D. R., Meredith, P. G., and Vinciguerra, S. (2010). Elastic moduli evolution and accompanying stress changes with increasing crack damage: implications for stress changes around fault zones and volcanoes during deformation. *Geophys. J. Int.* 183, 225–236. doi: 10.1111/j.1365-246X.2010.04726.x

28. Heap, M. J., Lavallée, Y., Laumann, A., Hess, K.-U., Meredith, P. G., and Dingwell, D. B. (2012). How tough is tuff in the event of fire. *Geology* 40, 311–314. doi: 10.1130/G32940.1

29. Heap, M. J., Lavallée, Y., Meredith, P. G., Dingwell, D. B., Huismann, S., and Weise, F. (2013b). The influence of thermal-stressing on the physical, mechanical and chemical properties of siliceous-aggregate high-strength concrete. *Constr. Build. Mater.* 42, 248–265. doi: 10.1016/j.conbuildmat.2013.01.020

30. Heap, M. J., Lavallée, Y., Petrakova, L., Baud, P., Reuschlé, T., Varley, N. R., et al. (2014a). Microstructural controls on the physical and mechanical properties of edifice-forming andesites at Volcán de Colima, Mexico. *J. Geophys. Res. Solid. Earth* 119, 2925–2963. doi: 10.1002/2013JB010521

31. Heap, M. J., Mollo, S., Vinciguerra, S., Lavallée, Y., Hess, K.-U., Dingwell, D. B., et al. (2013a). Thermal weakening of the carbonate basement under Mt. Etna volcano (Italy): implications for volcano instability. *J. Volcanol. Geotherm. Res.* 250, 42–60. doi: 10.1016/j.jvolgeores.2012.10.004

32. Heap, M. J., Vinciguerra, S., and Meredith, P. G. (2009). The evolution of elastic moduli with increasing crack damage during cyclic stressing of a basalt from Mt. Etna volcano. *Tectonophysics* 471, 153–160. doi: 10.1016/j.tecto.2008.10.004

33. Heap, M., Xu, T., and Chen, C.-F. (2014b). The influence of porosity and vesicle size on the brittle strength of volcanic rocks and magma. *Bull. Volcanol.* 76, 1–15. doi: 10.1007/s00445-014-0856-0

34. Hoek, E. (1994). Strength of rock and rock masses. *ISRM News J.* 2, 4–16.

35. Hürlimann, M., Mart, J., and Ledesma, A. (2000). Mechanical relationship between catastrophic volcanic landslides and caldera collapse.*Geophys. Res. Lett.* 27, 2393–2396. doi: 10.1029/2000GL011564

36. Iverson, R. (1995). Can magma-injection and groundwater forces cause massive landslides on Hawaiian volcanoes? *J. Volcanol. Geotherm. Res.* 66, 295–308.

37. Jaeger, J. C., and Cook, N. G. W. (1976). *Fundamentals of Rock Mechanics*. London: Chapman and Hall.

38. Jones, C., Keaney, G., Meredith, P. G., and Murrell, S. A. F. (1997). Acoustic emission and fluid permeability measurements on thermally cracked rocks. *Phys. Chem. Earth* 22, 13–17.

39. Kendrick, J. E., Lavallée, Y., Hess, K. U., Heap, M. J., Gaunt, H. E., Meredith, P. G., et al. (2013b). Tracking the permeable porous network during strain-dependent magmatic flow. *J. Volcanol. Geotherm. Res.* 260, 117–126. doi: 10.1016/j.jvolgeores.2013.05.012

40. Kendrick, J. E., Smith, R., Sammonds, P., Meredith, P. G., Dainty, M., and Pallister, J. D. (2013a). The influence of thermal and cyclic stressing on the strength of rock from Mount St. Helens, Washington. *Bull. Volcanol.* 75, 1–12. doi: 10.1007/s00445-013-0728-z

41. Kitamura, S., and Matías Gomez, O. (1995). *Tephra Stratigraphic Approach to the Eruptive History of Pacaya Volcano, Guatemala.* Science Reports–Tohoku University, Seventh Series: Geography, 45, 1–41.

42. Kono, Y., Miyake, A., Ishikawa, M., and Arima, M. (2008). Temperature derivatives of elastic wave velocities in plagioclase (An51±1) above and below the order-disorder transition temperature. *Am Mineral.* 93, 558–564. doi: 10.2138/am.2008.2591

43. Kueppers, U., Scheu, B., Spieler, O., and Dingwell, D. B. (2006). Fragmentation efficiency of explosive volcanic eruptions: a study of experimentally generated pyroclasts. *J. Volcanol. Geotherm. Res.* 153, 125–135. doi: 10.1016/j.jvolgeores.2005.08.006

44. Lavallée, Y., Hess, K. U., Cordonnier, B., and Dingwell, D. B. (2007). Non-Newtonian rheological law for highly crystalline dome lavas. *Geology*35, 843–846. doi: 10.1130/G23594A.1

45. Lavallée, Y., Meredith, P. G., Dingwell, D. B., Hess, K. U., Wassermann, J., Cordonnier, B., et al. (2008). Seismogenic lavas and explosive eruption forecasting. *Nature* 453, 507–510. doi: 10.1038/nature06980

46. Lindholm, U. S., Yeakley, L. M., and Nagy, A. (1974). The dynamic strength and fracture properties of dresser basalt. *Int. J. Rock Mech. Min.* 11, 181–191. doi: 10.1016/0148-9062(74)90885-7

47. Loaiza, S., Fortin, J., Schubnel, A., Gueguen, Y., Vinciguerra, S., and Moreira, M. (2012). Mechanical behavior and localized failure modes in a porous basalt from the Azores. *Geophys. Res. Lett.* 39, 6. doi: 10.1029/2012gl053218

48. Matías Gómez, R. O. (2009). *Volcanological Map of the 1961–2009 Eruption of Volcán de Pacaya, Guatemala.* M.S. thesis, Michigan Technological University, Guatemala.

49. Matías Gómez, R. O., Rose, W. I., Palma, J. L., and Escobar-Wolf, R. (2012). Notes on a map of the 1961–2010 eruptions of Volcán de Pacaya, Guatemala. *Geol. Soc. Am. Digit. Map Chart Ser.* 10, 10. doi: 10.1130/2012.DMCH010

50. Meredith, P. G., and Atkinson, B. K. (1982). High-temperature tensile crack propagation in quartz:

experimental results and application to time-dependent earthquake rupture, *Earthquake. Predic. Res.* 1, 377–391.

51. Merriam, R., Rieke, H. H., and Kim, Y. C. (1970). Tensile strength related to mineralogy and texture of some granitic rocks. *Eng. Geo.* 4, 155–160. doi: 10.1016/0013-7952(70)90010-4
52. Mueller, S., Melnik, O., Spieler, O., Scheu, B., and Dingwell, D. B. (2005). Permeability and degassing of dome lavas undergoing rapid decompression: an experimental determination. *Bull. Volcanol.* 67, 526–538. doi: 10.1007/s00445-004-0392-4
53. Okubo, C. H. (2004). Rock mass strength and slope stability of the Hilina slump, Kilauea volcano, Hawaii. *J. Volcanol. Geotherm. Res.* 138, 43–76. doi: 10.1016/j.jvolgeores.2004.06.006
54. Paterson, M. S., and Wong, T. F. (2005). *Experimental Rock Deformation-the Brittle Field.* Berlin; Heidelberg: Springer Science and Business Media.
55. Přikryl, R. (2001). Some microstructural aspects of strength variation in rocks. *Int. J. Rock Mech. Min.* 38, 671–682. doi: 10.1016/S1365-1609(01)00031-4
56. Reid, M. E., Christian, S. B., and Brien, D. L. (2000). Gravitational stability of three-dimensional stratovolcano edifices. *J. Geophys. Res.* 105, 6043–6056. doi: 10.1029/1999JB900310
57. Rocchi, V., Sammonds, P. R., and Kilburn, C. R. (2003). Flow and fracture maps for basaltic rock deformation at high temperatures. *J. Volcanol. Geotherm. Res.* 120, 25–42. doi: 10.1016/S0377-0273(02)00343-8
58. Rocchi, V., Sammonds, P. R., and Kilburn, C. R. J. (2004). Fracturing of Etnean and Vesuvian rocks at high temperatures and low pressures. *J. Volcanol. Geotherm. Res.* 132, 137–157. doi: 10.1016/S0377-0273(03)00342-1
59. Rose, W. I., Palma, J. L., Escobar Wolf, R., and Matías Gomez, R. O. (2013). A 50 yr eruption of a basaltic composite cone: Pacaya, Guatemala.*Geol. S. Am. S.* 498, 1–21. doi: 10.1130/2013.2498(01)
60. Sammis, C. G., and Ashby, M. F. (1986). The failure of brittle porous solids under compressive stress states. *Acta. Metall.* 34, 511–526. doi: 10.1016/0001-6160(86)90087-8
61. Schaefer, L. N., Oommen, T., Corazzato, C., Tibaldi, A., Escobar-Wolf, R., and Rose, W. I. (2013). An integrated field-numerical approach to assess slope stability hazards at volcanoes: the examples of Pacaya Volcano, Guatemala. *Bull. Volcanol.* 75, 1–18. doi: 10.1007/s00445-013-0720-7
62. Scheu, B., Kueppers, U., Mueller, S., Spieler, O., and Dingwell, D. B. (2008). Experimental volcanology on eruptive products of Unzen volcano. *J. Volcanol. Geotherm. Res.* 175, 110–119. doi: 10.1016/j.jvolgeores.2008.03.023
63. Scholz, C. H. (1968). Microfracturing and the inelastic deformation of rock in compre ssion. *J. Geophys. Res.* 73, 1417–1432. doi: 10.1029/JB073i004p01417
64. Simmons, G., and Cooper, H. W. (1978). Thermal cycling cracks in three igneous rocks. *Int. J. Rock Mech. Min.* 15, 145–148. doi: 10.1016/0148-9062(78)91220-2
65. Smith, R., Sammonds, P. R., and Kilburn, C. R. (2005). "Experimental studies of lava dome fracture," in *AGU Fall Meeting Abstracts* (San Francisco, CA), 1.
66. Song, H., Zhang, H., Kang, Y., Huang, G., Fu, D., and Qu, C. (2013). Damage evolution study of sandstone by cyclic uniaxial test and digital image correlation. *Tectonophysics* 608, 1343–1348. doi: 10.1016/j.tecto.2013.06.007
67. Sousa, J., and Voight, B. (1995). Multiple-pulsed debris avalanche emplacement at Mount St. Helens in 1980: evidence from numerical continuum flow simulations. *J. Volcanol. Geotherm. Res.* 66, 227–250. doi: 10.1016/0377-0273(94)00067-Q
68. Spieler, O., Kennedy, B., Kueppers, U., Dingwell, D. B., Scheu, B., and Taddeucci, J. (2004). The fragmentation threshold of pyroclastic rocks.*Earth Planet Sci. Lett.* 226, 139–148. doi: 10.1016/j.epsl.2004.07.016
69. Tapponnier, P., and Brace, W. F. (1976). Development of stress-induced microcracks in Westerly granite. *Int. J. Rock Mech. Min.* 13, 103–112. doi: 10.1016/0148-9062(76)91937-9
70. Thomas, M. E., Petford, N., and Bromhead, E. N. (2004). Volcanic rock-mass properties from Snowdonia and Tenerife: implications for volcano edifice strength. *J. Geol. Soc. Lond.* 161, 939–946. doi: 10.1144/0016-764903-166
71. Vallance, J. W., Siebert, L., Rose, W. I., Raul Girón, J., and Banks, N. G. (1995). Edifice collapse and related hazards in Guatemala. *J. Volcanol. Geotherm. Res.* 66, 337–355. doi: 10.1016/0377-0273(94)00076-S
72. Vinciguerra, S., Trovato, C., Meredith, P. G., and Benson, P. M. (2005). Relating seismic velocities, thermal cracking and permeability in Mt. Etna and Iceland basalts. *Int. J. Rock Mech. Min. Sci.* 42, 900–910. doi: 10.1016/j.ijrmms.2005.05.022
73. Voight, B., Janda, R. J., Glicken, H., and Douglass, P. M. (1983). Nature and mechanics of the Mount St Helens rockslide-avalanche of 18 May 1980. *Géotechnique* 33, 243–273. doi: 10.1680/geot.1983.33.3.243
74. Wang, X. Q., Schubnel, A., Fortin, J., Guéguen, Y., and Ge, H. K. (2013). Physical properties and brittle strength of thermally cracked granite under confinement. *J. Geophys. Res. Solid Earth* 118, 6099–6112. doi: 10.1002/2013JB010340
75. Wunderman, R. L., and Rose, W. I. (1984). Amatitlán, an actively resurging cauldron 10 km south of Guatemala City. *J. Geophys. Res. Solid Earth*89, 8525–8539. doi: 10.1029/JB089iB10p08525
76. Zhang, L., Mao, X., and Lu, A. (2009). Experimental study on the mechanical properties of rocks at high temperature. *Sci. China Ser. E* 52, 641–646. doi: 10.1007/s11431-009-0063-y
77. Zimbelman, D. R., Watters, R. J., Firth, I. R., Breit, G. N., and Carrasco-Nunez, G. (2004). Stratovolcano stability assessment methods and results from Citlaltépetl, Mexico. *Bull. Volcanol.* 66, 66–79. doi: 10.1007/s00445-003-0296-8

CHAPTER 2

GEOMECHANICAL ASPECTS AND SUITABILITY OF THE LIMESTONE (SULAIY LIMESTONE FORMATION) FOR FOUNDATION BEDROCK, SULAIY REGION, SAUDI ARABIA

Abd El-Aal AK[1,2]

[1]Civil Engineering Department, Faculty of Engineering, Najran University, Najran, Saudi Arabia

[2]Geology Department, Faculty of Science, Al Azhar University, Assuit Branch, Cairo, Egypt

ABSTRACT

The present study is a geomechanical study on the Sulaiy Limestone Formation Riyadh region, KSA which the limestone is the main rocks of the subsurface strata in this area. Due to high strength of rock formation, most of the buildings are designed on shallow foundations. Limestone strata may have cavities through its formation in some places which reduces the rock strength and increases its settlement. In our study, the geomechanical and physical analyses included point load strength (Is), uniaxial compressive strength (UCS), Schmidt hummer, dry density, porosity, permeability tests, (TCR), (SCR) and (RQD). The point load strength values are between 2.40 and 6.90 (Mpa). UCS of the different samples shows difference in values (55 to 143 MPa). Dry density shows slight differences in value (1.7 to 2.3 gm/cm3) with an average of 2.32 gm/cm3. Percentage of porosity is noticed by about (12%) and the lowest is (2%). The permeability is 22 milli Darcy (mD) (fair) and 202 milli Darcy (mD) (good). Schmidt hammer tests values are between 33 to 54. TCR is between 15-90%, SCR and RQD are between 20-90%. Then the allowable bearing stress is 18.90 kg/cm2. We can consider only 3.00 kg/cm2. The expected settlement is (Se=1.12 cm). Foundation level for building at 1.50 m below the lowest point on the ground surface. Finally, grouting should be added to fill the open cavities of limestone rock formation (if any) to the significant depth below footings before placing the foundations to gain more strength with reducing the limestone settlement.

INTRODUCTION

The purpose of this study is to investigate and provide reliable, specific and detailed information about the physical and mechanical properties, Such as point load strength (Is), uniaxial compressive strength (UCS), Schmidt hummer, dry density, porosity, permeability tests, (TCR), (SCR) and (RQD), site geology and any installations within the investigated area.

Foundation is an important part of every building, which interfaces the superstructures to the adjacent soil or rock below it. The superstructure loads will be transferred to the underlying soil or rock. Without a proper design and construction of foundation, problems such as cracking, settlement of building may occur and even to the extent, the whole building may collapse within its design life. Therefore, a proper foundation system is required to maintain the safeness of a building.

Citation: El-Aal AK Abd (2015) Geomechanical Aspects and Suitability of the Limestone (Sulaiy Limestone Formation) for Foundation Bedrock, Sulaiy Region, Saudi Arabia. J Geol Geophys 4:230. doi:10.4172/2381-8719.1000230

At the northwest part of Riyadh city, there are two units of limestone bedrock which have high potentiality of karst caves, sinkholes and open fractures. The first unit is exposed as NW-SE limestone belt at the eastern side of Riyadh city and is named as the Sulaiy Limestone Formation of Cretaceous. The Sulaiy Formation is typically composed of compacted limestone with few thin calcarenite beds. In outcrop, this formation shows slumping features in its lower beds exactly like those features which found in the Arab Formation. The higher beds, however, are unaffected by slumping and are moderately strong, forming erosion resistant, well-defined steep scarp slopes. Cavities and sinkholes are likely formed in the lower beds of this formation at the contact with the Arab Formation rather than the in upper beds as they made up mainly of compacted limestone [1] (Figure 1).

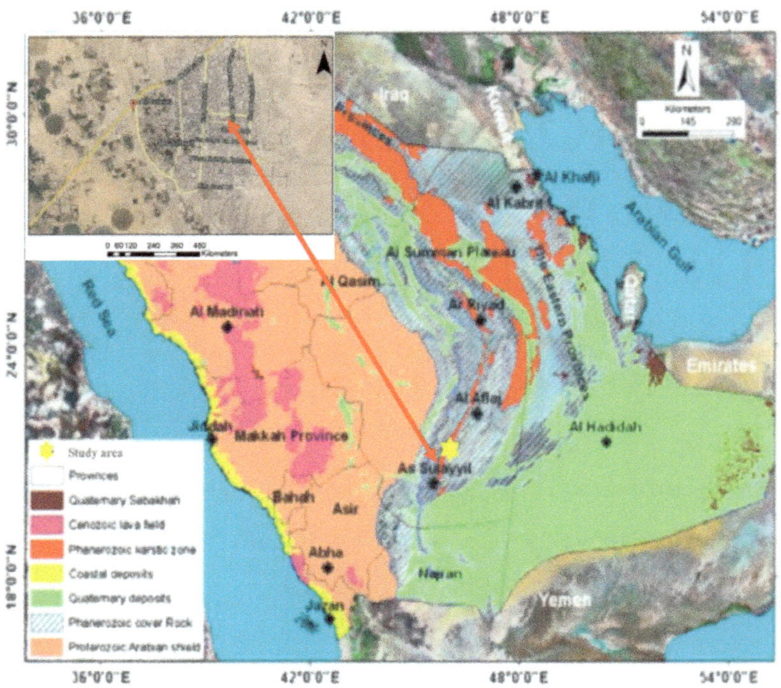

Figure 1: Geological map showing the locations of studied area, (modified after [2]), the corner is a Google Earth Satellite Image showing the locations of studied area.

GEOLOGICAL SETTING

The geological sequence in the Kingdom of Saudi Arabia was classified into three main groups; 1) Precambrian igneous and metamorphic rocks of the Arabian Shield located on the western part of the country, 2) sedimentary rocks surrounding the shield to the east assigned to the Arabian Platform, and 3) narrow strip of the younger rocks mainly of Cenozoic age outcrops in the coastal area west of the Shield [2,3]. Limestone is the main constituent of the sedimentary rocks of the Arabian platform and appears either as outcrops or underlies surface soil or sand dune fields (Figure 2). The distribution of the Phanerozoic karastic zone in the Kingdom of Saudi Arabia is shown in Figure 1. Many sinkholes have been discovered throughout the Phanerozoic karst zone in Ar Riyadh area and the Eastern Province and were reported in different localities such as the Al Summan Plateau northeast of Ar Riyadh, in the terrain south of Ar Riyadh as far as As Sulaiy, and in the Eastern Province and Northern Provinces. Similar sinkhole occurrences were reported in similar geological conditions in Qatar, Kuwait, and Jordan.

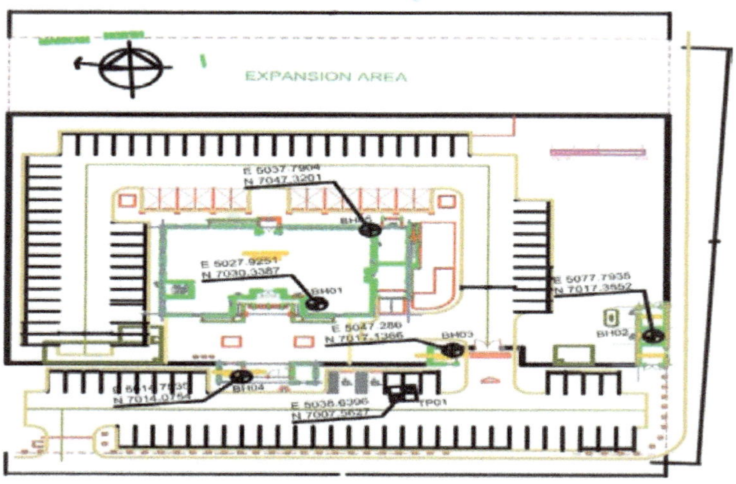

Figure 2: Schematic plan shoeing the location of some boreholes in study area.

PURPOSE OF THE STUDY

The purpose of this site investigation is to determine the existing engineering characteristics of the Sulaiy Limestone Formation and subsurface conditions at the site and to provide the designer with comments on the following:

- Suitable footing types, founding depths and geotechnical design parameters which will be required for a safe and economic design and excavation of the engineering works, such as the soil bearing capacity, expected foundation settlement at the site and other special recommendation which depends on the site nature.
- Methods of construction of foundation and footings, site characters, groundwater conditions, quality control requirements and outdoor subgrade and soil retaining parameters.
- Performing all necessary field and laboratory tests to obtain physical mechanical properties of the subsurface soil. This leads to the geological description of the obtained materials.
- Developing conclusions and recommendations concerning design and construction of the most safe and economical foundations, site preparation, road and parking areas and retaining walls (if any).

SITE INVESTIGATION

The investigation included 6 boreholes, 1 borehole with 15.0 m deep, 5 with 10.0 m and 1 test bit. The subsurface investigation revealed that the material at the site can be described as "a layer of fill material with variable depth (6.0-10.0) m, then a layer of very pale brown fractured weak lime stone up to the end of borings" as shown in the logs of boring in (Figure 2 and 3A-B) The ground water was not encountered within the drilled depth.

Figure 3: A- view of the site under study, B-the Aker ACKER mobile rig with rotary drilling system to take to core samples.

GEOTECHNICAL FIELD WORK

The site was geotechnically investigated by (6) boreholes, and two (1) test pit, of depths 0.50 and 0.75 m (Figure 4A). The boreholes were drilled by ACKER mobile rig using rotary drilling system. The drilling was started at the existing natural ground surface at the location of each borehole.

Figure 4: A-the JBC drilling the test bit, B- the depth of test bit.

During drilling, rock cores were extracted each 1.50 m of drilling depth. For each core, the coefficients of recovery (TCR), (SCR) and quality (RQD) were measured and the results were presented in the boreholes logs (Figure 4A-B).

PHYSICAL AND GEOMECHANICAL PROPERTIES

The geomechanical and physical analyses included point load strength (Is), uniaxial compressive strength (UCS), Schmidt hummer, dry density, porosity, and permeability tests. A total of 10, one-inch diameter specimens were selected for determining UCS, PLT, TCR, RQD and SCR. Schmidt hammer rebound hardness tests were also conducted on outcrop sections of the Sulaiy Formation. The values of geomechanical and physical properties for the Sulaiy Limestone Formation are shown in Table 1.

Rock Quality Designation Index (RQD)

The Rock Quality Designation index (RQD) was developed by Deere [4] to provide a quantitative estimate of rock mass quality from drill core logs. RQD is defined as the percentage of intact core pieces longer than 100 mm (4 inches) in the total length of core. The core should be at least (54.7 mm or 2.15 inches in diameter) and should be drilled with a double-tube core barrel.

RQD is intended to represent the rock mass quality in situ. When using diamond drill core, care must be taken to ensure that fractures, which have been caused by handling or the drilling process, are identified and ignored when determining the value of RQD [5]. In the present study the RQD is between 20% : 90%.

Total Core Recovery *(TCR)*

Total core recovery (TCR) is the borehole core recovery percentage.

TCR is defined as the quotient:

$$TCR = (L_{sum\ of\ pieces} / L_{tot\ core\ run}) \ X\ 100\% \qquad (1)$$

$L_{sum\ of\ pieces}$ =Sum of length of core pieces

$L_{tot\ core\ run}$ =Total length of core run.

The TCR is commonly between 15% to 90%, (Figure 5).

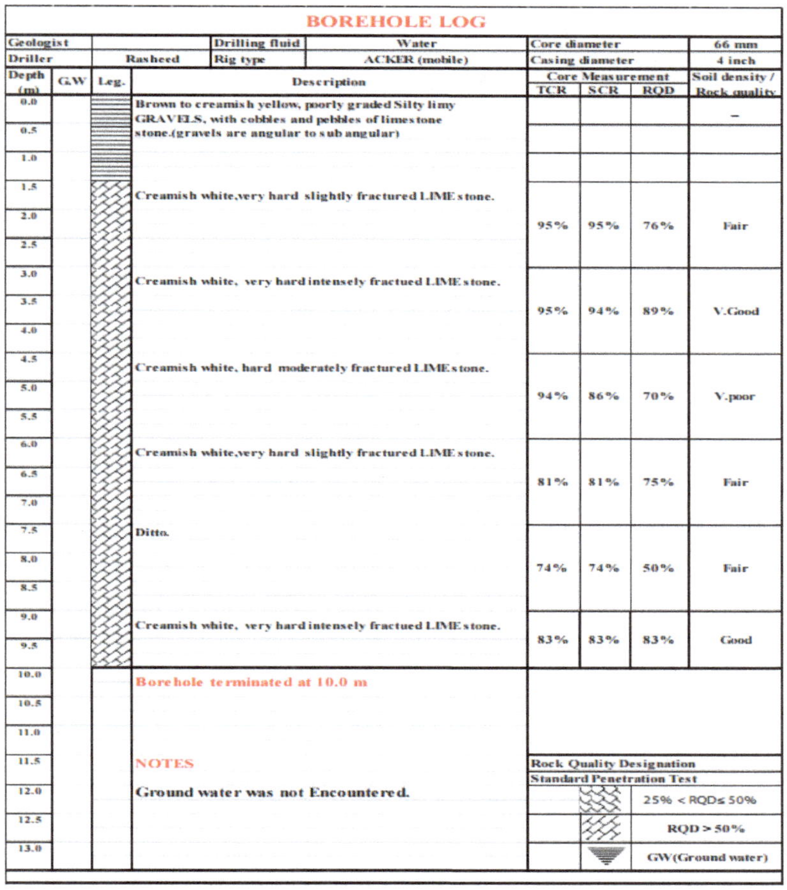

Figure 5: boreholes log represents the rock cores were extracted each 1.50 m of drilling depth which (TCR), (SCR) and (RQD) can be measured.

Solid core recovery (SCR) is the borehole core recovery percentage of solid, cylindrical, pieces of rock core.

SCR is defined as the quotient:

$$SCR=\left(L_{sum\ of\ solid\ core\ pieces} / L_{tot\ core\ run}\right) X\ 100\% \quad (2)$$

$L_{sum\ of\ solid\ core\ pieces}$= Sum of length of solid, cylindrical, core pieces

$L_{total\ core\ run}$ =Total length of core run.

The (SCR) is ranging between (20%) to (90%) (Figure 5).

The Point Load Test (PLT)

Point load testing is used to determine rock strength indexes in geotechnical practice. The point load test apparatus and procedure enables economical testing of core or lump rock samples in either a field or laboratory setting. The point load test (PLT) is an accepted rock mechanics testing procedure used for the calculation of a rock strength index. This index can be used to estimate other rock strength parameters. The PLT is an attractive alternative to the UCS because it can provide similar data at a lower cost. The PLT has been used in geotechnical analysis for over thirty years [6]. The PLT involves the compressing of a rock sample between conical steel plates until failure occurs. The apparatus for this test consists of a rigid frame, two point load platens, a hydraulically activated ram with pressure gauge and a device for measuring the distance between the loading points. The pressure gauge should be of the type in which the failure pressure can be recorded.

The point load strength values for the Sulaiy Limestone Formation are 2.40 and 6.90 megapascals (Mpa).

The Uniaxial Compressive Strength Test

The UCS is undoubtedly the geotechnical property that is most often quoted in rock engineering practice. It is widely understood as a rough index, which gives a first approximation of the range of issues that are likely to be encountered in a variety of engineering problems including roof support, pillar design, and excavation technique [6]. For most coal mine design problems, a reasonable approximation of the UCS is sufficient. This is due in part to the high variability of UCS measurements. Moreover, the tests are expensive, primarily because of the need to carefully prepare the specimens to ensure that their ends are parallel. UCS of the different types of limestone rocks of Sulaiy Formation in the study area shows difference in values (55 to 143 MPa).

Dry Density

Dry density is defined as a mass per unit volume. The mass of a unit volume of rock in its natural state is different from the mass of the same volume of rock containing only its solid phase. The statistical calculation of volumetric weight are given in Table 1. Eight specimens with different dimensions (L × W × H) of the limestone rocks from the study area were prepared by disc saw machine; then dried in oven at 105°C for 24 hr. Volumetric weight of the different types of limestone rocks of Sulaiy Formation in the study area shows slight differences in value (1.7 to 2.3 gm/cm^3) with an average of 2.32 gm/cm^3.

Table 1: Geomechanical and Physical Properties of Sulaiy Formation

Sample No.	Permeability	Porosity	UCS	PLT	Density	SHT
1	202	12	143	6.9	2.2	45
2	24	10	65	2.4	2.0	23
3	135	2	77	3.9	1.7	54
4	198	6	61	4.5	1.9	51
5	22	2	76	3.2	2.1	36
6	68	5	59	4.6	1.6	33
7	92	6	64	6.1	1.8	43
8	186	5	55	5.8	2.3	49

Porosity

In general, porosity is the property of solid to have free spaces, or pores, between individual particles not filled by the structural materials. Quantitatively, porosity is defined by the volume of pores per unit volume of a material [7].

To determine the porosity of a rock, the specimens were dried at 105°C for 24 hr and then weighed (W_1). Then, immersed completely in distilled water at 20i°C ± 2°C for 24 hr. The samples were removed from water, their surface dried by a damp cloth and weighed to the nearest 0.0001 gm. (W_2). The porosity (**N%**) by polished surface area were calculated as follows (**ASTM**, [8]).

$V_P=W_2 - \mathbf{W}_1$....................(4)

$N = V_{P/} V_b * 100\%$..................(5)

W_1=Weight of the dry sample, gm.

W_2=weight of specimen after immersion, gm.

V_b =Volume of the bulk sample, cm^3.

V_P=Volume of the pore space in the sample, cm^3.

N= Total porosity of the rock%

The results of porosity of the studied samples are given in Table (1). The porosity of the studied limestone samples is widely varied, the highest percentage of porosity is noticed by about (12%) and the lowest is (2%).

Permeability

The average permeability in of the study limestone of Sulaiy Formation is 22 mille Darcy (mD) (fair) and 202 mille Darcy (mD) (good).

The Schmidt Rebound Hammer

The Schmidt rebound hammer is principally a surface hardness tester. It works on the principle that the rebound of an elastic mass depends on the hardness of the surface against which the mass impinges. There is little apparent theoretical correlation between the strength of concrete and the rebound number of the hammer. However, within limits, empirical correlations have been established between strength properties and the rebound number, (International atomic energy agency, [9]).

Schmidt hammer tests indicate that the values are between 33 to 54 Schmidt hammer rebound values (SHV).

BEARING CAPACITY AND SETTLEMENT CALCULATIONS

Footing on (Sulaiy Formation) Limestone Rock

As per Canadian Geotechnical Society, Canadian Foundation Engineering Manual – 3rd edition, we can use the following equation for estimating the allowable bearing capacity of a footing on rock [10]:

$$q_{all}=k_{sp}\ \textbf{\textit{(UCS)}} \qquad (6)$$

where,

- (k_{sp}) is an influence factor depending on the nature of the rock, for intensely fractured rock *(k_{sp}=0.10)*.
- (UCS) is the unconfined strength of the rock. In the current case the minimum value is *(UCS=189.2 kg/cm^2)*.

Then the allowable bearing stress is 18.90 kg/cm^2. We can consider only *3.00 kg/cm^2*.

Estimating the Settlement

Willie [11] summarizes settlements of foundation on rocks are as following two different types. First, elastic settlements result from a combination of strain of the intact rock, slight closure and movement of fractures and compression of any minor clay seams (less than a few millimeters). Elastic theory can be used to calculate this type of settlement, AASHTO [11,12]. Second, settlements result from the movement of blocks of rock due to shearing of fracture surfaces. This occurs when foundations are sitting at the top of a steep slope and unstable blocks of rocks are formed in the face. The stability of foundations on rock is influenced by the geologic characterization of rock blocks. The information required on structural geology consists of the orientation, length and spacing of fractures, and their surface and infilling materials. Procedures have been developed for identifying and analyzing the stability of sliding blocks, stability of wedge blocks, stability of toppling blocks, or three-dimensional stability of rock blocks.

In the other hand, the settlement due to mat foundation placed on limestone formation with closed cavities reduced the settlement up to three times compared to that with open cavities. Therefore, grouting should be recommended to fill the open cavities of limestone rock formation to the significant depth below footings before placing the foundations to gain more strength with reducing the limestone settlement.

Recommendation of the study are given for geotechnical investigation engineers to expect the maximum depth for cavity probing search through any project depending on the loads and widths of the shallow foundation. The following equation can be used for estimating the elastic settlement (Se) under rigid foundation [13]:

$$S_e = \frac{qB}{E}\left(1-\mu^2\right)\alpha_r \qquad (7)$$

In the current case:

- q=3.00 kg/cm^2
- B=500 cm
- E=1000 kg/cm^2
- μ^2 =0.30
- a_r=0.82 (for L/B=1), i.e. a square rigid footing

Then, the expected settlement is (***Se***=1.12 cm).

CONCLUSION AND RECOMMENDATIONS

This paper is prepared to study and investigate the physical and mechanical properties, Such as point load strength (Is), uniaxial compressive strength (UCS), Schmidt hummer, dry density, porosity, permeability tests, (TCR), (SCR) and (RQD), site geology and any installations within the investigated area.

Based on the above mentioned rock properties and the expected structures, suitable footing types, founding depths and geotechnical design parameters which will be required for a safe and economic design and excavation of the engineering works, such as the bearing capacity, expected foundation settlement at the site and other special recommendation which depends on the site nature are also calculated as flows:

- The suitable foundation type, for buildings and fences, is reinforced concrete isolated Footings.
- Foundation level for buildings at 1.50 m below the lowest point on the ground surface at each building, and for fences at 1.50 m below the lowest point on ground surface along the fence.
- The allowable net bearing stress is 3.00 kg/cm^2. Under this stress, the expected settlements will not exceed 25 mm, and the angular distortion between any adjacent columns should not exceed (1 vertical: 500 horizontal).
- A plain concrete layer of minimum thickness 10 cm should be placed under each footing.
- Finally, it is recommended to inject cement-based grout into subsurface cavities in order to improve strength and reduce permeability.
- Recommendation of the study are given for geotechnical investigation engineers to expect the maximum depth for cavity probing search through any project depending on the loads and widths of the shallow foundation.

REFERENCES

1. Abdeltawab (2013) Karst limestone foundation geotechnical problems, detection and treatment: Case studies from Egypt and Saudi Arabia. International Journal of Scientific & Engineering Research 4: 376-387.
2. Youssef A, El-Kaliouby H,Zabramawi Y (2012) Integration of remote sensing and electrical resistivity methods in sinkhole investigation in Saudi Arabia. Journal of Applied Geophysics 87: 28-39.
3. ShehataW, Roobol JM, Stewart I, Khiyami H, Al Khammash A, et al. (2007) Evaluation of a subsidence hazard in the Al Khafji area. Confidential report Saudi Geological Survey.
4. Deere DU, Hendron AJ, Patton FD, Cording EJ (1967) Design of surface and near surface construction in rock, 8th US Symposium on Rock Mechanics, Minnesota.
5. Deere DU (1989) "Rock quality designation (RQD) after twenty years", US Army Corps of Engineers Contract Report GL-89-1, Waterways Experiment Station, Vicksburg.
6. (ISRM) International Society for Rock Mechanics (1985) Suggested method for determining point load strength: ISRM Common testing methods. Int J Rock Mech Min SciGeomechAbstr 22:112.
7. Hoek E (1977) Rock Mechanics Laboratory Testing in the Context of a Consulting Engineering Organization. Int J Rock Mech Min Sci and Geomech Abstract 14: 93-101.
8. ASTM (1958) Standers of Building stones. Part 5: 9-35.
9. International atomic energy agency (2002) Guidebook on non-destructive testing of concrete structures, Vienna, Austria.
10. Canadian Geotechnical Society (1992) "Canadian Foundation Engineering Manual (3rd edn)", BiTech Publishers, Canada.
11. Hoek E, Diederichs MS "Empirical estimation of rock mass modulus," International Journal of Rock mechanics and Mining Sciences43: 203-215.
12. AASHTO (1997) LRFD Bridge System Design Specification (Interim Revisions), American Association of State Highway and Transportation Officials, Washington, DC.
13. Das B (1995)Principles of soil dynamics, PWS Kent, MA.

CHAPTER 3

GEOMECHANICAL PROPERTIES OF UNCONVENTIONAL SHALE RESERVOIRS

Mohammad O. Eshkalak,[1] Shahab D. Mohaghegh,[2] and Soodabeh Esmaili[3]

[1]Department of Petroleum and Geosystems Engineering, The University of Texas at Austin, Austin, TX 78712, USA

[2]Department of Petroleum and Natural Gas Engineering, West Virginia University, Morgantown, WV 26506, USA

[3]Occidental Petroleum Corporation, Bakersfield, CA 93311, USA

ABSTRACT

Production from unconventional reservoirs has gained an increased attention among operators in North America during past years and is believed to secure the energy demand for next decades. Economic production from unconventional reservoirs is mainly attributed to realizing the complexities and key fundamentals of reservoir formation properties. Geomechanical well logs (including well logs such as total minimum horizontal stress, Poisson's ratio, and Young, shear, and bulk modulus) are secured source to obtain these substantial shale rock properties. However, running these geomechanical well logs for the entire asset is not a common practice that is associated with the cost of obtaining these well logs. In this study, synthetic geomechanical well logs for a Marcellus shale asset located in southern Pennsylvania are generated using data-driven modeling. Full-field geomechanical distributions (map and volumes) of this asset for five geomechanical properties are also created using general geostatistical methods coupled with data-driven modeling. The results showed that synthetic geomechanical well logs and real field logs fall into each other when the input dataset has not seen the real field well logs. Geomechanical distributions of the Marcellus shale improved significantly when full-field data is incorporated in the geostatistical calculations.

INTRODUCTION

Shale gas reservoirs, which are also called source rock reservoirs (SRR), have some unique attributes that make hydraulic fracturing an essential option in order to commence an economic level of the natural gas production. Unlike conventional gas reservoirs, insufficient permeability, ultra-low porosity of shale rock, and limited reservoir contact area, but vastly organic-rich formation, cannot offer production in a commercial value without stimulation processes. Many studies are conducted from shale pore-scale level to field scale reservoir simulations to improve the understanding of complex flow behavior that are developed and discussed through numerical, analytical, and semianalytical reservoir models for unconventional reservoirs [1–10]. However, in order to predict the performance of a shale gas reservoir, implementing accurate shale rock properties is essential for developing a geologic model for the entire asset. Hence, it is very critical to access more data while working on an unconventional

Citation: Mohammad O. Eshkalak, Shahab D. Mohaghegh, and Soodabeh Esmaili, "Geomechanical Properties of Unconventional Shale Reservoirs," Journal of Petroleum Engineering, vol. 2014, Article ID 961641, 10 pages, 2014. doi:10.1155/2014/961641.

reservoir. In this study, synthetic data are generated using artificial intelligence and data mining techniques (AI&DM).

Principal stress profile of an oil and gas reservoir depends highly on the rock geomechanical properties. Geomechanical properties of reservoir rock include Poisson's ratio, total minimum horizontal stress, and bulk, Young, and shear modulus. These properties play significant role in current development plans of shale assets compared to conventional reservoirs that have established sufficient information available. Moreover, having access to geomechanical data can assist engineers and geoscientists during geomechanical modeling, hydraulic fracture treatment design, and reservoir simulation in shale gas fields across the U.S. and worldwide. Geomechanical well logs are one of the sources that secure such data. Running geomechanical well logs (in all wells in a shale asset) is not common practice among operators and the reason is attributed to the cost associated with running such logs. In this paper, data-driven models are developed to accurately determine the Marcellus shale rock properties.

Artificial intelligence and data mining have been used within last 20 years in reservoir modeling and characterization to perform analysis on formation of interest [11, 12]. Also, some studies indicated that artificial neural network (ANN) is a powerful tool for pattern recognition and system identification such as methodology developed by Mohaghegh et al. in 1998 [13] to generate synthetic magnetic resonance imaging (MRI) logs using conventional logs such as SP, GR, and resistivity. Their methodology incorporated an artificial neural network as its main tool to generate the target variable. The synthetic magnetic resonance imaging logs were generated with a high degree of accuracy even when the model developed used data not employed during model development. Moreover, Basheer and Najjar demonstrated that ANN is suitable to predict and classify soil compaction and rocks characteristics as well as determining some mechanical parameters such as Young's modulus, Poisson's ratio [14]. They mainly investigated the neural network capability in solving geotechnical engineering problems and they provided a general view of some neural network application in their field of research.

In this study, it is demonstrated that AI&DM technology is able to develop data-driven models for generating rock geomechanical properties. The overall work-flow includes development of synthetic geomechanical well logs from conventional logs such as gamma ray and bulk density that are commonly available. These data-driven models used around 30 percent of data (coming from geomechanical logs) of the entire asset, which were available to expand them for the rest of the field with conventional logs but no geomechanical logs. Data-driven models have been validated with blind wells. Blind wells are wells with actual data which are selected due to different locations in the asset of 100 horizontal gas wells. Moreover, the logs generated from data-driven models are used to build an integrated field-wide geomechanical distribution (maps and volumes) for rock geomechanical properties. In this work, the ultimate purpose is meant to propose a technique in order to omit the necessity of running geomechanical logs for the entire asset once such logs are obtained for some portion of the asset to be used in the data-driven models. The number of well logs required to run the data-driven models that leads to an accurate result is determined to be 30 percent of the wells in an unconventional asset. In this study, just 30 out of 100 horizontal wells in the Marcellus shale asset in southern Pennsylvania own actual geomechanical well logs that are provided by a major service company.

METHODOLOGY

In order to accomplish the objectives of this study, a methodology and thorough procedure are required to be defined. The methodology used to accomplish the objectives of this study includes four steps as indicated below. A detailed description of each step is explained afterwards:

- data preparation,
- data-driven model development using AI&DM,
- validation of data-driven models,
- geomechanical property distribution.

Step (a): Data Preparation

Data preparation is the most important step in developing data-driven models due to the fact that all the other steps are using the data prepared in this step. Data preparation involves checking the data for accuracy, entering data in a right format in a computer file, and developing and documenting a database structure that integrated the various properties used in the next steps. In this study, a dataset form the available well logs are required to be prepared that portrays specific property of the rock versus depth. First, production pay zone must be identified from the conventional well logs along with the horizontal wells trajectory. After identifying the depth of the producing zones for Marcellus shale from conventional well logs such as gamma ray, available information of each individual well is extracted from well logs in every one or half a foot according to its log characteristics and tools used. Also, in order for the models to understand the differences between production pay zones and while using these datasets, it is essential to specify the contrast between pay zones (upper Marcellus, Purcell, lower Marcellus, and Onondaga) and the adjacent rocks, nonshale. To account for this contrast, 50 feet depth of log data located above and below the pay zones of interest is also added to the main dataset.

The prepared dataset contains rows and columns consisting of following data that are recorded versus depth: the wells name, the well coordinates, the values for gamma ray (GR), bulk density (BD), sonic porosity, bulk modulus (BM), shear modulus (SM), Young's modulus (YM), Poisson's ratio (PR), and total minimum horizontal stress (TMHS) for each horizontal shale well. It must be emphasized that not all the wells include geomechanical well logs; thus geomechanical values are only recorded for the wells that have such real data, 30 wells.

Step (b): Data-Driven Models Development

In this step, the prepared dataset was processed using backpropagation algorithm of neural network into two main parts.

Part 1 (Conventional Models). In order to have consistent conventional logs data for all wells for the shale production pay zone, part 1 is defined and the conventional logs are generated for those wells that missed some conventional logs. As it is shown in Figure 1, the bulk density and sonic porosity for 30 wells were produced by using two different data-driven models. First model (neural network model 1) used gamma ray, depth, location, and well coordinates as an input to develop training, calibration, and verification segments for generating the bulk density of around 30 wells in the asset where bulk density and sonic data were missing. Second model (neural network model 2) used also bulk density as input beside inputs of first step to generate sonic porosity for the part of asset without this property (around 30 wells also used in this part).

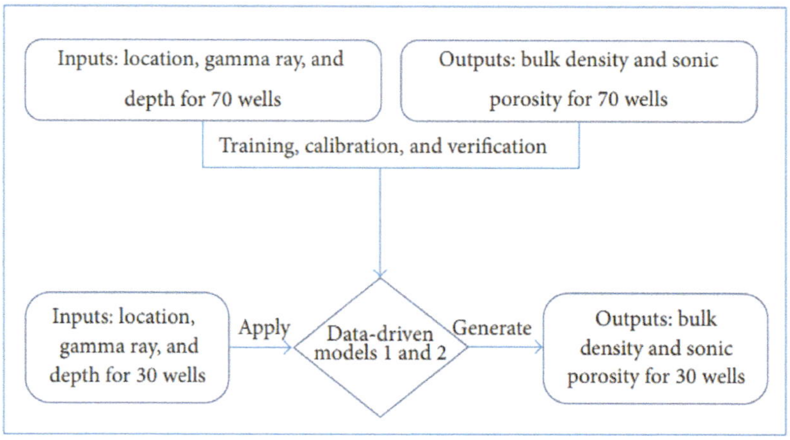

Figure 1: Part 1—data-driven model for conventional logs.

At the end of this step, all existing wells in the asset have the required conventional well log properties to be used in part 2.

Part 2 (Geomechanical Models). After completing the missing data for conventional logs from part 1 for all horizontal wells, another neural network structure is used to develop five different data-driven models (models 3 to 7) as shown in Figure 2 to generate the geomechanical well logs for all wells. As it is shown in Figure 2, this step consists of five neural network models in which the inputs were completed in each step by using the generated geomechanical property in the previous step. In detail, for each of these neural networks, the same conventional logs of 30 wells have been provided as the input and one geomechanical property was generated at a time. Then each generated geomechanical property was used as an input for the next neural network model and the process continued until all five geomechanical properties of interest were generated for the entire Marcellus shale asset.

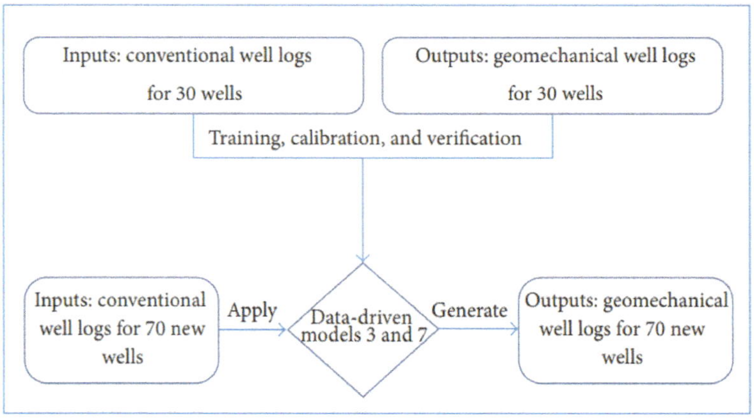

Figure 2: Part 2—data-driven models for geomechanical logs.

All models are multilayer networks that are trained using a backpropagation method of neural network technology. In order to achieve the least error in backpropagation method, a different percentage of datasets are incorporated in the models and finally it is concluded that considering 80% of data used in the training process, and by considering 20% for the calibration process and the remaining for the last step, verification (10% for each), the universal backpropagation error is minimized.

In the next section, two different methods for error calculation S are explained.

Error Analysis. The mean absolute percentage error (MAPE), also known as mean absolute percentage deviation, is a measure of accuracy of a method for constructing fitted time series values in statistics, specifically in trend estimation. It usually expresses accuracy as a percentage and is defined by

$$\text{MAPE} = \frac{100\%}{n} \sum_{t=1}^{n} \left| \frac{At - Ft}{At} \right|, \tag{1}$$

where At is the actual value and Ft is the forecast value.

The difference between At and Ft is divided by the actual value At again. The absolute value in this calculation is summed for every fitted or forecasted point in time and divided again by the number of fitted points n. Multiplying by 100 makes it a percentage error. Also, R-squared is defined and determined in (2) for all three steps, training, calibration, and verification. The higher the R-squared, the closest the results to the actual values:

$$R\text{-squared} = 1 - \frac{\text{Sum of squared distances between the actual and predicted values}}{\text{Sum of squared distances between the actual and their mean values}}. \tag{2}$$

In our study the highest achieved R-squared was around 98 percent and the lower one in some cases around 89 percent, and in both situations, the results presented are highly acceptable. A higher level of R-squared reflects, in all three stages of training, calibration, and verification, a reliable correlation between actual and generated data. It is also important to mention that during the initial training of datasets; the results obtained were with low R-squared. Unsuccessful behavior of models was understood because of having some wells with log data for each 0.5 ft., which is in contrast with the rest of the wells with every available 1 ft. log data. Once piece of data of 0.5 ft. turned to 1 ft., which is considered as discrepancy that misleads the data-driven models; the results came out properly and the data-driven models showed rapid improvements. Further, second issue that resulted in smoother behavior of the data-driven models was related to the removal of nonshaly thin intervals log data from the pay zones that exists within the upper Marcellus, Purcell, lower Marcellus, and Onondaga. Once these layers were removed, the models converged much faster with very low backpropagation error.

Step (c): Data-Driven Model Validation

This step explains a robust method to analyze the accuracy and validity of the data-driven model's results, although the universal error of all the models was very low according to previous section. To examine the models validity, the well log data of some wells (which have both real conventional and geomechanical logs) was removed from the training dataset and it was attempted to regenerate the geomechanical logs. These removed wells are so called blind wells. Blind wells have been chosen from different location in the Marcellus shale asset under study. Then, data-driven models used this new dataset to be trained to generate geomechanical properties for blind wells. Data-driven models number 3 to 7 was separately validated to generate geomechanical properties. In each step, the generated property compared and plotted against the actual values which had been removed from main dataset. The results of this step are explained in Results and Discussions section.

Figures 4 through 8 demonstrate the actual well logs and generated logs for 5 blind wells that are chosen form different location in the asset as illustrated in Figure 3. To compare the results, both actual and generated properties are plotted in the same figure like an actual well log. Properties such as bulk modulus, Young modulus, Poisson's ratio, shear modulus, and total minimum horizontal stress are presented, respectively. Blue line shows the actual value and the red line is for generated values by data-driven models. For well #1 to well #4, there is a perfect match between blue and red lines that shows the models have generated the exact actual data. These wells are in proximity of wells with actual geomechanical properties according to their locations and depths. As it was expected, results shown for these wells are accurate which demonstrate data-driven model's capability and accuracy in generating geomechanical properties of Marcellus shale.

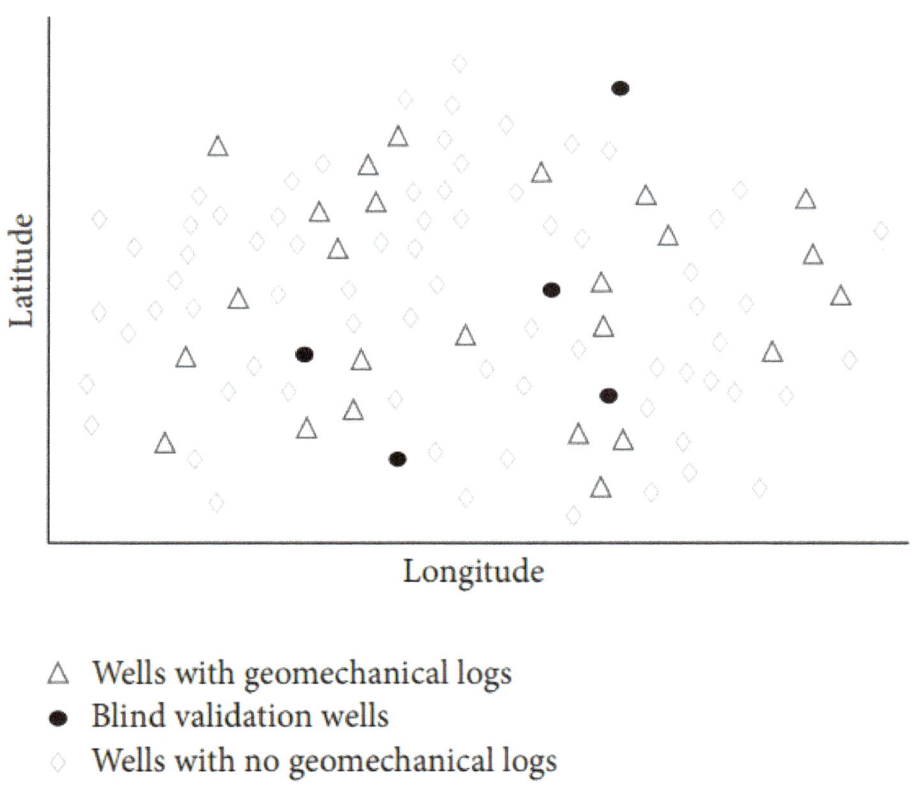

Figure 3: Marcellus shale gas field, southern Pennsylvania.

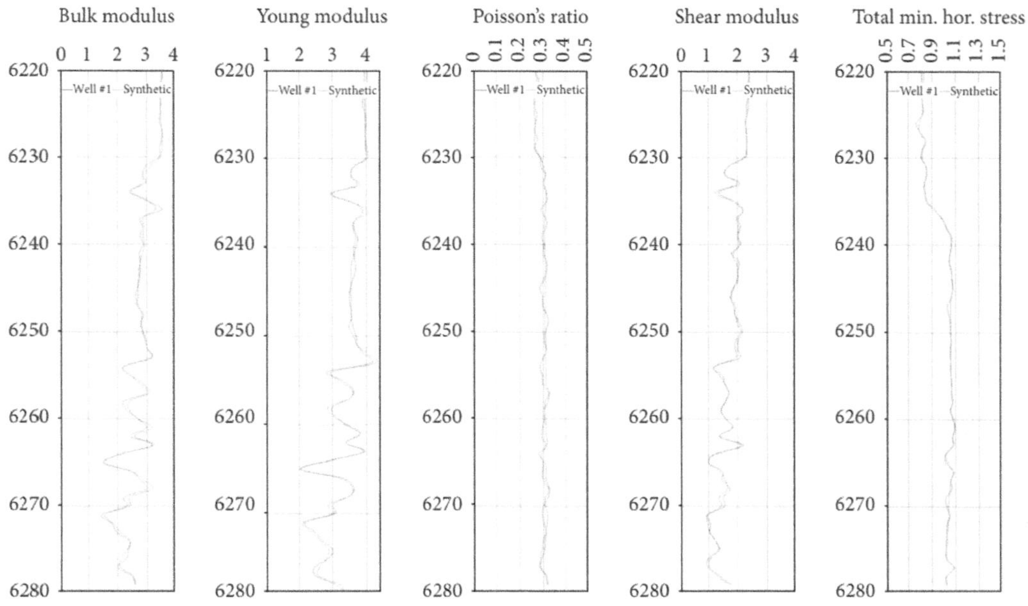

Figure 4: Well #1, actual versus generated well logs.

Step (d): Geomechanical Property Distribution

The first objective of this paper is accomplished in the previous section and the geomechanical properties are generated for all existing wells in the Marcellus shale asset. To accomplish this step, different geostatistical methods, from Petrel commercial software, are considered to create geomechanical property distribution for the Marcellus shale field. Further, geomechanical well logs generated from the data-driven models are coupled with a commercial reservoir simulator in order to create geomechanical distributions for properties such as total minimum horizontal stress, Poisson's ratio, and Young, shear, and bulk modulus.

Sequential Gaussian simulation (SGS) is finally used to create distribution according to well locations for the entire field due to its very smooth and consistent surfaces and distributions (maps) obtained compared to other methods. Two types of maps were created. First map is only incorporated with 30 wells which already had actual geomechanical logs. The second map is related to entire field (70 wells with generated property and 30 wells with actual data). With comparing these two maps, significant difference between geomechanical property distribution with and without having full-field data is observed as shown in Figures 9 through 13. Ten maps that show distribution of five rock geomechanical properties in the Marcellus shale asset were created.

RESULTS AND DISCUSSIONS

The Marcellus shale under study consists of 100 multifractured horizontal wells. Figure 3 depicts the distribution of existing wells in the asset that is used in this study. Table 1 shows the information and number of wells that were used to develop data-driven models in different steps as well as the validation purpose in step (c).

Table 1: Information used for databases for developing data-driven models

Well identifier	Description	Conventional well logs	Geomechanical well logs	Number of wells
Circle	Blind validation wells	Yes	Yes	5
Triangle	Wells used for validation	Yes	Yes	25
Diamond	Wells with no geomechanical logs	Yes	No	70

In this study, a multilayer neural networks or multilayer perceptions are considered to develop the data-driven models. These networks are most suitable for pattern recognition specially in nonlinear problems neural network that have one hidden layer with different number of hidden neurons that are selected based on the number of data records available and the number of input parameters selected in each training process.

The training process of the neural networks is conducted using a backpropagation technique. In the training process, the dataset is partitioned into three separate segments. This is done in order to make sure that the neural network will not be trapped in the memorization phase. Moreover, the intelligent partitioning process allows the network to adapt to new data once it is being trained. The first segment, which includes the majority of the data, is used to train the model. In order to prevent the memorizing and overtraining effect in the neural network training process, a second segment of the data is taken for calibration that is blind to the neural network and at each step of training process, the network is tested for this set. If the updated network gives better predictions for the calibration set, it will replace the previous neural network; otherwise, the previous network is selected. Training will be continued once the error of predictions for both the calibration and training dataset is satisfactory. This will be achieved only if the calibration and training partitions are showing similar statistical characteristics. Verification partition is the third and last segment used for the process that is kept out of training and calibration process and is used only to test the precision of the neural networks. Once the network is trained and calibrated, then the final model is applied to the verification set. If the results are satisfactory then the neural network is accepted as part of the entire prediction system [15, 16].

Figures 4 to 8 show the actual well logs and generated logs for 5 blind wells shown as black circles in Figure 3. To compare the results, both actual and generated properties are plotted in the same figure similar to an actual well log. In these plots, properties such as bulk modulus, Young modulus, Poisson's ratio, shear modulus, and total minimum horizontal stress are presented, respectively. Blue line shows the actual value and the red line is for generated values by data-driven models. For well #1 to well #4 (Figures 5, 6, and 7), there is perfect match between blue and red lines. These wells are in proximity of wells with actual geomechanical properties according to their locations and depths. As it was expected, results shown for these wells are accurate which demonstrate data-driven models capability in predicting geomechanical properties.

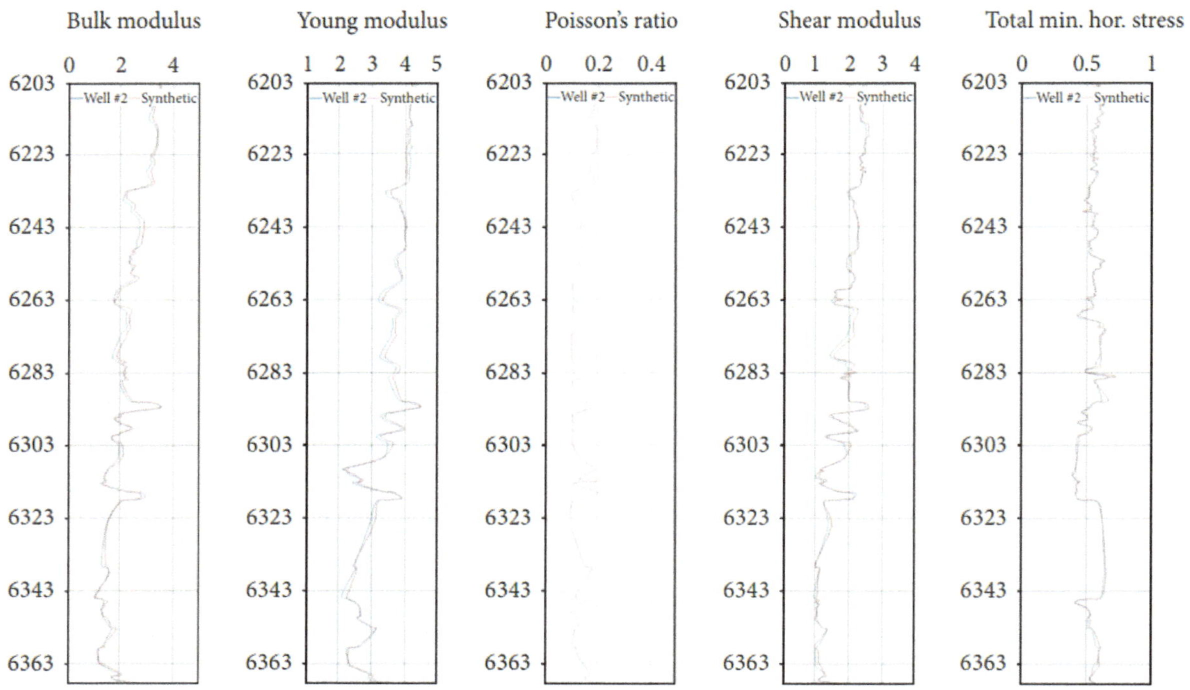

Figure 5: Well #2, actual versus generated well logs.

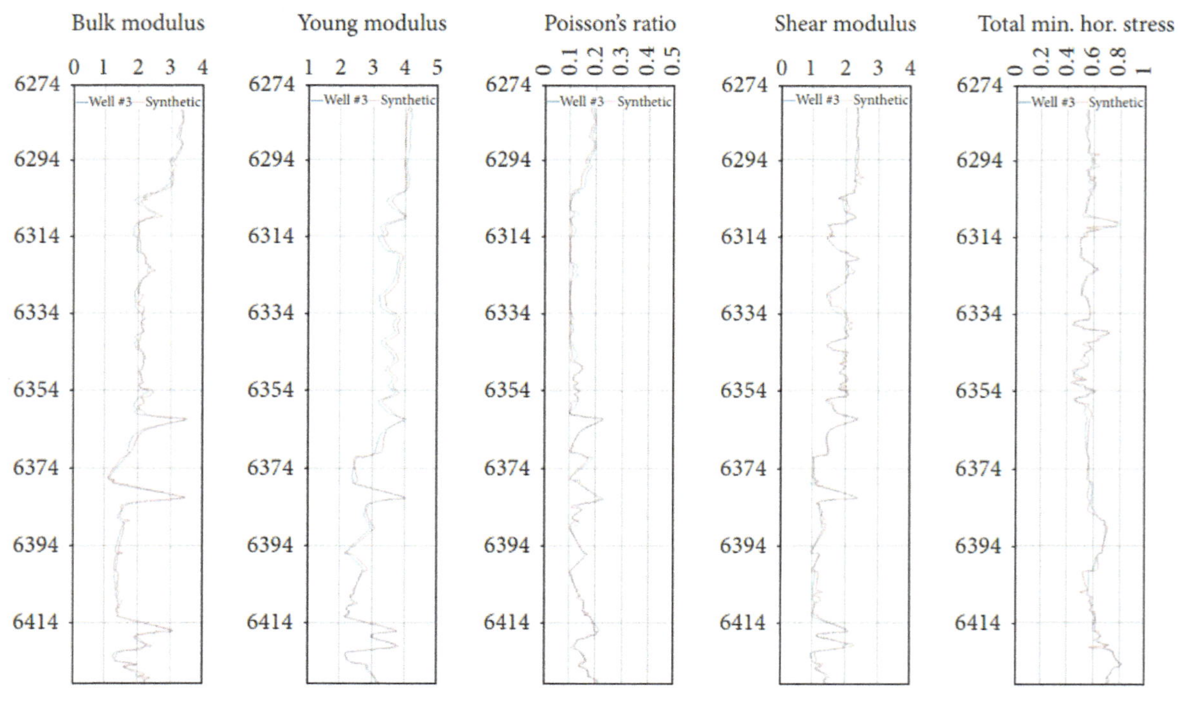

Figure 6: Well #3, actual versus generated well logs.

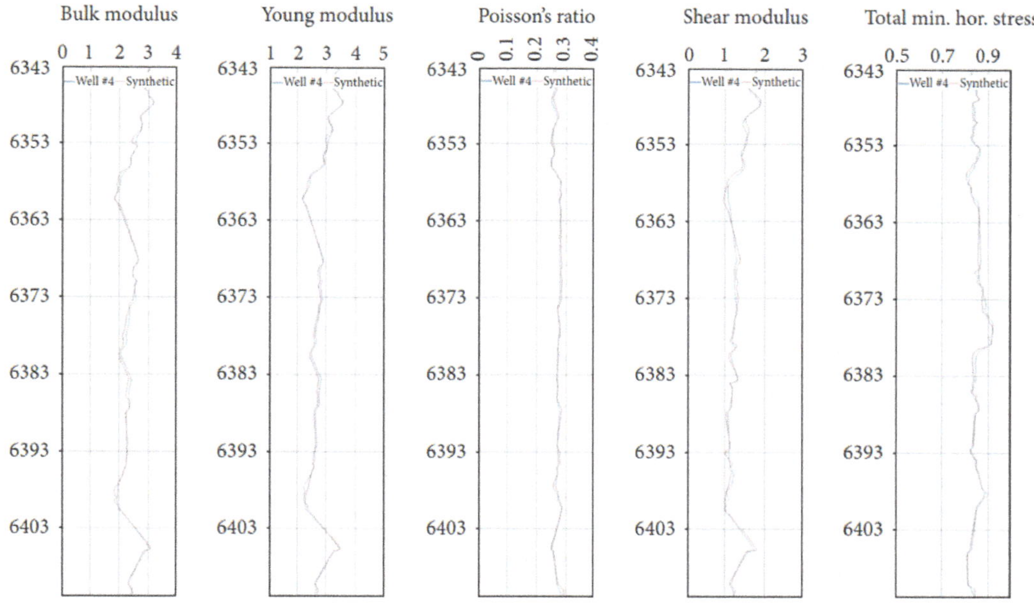

Figure 7: Well #4, actual versus generated well logs.

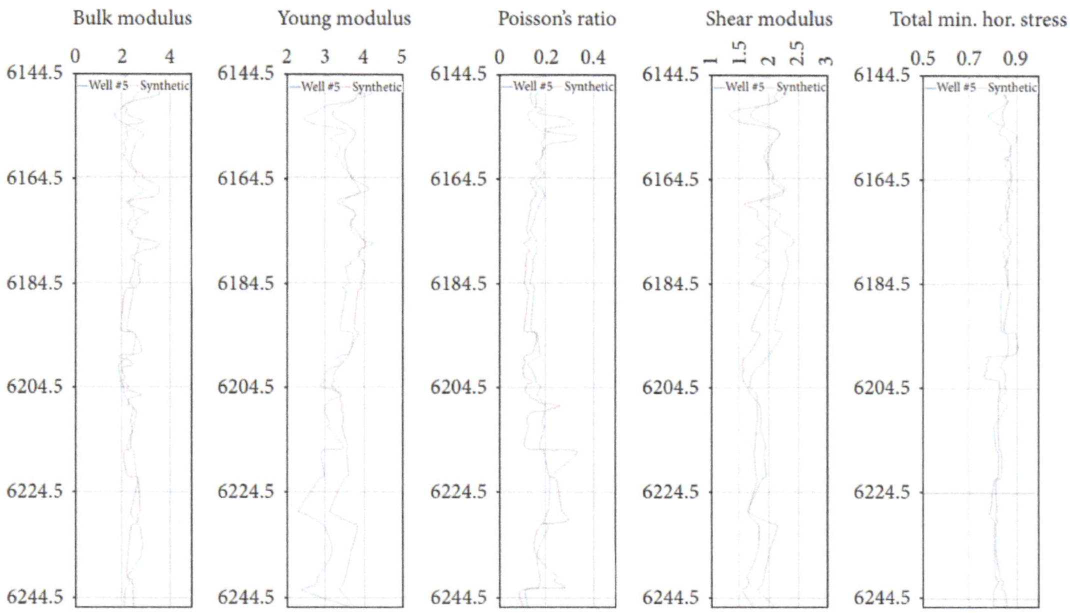

Figure 8: Well #5, actual versus generated well logs.

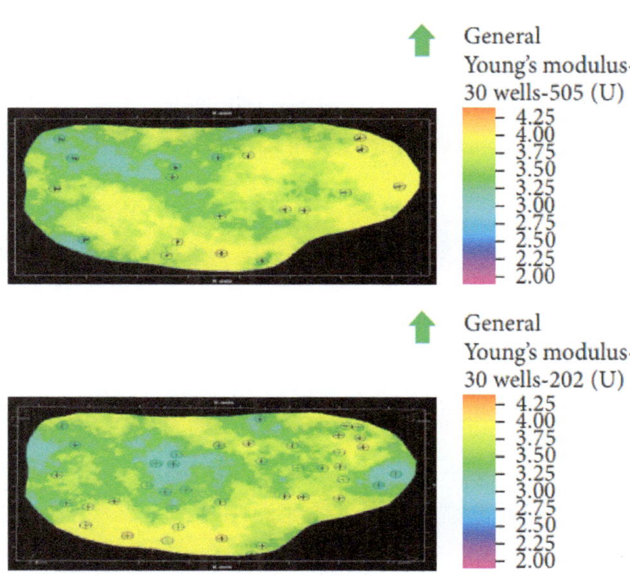

Figure 9: Young modulus.

For well #5 in Figure 8, the generated data is not in agreement with the actual logs and it is because of the location of the well that is far from (upper side of the asset in the field—Figure 2) the rest of wells in the asset that we have used for the training purposes. This fact indicates that the models could not predict the behavior of outlier wells and most importantly emphasizes on the fact that data-driven modeling is perfect for interpolations and not accurate for the extrapolation as it is in agreement with the neural network literature. Moreover, it is found that the depth of producing pay zone of this well #5, compared to other four blind wells, is different (out of range) and it might be another reason related to the fact that models could not capture the behaviors very well.

Figures 9, 10, 11, 12, and 13 are showing distributions and maps for the five geomechanical rock properties of interest in this study. For each property, there are two distributions: one that is generated by using the actual data and the second that considered the information of both generated and actual data (full-field data). A comparison between maps for each property demonstrates that more reasonable and accurate distribution is achieved using more data for the asset. The sequential Gaussian simulation (SGS) algorithm was used in order to generate these maps. In the top distribution map, plus signs represent the wells with actual data which have been used in dataset for training, calibration, and verification during data-driven model development.

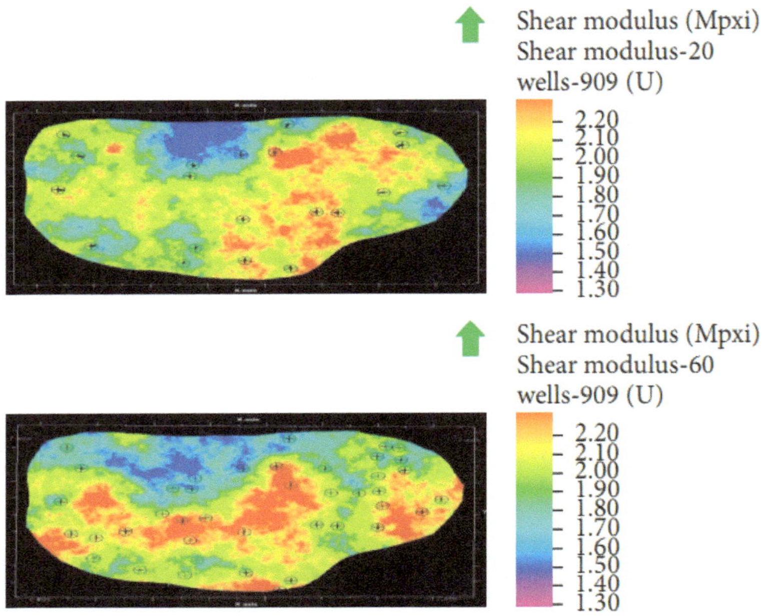

Figure 10: Shear modulus.

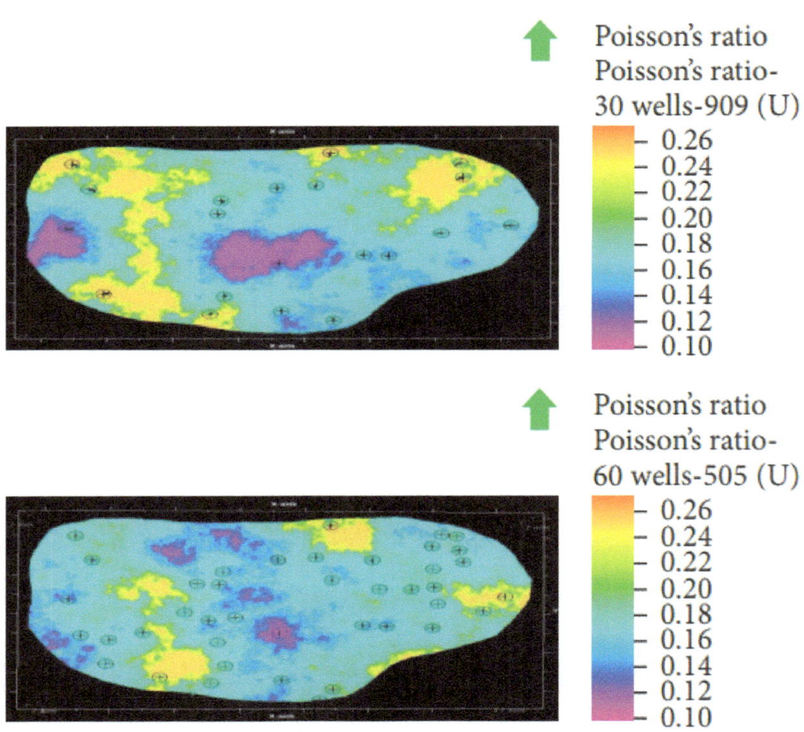

Figure 11: Poisson's ratio.

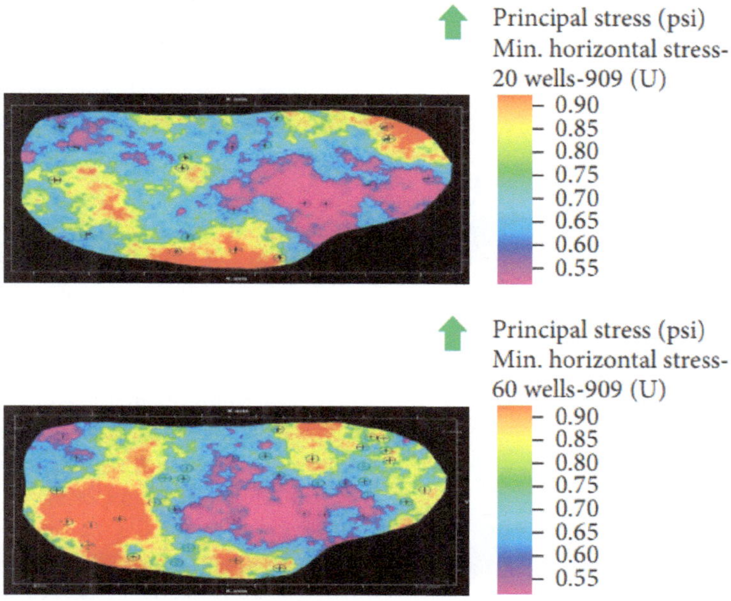

Figure 12: Total minimum horizontal stress.

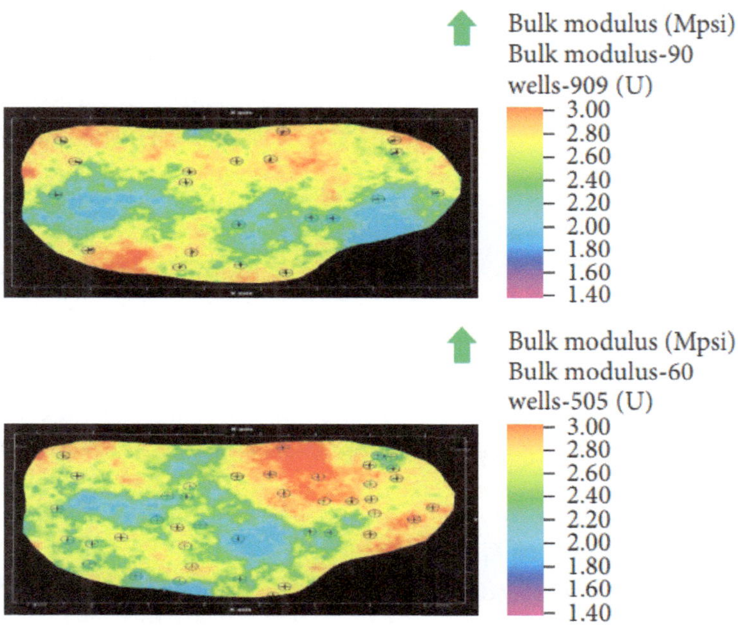

Figure 13: Bulk modulus.

CONCLUSIONS

In this study, it is demonstrated that the data-driven modeling using AI&DM technology is a reliable and robust tool to obtain accurate results for generating synthetic geomechanical logs for unconventional shale resources. In simple terms, we used conventional well logs to generate field-wide geomechanical properties and distribution maps of geomechanical properties for the entire asset, Marcellus shale in southern Pennsylvania.

Five data-driven models were designed, trained, and validated to predict five geomechanical properties of interest for Marcellus shale unconventional reservoir. First, data mining issue in this study, removing nonshaly intervals and adding 50 feet contrast zone, was successfully managed to lead a reliable prediction with least error calculated in backpropagation method. Also, second validation process, the use of 5 blind wells, was performed to show the robustness and accuracy of data-driven models for predicting Young modulus, Poisson ratio, bulk modulus, shear modulus, and total minimum horizontal stress.

Geomechanical property distribution maps of the entire asset illustrated a significant difference between distributions when there are just a few available pieces of actual data rather than having access to the full-field data. These synthetic geomechanical logs and property distributions for Marcellus shale exhibit a great deal of assistance to better performing reservoir modeling, characterization and the optimization of hydraulic fracturing issues related to the current Marcellus shale development process. Authors expect these models will conclude also accurate results in other unconventional shale resources.

NOMENCLATURE

X_j: Input vector for unit j

W_j: Weight vector for unit j

Z_j: Weighted sum of inputs for unit j

O_j: Output of unit j

T_j: Laplace transform parameter

E: Calculated error for each unit

MAPE: Mean absolute percentage error

A_t: Actual value

F_t: Predicted value by data-driven model.

HIGHLIGHTS

Advanced artificial intelligence and data mining technique is used to develop data-driven models in order to generate synthetic geomechanical well logs.

Highly accurate results from data-driven models are achieved that are validated against blind wells that have actual file data in the Marcellus shale asset.

The geomechanical distributions created with field-wide data demonstrate much better consistency and improvement compared to using just partial field data.

REFERENCES

1. C. L. Cipolla, E. P. Lolon, J. C. Erdle, and B. Rubin, "Reservoir modeling in shale-gas reservoirs," SPE Reservoir Evaluation & Engineering, vol. 13, no. 4, pp. 638–653, 2010. · ·
2. W. Yu and K. Sepehrnoori, "Simulation of gas desorption and geomechanics effects for unconventional gas reservoirs," in Proceedings of the SPE Western Regional/ Pacific Section AAPG Joint Technical Conference: Energy and the Environment Working Together for the Future, pp. 718–732, Monterey, Calif, USA, April 2013.
3. S. Esmaili, A. Kalantari-Dahaghi, and S. D. Mohaghegh, "Forecasting, sensitivity and economic analysis of hydrocarbon production from shale plays using artificial intelligence & data mining," in Proceedings of the Canadian Unconventional Resources Conference, Calgary, Canada, October-November 2012, paper SPE 162700.
4. T. W. Patzek, F. Male, and M. Marder, "Gas production in the Barnett Shale obeys a simple scaling theory," Proceedings of the National Academy of Sciences of the United States of America, vol. 110, no. 49, pp. 19731–19736, 2013. · ·
5. U. Aybar, M. O. Eshkalak, K. Sepehrnoori, and T. W. Patzek, "The effect of natural fracture›s closure on long-term gas production from unconventional resources," Journal of Natural Gas Science and Engineering, 2014. ·
6. M. O. Eshkalak, "Simulation study on the CO_2-driven enhanced gas recovery with sequestration versus the re-fracturing treatment of horizontal wells in the U.S. unconventional shale reservoirs," Journal of Natural Gas Science and Engineering, 2014. ·
7. M. O. Eshkalak, U. Aybar, and K. Sepehrnoori, "An integrated reservoir model for unconventional resources, coupling pressure dependent phenomena," in Proceedings of the the SPE Eastern Regional Meeting, pp. 21–23, Charleston, WV, USA, October 2014, paper SPE 171008. ·
8. M. O. Eshkalak, U. Aybar, and K. Sepehrnoori, "An economic evaluation on the re-fracturing treatment of the U.S. shale gas resources," in Proceedings of the SPE Eastern Regional Meeting, Charleston, WV, USA, October 2014, paper SPE 171009. ·
9. M. O. Eshkalak, S. D. Mohaghegh, and S. Esmaili, "Synthetic, geomechanical logs for marcellus shale," inProceedings of the SPE Digital Energy Conference and Exhibition, The Woodlands, Texas, USA, March 2013, paper SPE 163690.
10. M. Omidvar Eshkalak, Synthetic geomechanical logs and distributions for marcellus shale [M.S. thesis], West Virginia University Libraries, West Virginia University, Morgantown, WV, USA, 2013.
11. L. Rolon, S. D. Mohaghegh, S. Ameri, R. Gaskari, and B. McDaniel, "Using artificial neural networks to generate synthetic well logs," Journal of Natural Gas Science and Engineering, vol. 1, no. 4-5, pp. 118–133, 2009. · ·
12. S. Mohaghegh, R. Arefi, S. Ameri, K. Aminiand, and R. Nutter, "Reservoir characterization with the aid of artificial neural network," Journal of Petroleum Science and Engineering, vol. 16, pp. 263–274, 1996.
13. S. Mohaghegh, M. Richardson, and S. Ameri, "Virtual magnetic imaging logs: generation of synthetic MRI logs from conventional well logs," in Proceedings of the SPE Eastern Regional Meeting, pp. 223–232, Pittsburgh, Pa, USA, November 1998.
14. I. A. Basheer and Y. M. Najjar, "A neural network for soil compaction," in Proceedings of the 5th International Symposium on Numerical Models in Geomechanics, G. N. Pande and S. Pietruszczak, Eds., pp. 435–440, Balkema, Roterdam, The Netherlands, 1995.
15. A. J. Maren, C. T. Harston, and R. M. Pap, Handbook of Neural Computation Applications, Academic Press, San Diego, Calif, USA, 1990.
16. Y. Khazani and S. D. Mohaghegh, "Intelligent production modeling using full-field pattern recognition,"SPE Reservoir Evaluation & Engineering, vol. 14, no. 6, pp. 735–749, 2011.

CHAPTER 4

FACTORS INFLUENCING COLLECTION PERFORMANCE OF NEAR SURFACE INTERSEASONAL GROUND ENERGY COLLECTION AND STORAGE SYSTEMS

José Javier Muñoz-Criollo, Peter John Cleall, Stephen William Rees

Cardiff School of Engineering, Cardiff University, Cardiff, CF24 3AA, Wales, UK

ABSTRACT

The influence of surface boundary conditions, varying climatic conditions and engineering material parameters on the collection performance of near surface interseasonal ground energy collection and storage systems are investigated. In particular, the performance of a proposed design of an interseasonal heat storage system which has also been investigated by others as part of a full scale demonstration project is considered. A numerical model is developed and validated against field data. It is then applied to undertake a series of simulations with varying system parameters. It is found that (i) higher values of thermal conductivity of the storage layer result in increased storage of thermal energy and lower peak temperatures, (ii) system heat losses are strongly influenced by the performance of insulation layers, (iii) warmer climatic conditions provide more thermal energy available to be stored; however, changes in the amplitude of seasonal air temperature variations have an effect on the rate of collection of thermal energy and (iv) the use of correct surface boundary conditions is critical in modelling the dynamics of these systems.

Keywords

Ground energy; Thermal; Climate

INTRODUCTION

The use of the ground as a reservoir or source of thermal energy is long established. In recent times, systems utilizing modern engineering materials and technology have become more widespread, examples include ground source heating (e.g. Ref. [1]), shallow energy piles (e.g. Ref. [2]), passive heating and cooling of buildings (e.g. Refs.[3] and [4]) and inter-seasonal thermal energy storage (e.g. Ref. [5]). The performance of near surface ground energy collection and storage systems is highly dependent on the spatial and temporal variation in the amount of energy present in the near-surface region of the soil as well as the specific design and operation characteristics of the system. Inter-seasonal heat storage systems are of use

Citation: José Javier Muñoz-Criollo, Peter John Cleall, Stephen William Rees, Factors influencing collection performance of near surface interseasonal ground energy collection and storage systems, Geomechanics for Energy and the Environment, Volume 6, June 2016, Pages 45-57, ISSN 2352-3808, http://dx.doi.org/10.1016/j.gete.2016.04.001.

in applications that have a cyclical annual thermal energy demand typically driven by energy demands in the winter that may be met by using excess heat energy stored in the summer. In some cases, waste heat may be captured from heating and ventilation systems. Applications include heating of buildings, winter maintenance of highways and minimizing ice formation at aircraft stands.

The ability to model such facilities offers potential benefits at the design stage, in particular for scenario testing to help optimize the system. However, for a model to be representative it must be capable of simulating the transient temperature regime in the surrounding soil with reasonable precision. To this end, a significant body of research has appeared in the literature. For example, Ma et al. [6] implemented a 2D heat transfer FEM model to obtain soil temperature profiles suitable to be used as initial conditions in problems involving heat and mass transfer in soils. Quin et al. [7] presented a detailed algorithm for the computation of soil surface heat fluxes and temperature changes using a complete description of the surface energy balance under bare soil conditions. Their model was validated against experimental measurements using publicly available meteorological data for a desert in southern Israel. Rajeev et al. [8] developed a 1D numerical model to describe the ground–atmosphere interaction and soil moisture and temperature profiles to a depth of 2 m below ground surface over a period of 2 years. Their model was validated with data measured at a site in Melbourne, Australia and meteorological data was obtained from a station installed on-site. Liu et al. [9] used a simplified 1D model that estimated the soil surface temperature as a function of air temperature measured close to the soil surface. They predicted underground temperature profiles over a summer season in Nanjing, China.

In the context of the numerical analysis, the methods and assumptions employed vary depending on the specific physical characteristics of the problem in hand. For example, Yumrutaş et al. [10] developed a semi-analytic model to investigate the annual periodic performance of a cooling system which coupled a chiller with a spherical underground thermal energy storage element. Shang et al. [11] studied the temperature recovery of the ground surrounding a relatively deep (50 m vertical U-tubes) ground-source heat pump under intermittent operation. Several influencing factors were taken into account including the soil thermal conductivity, air temperature and solar radiation. Wu et al. [12] assessed the performance of a shallow (1.2 m depth) horizontal slinky ground source heat exchanger under UK (Oxfordshire) weather conditions. A 3D numerical model was validated with experimental measurements collected over a 2 month period using soil thermal properties measured from in situ undisturbed soil samples. The model was then used to study the impact of varying pipe diameters and slinky interval distances. Esen et al. [13] developed a 2D numerical model to evaluate a shallow (1 m depth) ground-coupled heat pump system designed for space heating. They validated the numerical model using experimental measurements [14] and measured soil thermal properties. Congedo et al. [15] performed a set of 3D numerical simulations to study the performance of shallow (1 m depth) horizontal ground heat exchangers under varying pipe distributions, heat carrier fluid velocities, pipe depths and soil thermal properties. It was concluded that most important parameter for the heat transfer performance of the system was the soil thermal conductivity with higher values (1, 2 and 3 W/m K considered by the authors) delivering a better performance. Ramírez-Dávila et al. [16] studied the performance of a relatively shallow earth-to-air heat exchanger (10 m depth) under three different types of weather in Mexico: desert (Cd. Juárez, Chihuahua), mild weather (México city) and hot-humid weather (Mérida,Yucatán). It was found that the performance of the system is dependent on the weather conditions and the season under which it operates.

For effective assessment and design of ground energy collection and storage systems, and in many geomechanical environmental and energy related applications, it is necessary to be able to correctly represent a number of key factors. These include (i) the transfer of heat between the ground surface and the atmosphere, (ii) the movement of heat within the engineered soil mass and (iii) the movement of heat energy within the collector and storage systems. This paper explores how the performance of interseasonal heat storage systems may be affected by meteorological conditions, surface flux boundary conditions, and the thermal properties of the storage materials. It appears that these factors have not been fully explored and so the objective of the paper is to investigate and quantify these three factors via a series of numerical analyses of a typical system. To this end, a numerical framework is developed and validated against a comprehensive dataset produced by others. [5] The impact on system performance is assessed in terms of soil temperature profiles and thermal energy stored.

MODELLING FRAMEWORK

The analysis of thermal problems usually requires the solution of the transient heat transfer equation for a 2D or 3D soil domain. This may be coupled with the solution of the transient heat advection equation (in 3D) or a suitable mathematical algorithm able to represent the transfer of heat in the soil–pipe–fluid system in 2D. In both cases, suitable boundary conditions are required to represent both energy balance at the soil surface and energy transfer between the soil mass and the pipes. In the present study, the mathematical framework for a 2D case is presented (full details are given in Ref. [17]). In order to simplify the analysis, only heat transfer by conduction is considered in the soil mass while other physical processes like convection and mechanical deformation are neglected.

Since the current model addresses the performance of a near-surface thermal device, the theoretical representation of heat transfer at the soil surface includes several terms that are typically not required in models dealing with systems buried deep into the ground and therefore comparatively insulated from soil–surface interactions. The relative contribution of these flux terms is of course problem dependent but in general terms solar radiation is the most significant followed by infrared radiation and then smaller convective and evaporative heat fluxes.

Heat Transfer

Transient heat transfer, when only conduction is considered, can be expressed as[18]

$$C_{p,b}\frac{\partial T_s}{\partial t} = \nabla \cdot (\lambda_b \nabla T_s) \quad (1)$$

where $C_{p,b}$ (J/m^3 K) is the volumetric heat capacity of the soil, λ_b (W/m K) is its thermal conductivity, T_s (°C) is the temperature of the soil, t is time, $\nabla\cdot$ is the divergence operator and ∇ is the gradient operator.

Surface Heat Fluxes

A general energy balance can be used to define the boundary condition at the soil surface. Taking solar radiation, infrared radiation, convective and evaporative heat fluxes into account the energy balance can be expressed as[19]

$$-\lambda_b \frac{dT_{ss}}{d\hat{n}} = (1-\alpha_S)R + \sigma\varepsilon_{ss}\left(\varepsilon_{sky}T_{a,K}^4 - T_{ss,K}^4\right) + h_E(q_a - q_{ss}) + h_C\,(T_a - T_{ss}) \quad (2)$$

where α_S (–) is the solar albedo, $\hat{n}$ is the unitary vector normal to the surface, R (W/m^2) is the magnitude of solar radiation, σ (W/m^2 K^4) is the Stefan–Boltzmann constant, ε_{ss} (–) is the infrared emissivity of the surface, ε_{sky} (–) is the emissivity of the sky,[19] T_a(°C) and $T_{a,K}$ (K) are the temperature and absolute temperature of the air respectively, $T_{ss,K}$ (°C) and $T_{ss,K}$ (K) are the temperature and absolute temperature of the soil surface respectively, q_a(kq/kg) and q_{ss} (kg/kg) are the specific humidity of air and soil surface respectively, h_E (W/m^2) is the evaporative heat transfer coefficient, and h_c (W/m^2 K) is the convective heat transfer coefficient.

In principle, the surface of the soil can be composed of many regions with different properties that correspond to different types of cover. In this paper three, particular surface types are considered: bitumen pavement, bare soil and soil partially covered by vegetation.

Paved Surface and Bare Soil

For bitumen pavement and bare soil cases, values for variables used and expressions for h_E and h_C in Eq. (2) are taken from Refs. [20] and [19] respectively.

$$h_E = \rho_a L_V(C_{fc}C_{sh}u + C_{nc}\Delta\theta_v^{0.33}) \quad (3)$$

$$h_C = \rho_a c_{p,a}(C_{fc}C_{sh}u + C_{nc}\Delta\theta_v^{0.33}) \quad (4)$$

where ρ_a (kg/m^3) is air density, $c_{p,a}$ (J/kg K) is the specific heat capacity of air, L_v (J/kg) is the latent heat of vaporization of water, $\Delta\theta_v$ (K) is the difference in virtual temperature between the air and the soil surface and u (m/s) is the surface wind velocity. The concept of virtual temperature arises in meteorology and is introduced when working with moist air as it allows use of the ideal gas law for dry air. [21], C_{nc} (m/s K$^{1/3}$), C_{fc} (–) and C_{sh} (–) are coefficients that weight the natural and forced convective processes and the sheltering caused by surrounding objects respectively. It is assumed that the paved surface is impervious and that any water will run-off, as such the contribution of evaporation has been neglected and only natural and forced convection are considered. Evaporation from the soil surface is considered and a soil surface water tension equal to the average of the saturation and wilting point values is assumed.

Soil Covered by Vegetation

Where the soil is covered by vegetation, the approach used to consider energy balance is modified to take into account the presence of a canopy cover using the model described in the works of Deardorff [22] and Best,[23] and applied as suggested by Herb et al. [19] who also provides suitable values for variables and coefficients present in Eq. (2). The consideration of the heat transfer interactions of the canopy layer results in an additional heat balance equation that is coupled to the heat balance equation for the soil surface. The latter is given as

$$\begin{aligned}-\lambda_b\frac{dT_{ss}}{dx} &= (1-\alpha_s)(1-\nu)R \\ &+ \left[\sigma\varepsilon_{ss}\left((1-\nu)\,\varepsilon_{sky}T_{a,K}^4 - T_{ss,K}^4\right) + \nu\varepsilon_c\sigma T_{c,K}^4\right] + h_E(1-C_e\nu)(q_a - q_{ss}) \\ &+ h_C(1-C_e\nu)(T_a - T_{ss})\end{aligned} \quad (5)$$

where v (–) is the canopy density, $T_{c,K}$ (K) is the absolute temperature of the canopy cover, ε_c (–) is the canopy emissivity and the constant C_e (–) establishes the level of soil evaporation for fully dense canopies. The heat balance equation for the canopy cover (which is assumed to have negligible heat capacity) is given by

$$\begin{aligned}0 &= (1-\alpha_c)\nu R + \nu\sigma\varepsilon_{ss}\varepsilon_{sky}T_{a,K}^4 \\ &- 2\nu\sigma\varepsilon_c T_{c,K}^4 + \nu\sigma\varepsilon_{ss}T_{ss,K}^4 \\ &+ \frac{\rho_a L_v \nu}{r_{a,c} + r_s}(q_a - q_{sat}) + \frac{\rho_a c_{p,a}\nu}{r_{a,c}}(T_a - T_c)\end{aligned} \quad (6)$$

where α_c (–) is the canopy solar albedo, T_c(°C) is the temperature of the canopy cover, q_{sat} (kg/kg) is the saturation specific humidity of the canopy cover,[21] $r_{a,c}$(s/m) is the aerodynamic resistance for the canopy cover and r_s(s/m) is the stomata resistance calculated as

$$r_{a,c} = \frac{1}{0.01\,(0.3 + u)} \quad (7)$$

$$r_s = 200\left(\frac{R_{max}}{R + 0.03R_{max}} + \left(\frac{\theta_{wp}}{\theta}\right)^2\right) \quad (8)$$

where R_{max} (W/m^2) is the maximum solar radiation, θ (m^3/m^3) is the moisture content and θ_{wp} (m^3/m^3) is the wilting point soil moisture.

Diurnal Shading

The effect of shading due to nearby surface features (e.g. trees or buildings) on the surface temperature of a paved surface has been taken into account by modifying the solar radiation term to include a factor to account for the impact of shading[24]:

$$R_d = RD_s \quad (9)$$

where R_d (W/m^2) is the effective solar radiation and D_s (–) is a diurnal shading factor which can be defined to be a function of both time and surface location.

Heat Transfer within Storage and Collector Pipes

In this paper, the case of a pipe running perpendicular to the 2D plane flow domain representing the soil is considered, as illustrated in Figure 1. Following the typical mathematical description used to describe heat exchangers (e.g. Ref. [18]), the energy transferred between a unit length of an idealized pipe perpendicular to the plane of a 2D plane flow domain and the surrounding soil, q_p (W/m^2), is calculated as

$$q_p=U_p(T_s-T_f) \quad (10)$$

where T_s (°C) is the temperature of the soil in direct contact with the pipe. U_p (W/m^2 K) is the overall heat transfer coefficient of the pipe and is a function of the pipe geometry, fluid velocity and material properties of the pipe and fluid. T_s is calculated as the average value of all the degrees of freedom at the soil–pipe boundary. T_f is estimated as the average of the inlet and outlet temperatures of the idealized pipe, in which heat transfer along the direction of the fluid flow can be described as

$$q_p = \varepsilon_p c_{p,f} \dot{m}_f (T_s - T_{f,i})/A_p \quad (11)$$

where $c_{p,f}$ (J/kg K) is the specific heat capacity of the pipe fluid, $\dot{m}_f$ (kg/s) is the mass flow inside the pipe, $T_{f,i}$ (°C) is the temperature at the inlet of the idealized pipe, A_p (m^2) is the area through which the heat is being transferred and ε_p is the heat exchanger thermal efficiency that can be defined as[18]

$$\varepsilon_p = 1 - \exp\left[-\frac{U_p A_p}{\dot{m}_f c_{p,f}}\right] \quad (12)$$

Eq. (11) implies that the soil temperature at any time step remains constant for the whole length of the assumed pipe, this assumption is a necessary limitation of the adopted 2D modelling approach. When two or more pipes perpendicular to the domain form a closed system, the estimated outlet temperatures of a pipe or set of pipes can be used as inlet temperatures for the remaining set of pipes closing the system and analyzed in an iterative manner.

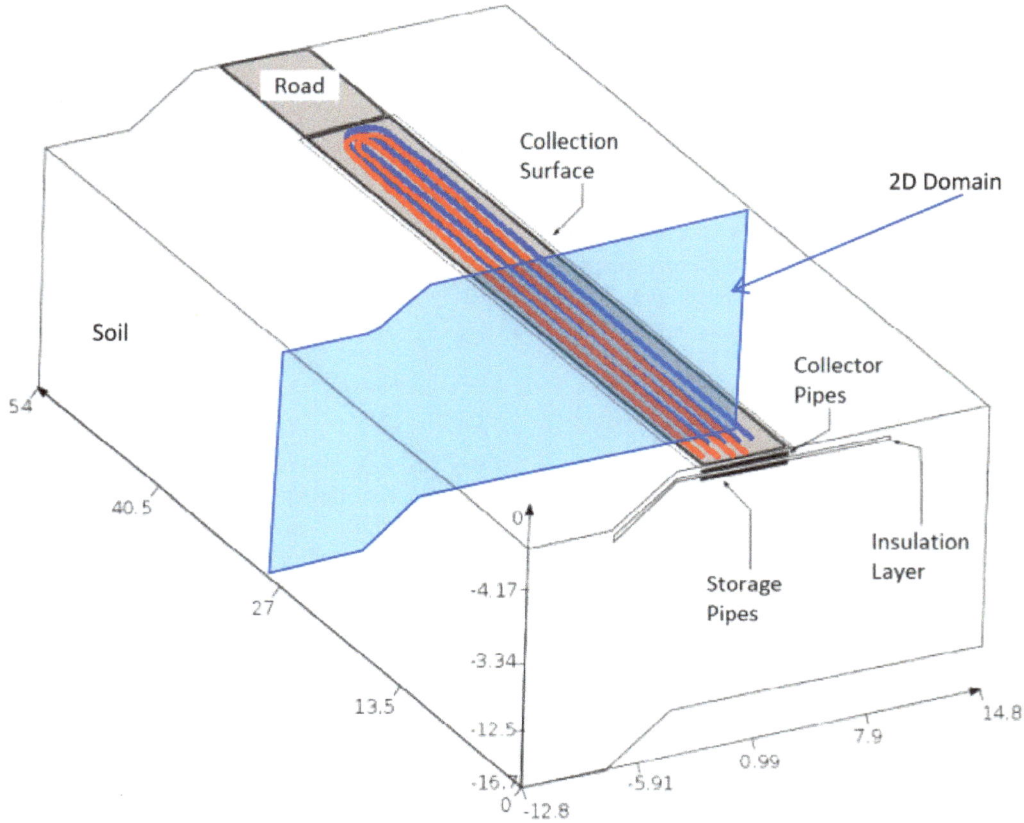

Figure 1. Schematic of interseasonal heat storage system, dimensions in (m).

Stored Thermal Energy

In this paper, the thermal energy collected per unit depth by an interseasonal heat storage facility is used as a parameter to compare its performance under varying conditions of selected variables. The additional thermal energy contained in a defined region E (W/m) is calculated as[25]

$$E = C_{p,b} \int_{sr} \Delta T_{sr} dA_{sr} \quad (13)$$

where ΔT_{sr} (°C) is the temperature difference in the region under consideration (e.g. *Storage Region*, Figure 2) due to the active system in comparison to a case without the activation of the system and A_{sr} (m^2) is the area of the region.

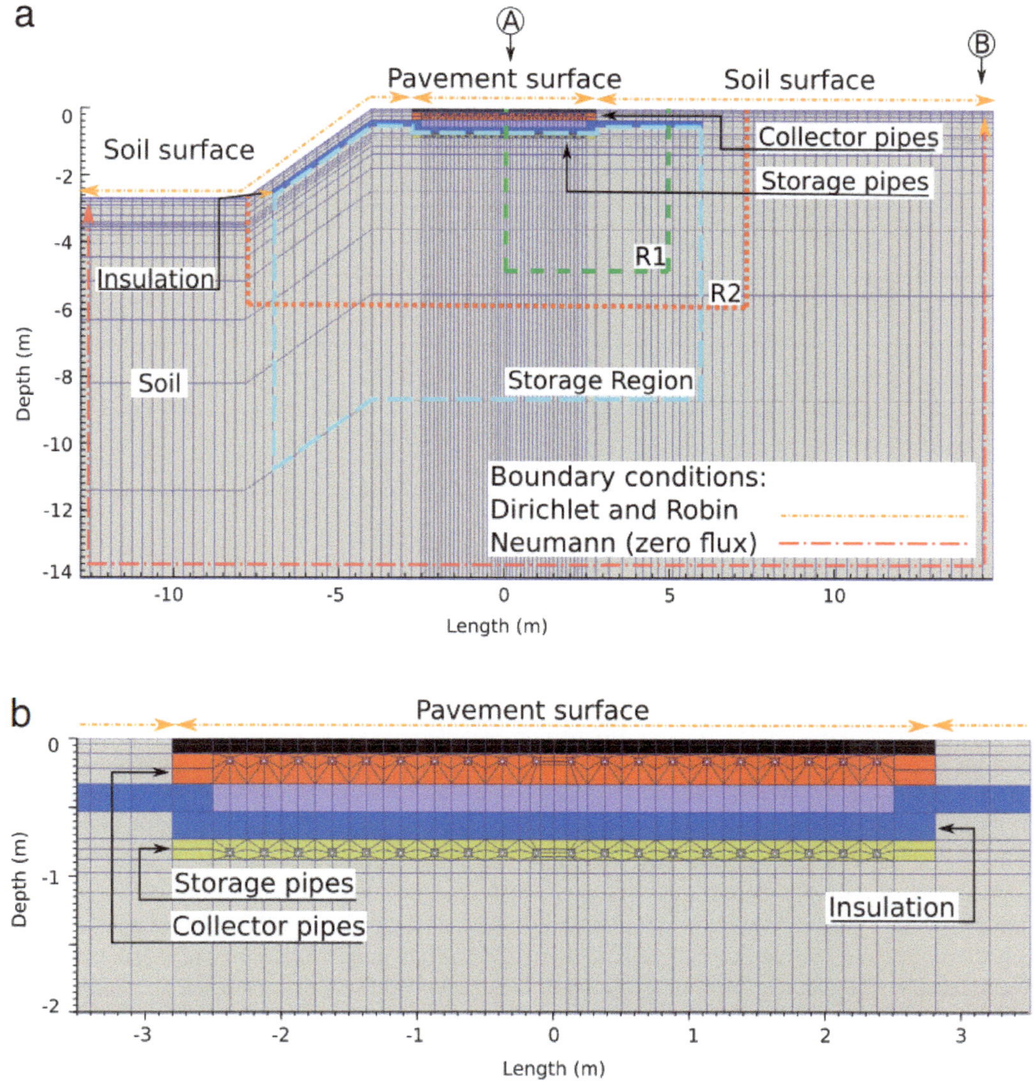

Figure 2. (a) Cross-section of interseasonal heat storage system. (b) Close up of collector and storage pipes region.

Numerical Solution

A numerical solution of the 2D model has been developed using the finite element method (Galerkin weighted residual) to solve the transient heat transfer (Eq. (1)) with boundary conditions as described in Section 2.2. Time discretization is achieved following Rothe's method via a Crank–Nicholson scheme. The open source and publicly available finite element library Deal.ii is used to implement the numerical method and solve the system of equations.[26]

In order to calculate the surface heat flux in each time step, the heat balance equations at the soil surface as defined by Eq. (2) for bare soil conditions and pavement and by Eqs.(5) and (6) for a soil covered with vegetation are solved using Newton's method, as suggested by Deardorff,[22] since they are non-linear due to the infrared and evaporative heat transfer fluxes. In particular for the case of vegetative cover, Eqs. (5) and (6) are solved iteratively by estimating a canopy temperature and then calculating the corresponding ground surface heat flux and solving Eq. (1).

An algorithm that takes into account two interconnected pipe heat exchangers is developed to estimate the temperature at the outlet of the first heat exchanger pipe using Eq. (11) that in turn is used as temperature at the inlet of the second pipe. The total amount of energy transferred between the two heat exchangers and the soil is calculated iteratively using Eq. (10). Full details and validation of this approach to modelling thermal soil–pipe–fluid interactions are available in Ref. [17].

CASE STUDY: INTERSEASONAL HEAT STORAGE SYSTEM

The design of the system considered in this paper is based on the interseasonal heat storage facility investigated by Carder et al. [5] as part of a full scale demonstration project. It consists of a collector pipe array (5 m wide and 30 m long, using 10 polyethylene pipes spaced at 0.25 m centers and arranged in horizontal U loops) placed at a depth of 0.1 m below a road surface and a second similar storage pipe array at a depth of 0.875 m, below a 0.2 m thick insulation layer, as illustrated in Figure 1 and Figure 2. The extent of the insulation layer and the position of collector and storage pipes are indicated. For clarity further material layers composing the road structure (detailed in Table 1) are not shown. Full details can be found in Ref. [5] and Ref. [17]. Additionally, three sub-regions labelled as R1, R2 and Storage Region and two points A and B used in the analyses presented in Sections 4 and 6 are shown. Figure 2(b) shows a close-up of the region containing the collector and storage pipes.

Table 1. Material parameters.

Material type	Density (kg/m^3)	Specific heat (J/kg K)	Thermal conductivity (W/m K)	Thickness (m)
Asphalt	2400	850	0.85	0.105
Concrete	2100	840	1.4	0.055
Silty clay	1960	840	1.21	14
Backfill	840	840	1.21	0.5
Polystyrene insulation	30	1130	0.034	0.2
Sand	2240	840	0.33	0.150
Near ideal insulation	30	1130	0.00034	0.2

All from Ref. [5], except for 'Near ideal insulation'.

The purpose of an interseasonal heat storage system is to take advantage of the difference in thermal energy available between seasons. One possibility is the collection of heat in the summer months to be used at appropriate times during winter months. Under these circumstances the system is activated (i.e. pumps switched on) whenever the temperature difference between the collection and storage pipes is higher than a certain value and is stopped when it drops below another (not necessarily the same) value. In this paper the criteria for activation and deactivation of the collector system follow those suggested by Carder et al. [5] who set the activation of the system when the temperature difference between the control sensors located at the collector and storage levels reached approximately 1.4 °C. Once active, the system was turned off when the temperature difference dropped to approximately 0.3 °C.

VALIDATION

Validation of the model is addressed here via simulation of the above demonstration project. The system design and activation and deactivation criteria used in the demonstration project are as described in Section 3. Meteorological data measured on site, provided by Carder et al.,[5] is used to consider the performance of the system during a 2.5 month collection period.

The simulation is undertaken in a series of sequential steps. Initially, a preliminary simulation consisting of 8 yearly cycles (each using meteorological data from August 2005 to August 2006) is undertaken to establish suitable initial conditions (Period 1). These initial conditions are then used to simulate from May 2005 to August 2005 after installation of the system but before its activation (Period 2), during this period the soil in the storage region is largely prevented from natural warming due to the presence of the insulation layer (for brevity results from this analysis are not shown here but are very similar to those presented in Fig. 9(a)). Finally, a collection period (Period 3) where the system was activated for a first collection stage (from the end of August 2005 until the beginning of November 2005) is simulated. The mesh used for the numerical analysis is shown in Fig. 2 and is made up of 2452 four node isoparametric linear elements (4 temperature degrees-of-freedom per element). After checks on both spatial and temporal convergence Periods 1 and 2 are simulated using hourly time steps, while in Period 3 a 15 min time step is used. Far field boundary conditions and bottom boundary conditions are assumed as insulated (zero flux). Boundary conditions at the surface are set using Eqs. (2)–(4) for the paved surface and Eqs. (5) and (6) for the soil surface, which is assumed to be partially covered by vegetation based on field observations. A schematic representation of the boundary conditions used for the problem is presented in Fig. 2. Analysis of a far field control borehole found a canopy density (v) of 0.85 and an average level of shading of 0.5 to be suitable.17 Reference values for thermal material properties considered in the domain are taken from Ref. [5] and listed in Table 1, coefficients and constants used in the theoretical description of the heat transfer process are given in Table 2. Pipe fluid velocity and composition are also based on typical values reported by Carder et al.5 Fig. 3(a) and (b) show the numerical results obtained compared with experimental measurements at collector pipes depth and storage pipes depth respectively. It can be seen that the temperatures calculated by the numerical model proposed are in good agreement with the measured data. Statistical analysis of the simulated results and the measured data for the whole collection period yields a root mean square error value of 1.09 °C in the collector region and 0.38 °C in the storage region. This validation of the modelling approach for heat collection and storage builds confidence to allow further exploration of the impact of various system parameters.

Table 2. Coefficients and constants used in analyses.

λ_b (W/m °C)	1.2	C_{nc} (m/s $K^{1/3}$)	0.0015
α_s	0.15	C_{sh}	1
α_c	0.15	C_{pf} (J/kg K)	4091.41
σ (W/m^2 K^4)	5.67E−8	$c_{p,a}$ (J/kg K)	1012
ε_{ss}	0.95	ν	0.85
ρ_a (kg/m^3)	1.2041	C_e	0
L_v (J/kg)	2.45E6	ε_c	0.95
C_{fc}	0.0015	R_{max} (W/m^2)	1000
θ	0.24	θ_{wp}	0.12
D_s	0.5	U_p (W/m^2 K)	192
A_p (m^2)	2.17	$\dot{m}_f$ (l/s)	0.155

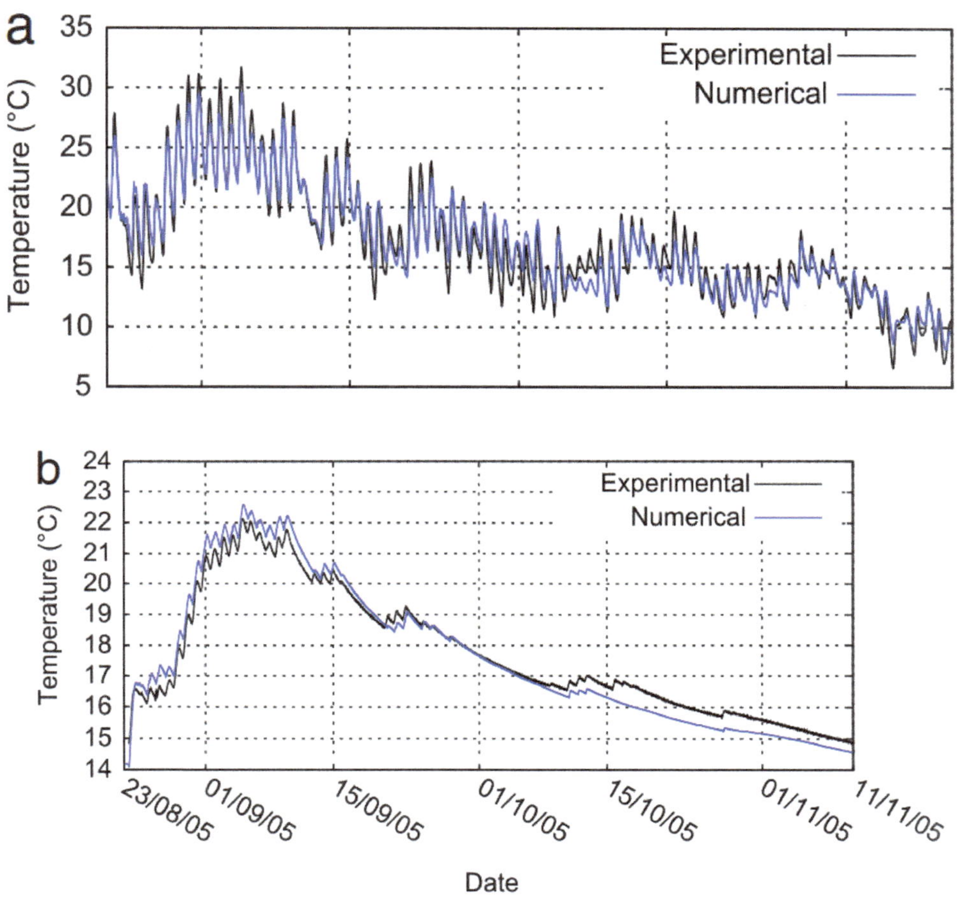

Figure 3. Experimental and numerical temperature transients at (a) collector depth (0.1 m) and (b) storage depth (0.875 m) for the first activation period reported by Carder et al. [5].

Table 3. Coefficients used in air temperature and solar radiation analytical expressions (14) and (17) for three different climates.

Coefficient	Mild weather	Cold weather	Hot weather
A	241.5 W/m^2	218.3 W/m^2	260.8 W/m^2
B	19.6 W/m^2	1.6 W/m^2	167.8 W/m^2
C	18.4 °C	11.2 °C	29.9 °C
D	6.4 °C	−0.2 °C	24.7 °C
E	9.8 °C	5.0 °C	8.2 °C
F	6.3 °C	9.8 °C	9.6 °C

NUMERICAL INVESTIGATION

The main objective of this paper is to explore the impact that the variation of certain factors have on the thermal performance of interseasonal heat storage devices, specifically, in terms of the thermal energy collected and the temperature profiles obtained after a collection period. As above, in the validation section, these simulations consist of three steps:

- an 8-yearly-cycle pre-analysis performed in order to obtain steady state initial conditions (Period 1),
- an analysis after installation of the system at the beginning of summer (Period 2), and
- an analysis of the active system with the system active for ~3 months (84 days) from the beginning of autumn (Period 3).

Three sets of simulations are performed, each exploring the influence of a particular system variable on performance. In each set of simulations the same base case analysis is performed using meteorological data generated for mild climatic conditions, a soil surface boundary condition that includes a vegetative cover and a thermal conductivity of 1.2 W/m K in the storage region under the insulation layer. The mesh, time steps and far-field and bottom boundary conditions are as described in the validation section. The specific simulations undertaken and details of the base case conditions are defined in the following sections.

Variation in Thermal Conductivity (Simulations 1–4)

The influence of varying the thermal conductivity values employed in the Storage Region (shown in Figure 2(a)) is explored via a simple parametric study. A base case value of 1.2 W/m K has been assumed (*Base case*/*Simulation* 1) and two additional values of 0.96 W/m K (*Simulation* 2) and 1.44 W/m K (*Simulation* 3) are also considered. These variations could arise in practice as a result of the process of installation of the system or from disrupting the natural cycles of wetting and drying since the insulation is likely to be relatively impermeable to water. A further analysis of the base case but with a near-ideal insulation layer is also considered (*Simulation* 4), material parameters used for this layer are given in Table 2. Simulations of Periods 1 and 2 in this section are run using the base case conditions, while in Period 3 the particular value of conductivity under consideration is employed. The change in soil thermal properties between Period 2 and 3 is done in order to simplify comparisons between the analyses by having common initial conditions, it is recognized however that this is a limitation and a potential source of error in the analysis due to an imbalance between the temperature field within the ground and its thermal properties.

Assumed Climatic Conditions (Simulations 5–9)

The meteorological conditions used for the numerical simulations performed in this section are generated using two analytical expressions for air temperature and solar radiation proposed by Cleall et al. [25] based on experimental observations as part of the development of analytical solutions for temperature profiles and thermal energy stored in the ground.

It is hypothesized that the performance of an interseasonal heat storage device is likely to be influenced by the type of weather conditions under which it operates. For this reason, three types of climatic conditions are explored:

- Mild (*Base case*/*Simulation* 5)—Obtained from the British Atmospheric Data Centre (BADC 2014) for the years 2004 and 2005 for Toddington, UK. [27]
- Cold (*Simulation* 6)—Obtained from the Icelandic Meteorological Office for the years 2004 and 2005 for the city of Reykjavík, Iceland.
- Hot (*Simulation* 7)—Obtained from the meteorological station of the School of Engineering of the Autonomous University of Yucatán for the year 2009 for the city of Mérida, México.

Using these datasets the values of the coefficients present in Eqs. (15) and (16), and in Eqs. (18)–(21) were calculated and are listed in Table 3. The meteorological data produced with Eqs. (14) and (17) were used for all simulation periods and are illustrated in Figs. 4 and 5 (half yearly cycles of daily maximum solar radiation, and maximum and minimum temperatures are shown). Values for wind speed and relative humidity were fixed at 1.14 m/s and 80.6% respectively for simplicity based on the meteorological data measured on site by Carder et al.5 in a location close to Toddington, UK. Two additional analysis are performed to investigate the impact of the amplitude of temperature variation by replacing coefficients E and F (which represent the mid-summer average amplitude, and mid-winter average amplitude respectively) for the mild climatic conditions with the corresponding coefficients of cold (Simulation 8) and hot conditions (Simulation 9).

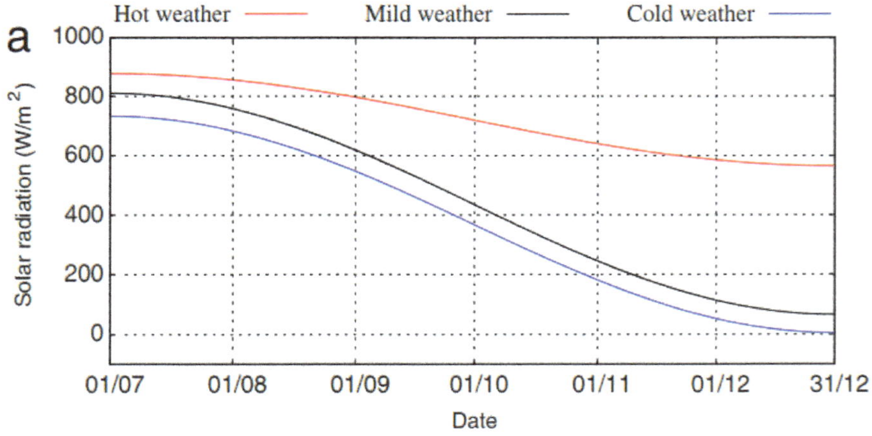

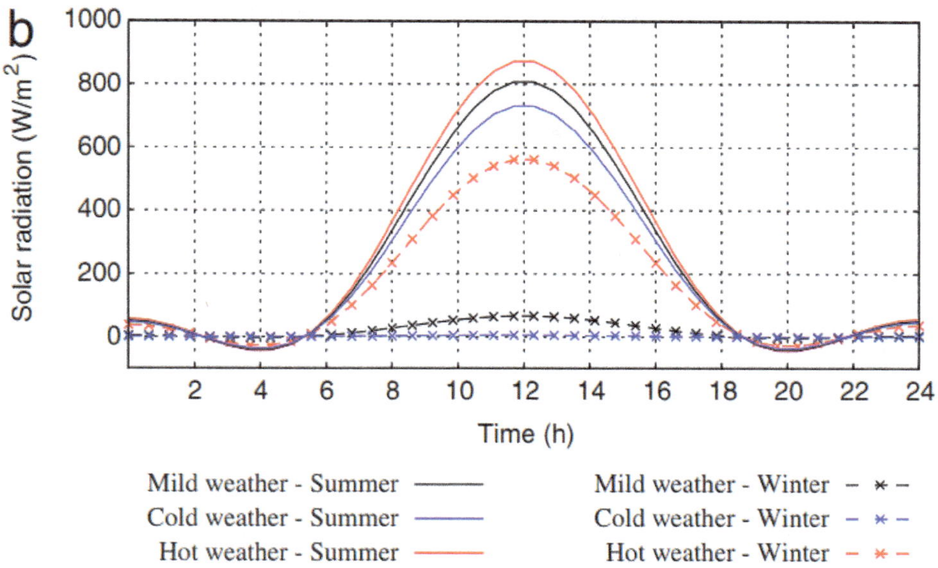

Figure 4. Solar radiation profiles obtained with Eq. (14) and coefficients listed in Table 3. (a) Seasonal variation of peak daily values from summer (July 1st) to winter (December 31st) for three climates considered. (b) Daily variation in summer (July 1st) and winter (December 31st).

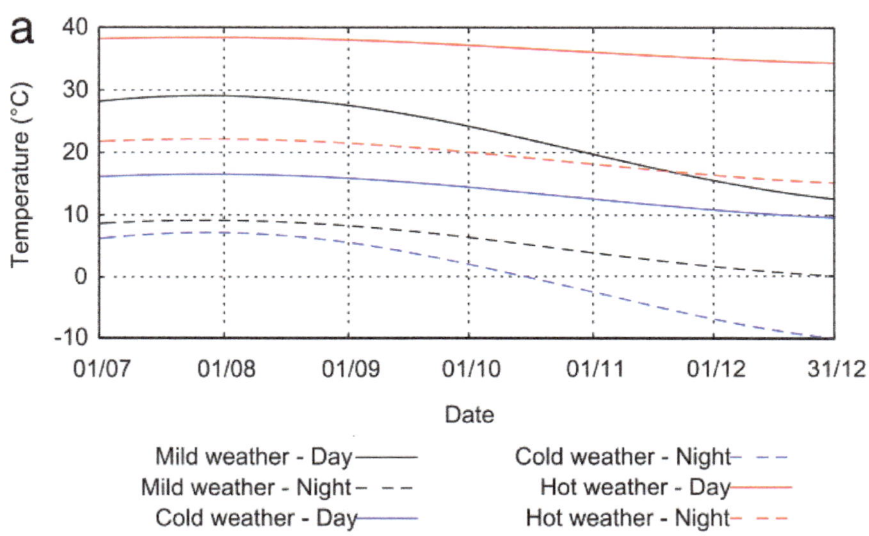

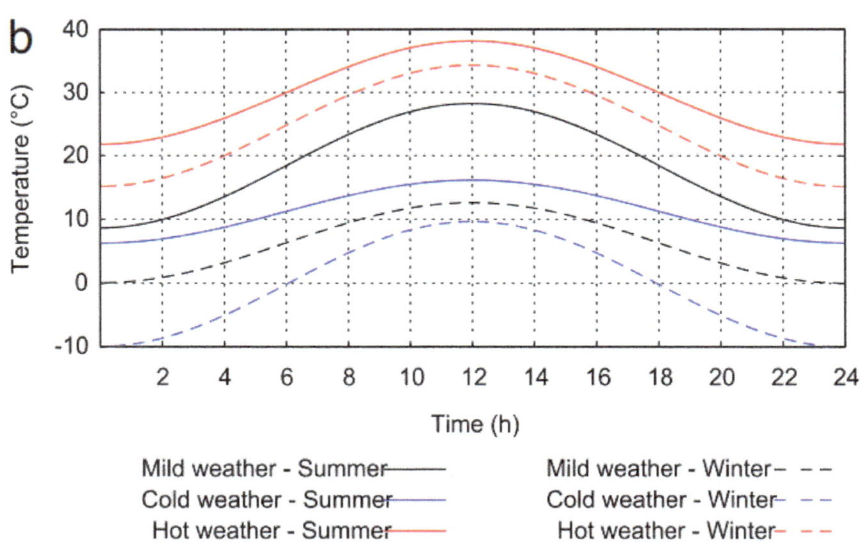

Figure 5. Air temperature profiles obtained with Eq. (17) and coefficients listed in Table 3. (a) Seasonal variation at day (noon) and night (midnight) from summer (July 1st) to winter (December 31st) for three climates considered. (b) Daily variation in summer (July 1st) and winter (December 31st).

Surface Conditions (Simulations 10–12)

Typically two different approaches are applied to consider soil–atmosphere thermal interactions. The simpler, in terms of application, is the use of a fixed (Dirichlet) boundary condition based on measured or estimated temperatures to define the variation of temperature of the soil surface (including any pavement) with time. In many cases, this is approximated to near surface air temperatures as these are often readily available. In this study, the air temperature defined by Eq. (17) is used (*Simulation* 10).

A more complex, but more physically representative, approach is to apply a third kind (Robin) boundary condition which allows representation of the surface fluxes defined in Section 2.2. In this study, two sets of boundary conditions have been utilized for the soil surface (not covered by pavement):

- Bare soil (Simulation 11). Eq.(2)is used to define the surface energy balance with Eqs. (3) and (4) representing the heat transfer coefficients for evaporation and convection respectively
- Soil covered by vegetation (Base case/Simulation 12). Eqs. (5) and (6) are used to represent energy balance between the soil and canopy cover. The heat transfer coefficients for evaporation and convection are as defined by Eqs. (3) and (4).

The flux boundary condition applied to the soil surface covered by pavement in *Simulations* 11 and 12 is given by Eq. (2), using Eq. (4) to define the heat transfer coefficient for convection.

RESULTS AND DISCUSSION

Variations in thermal conductivity

Simulations 1–3 study the effect of varying the thermal conductivity in the storage region under the insulation layer. To observe the influence of the active system, temperatures for Period 3 are obtained both with and without collector system activation. The difference in these temperatures indicates the area of influence of the system in isolation from temperature variations due to meteorological and surface conditions. Figure 6 shows the temperature differences obtained in the region marked as R2 in Figure 2(a) at the end of Period 3 for *Simulations* 1–3. It can be seen that higher temperature differences near the storage pipes are obtained for the case with low thermal conductivity while in turn a wider area of influence approaching the edge of the insulation layer is obtained with higher thermal conductivities.

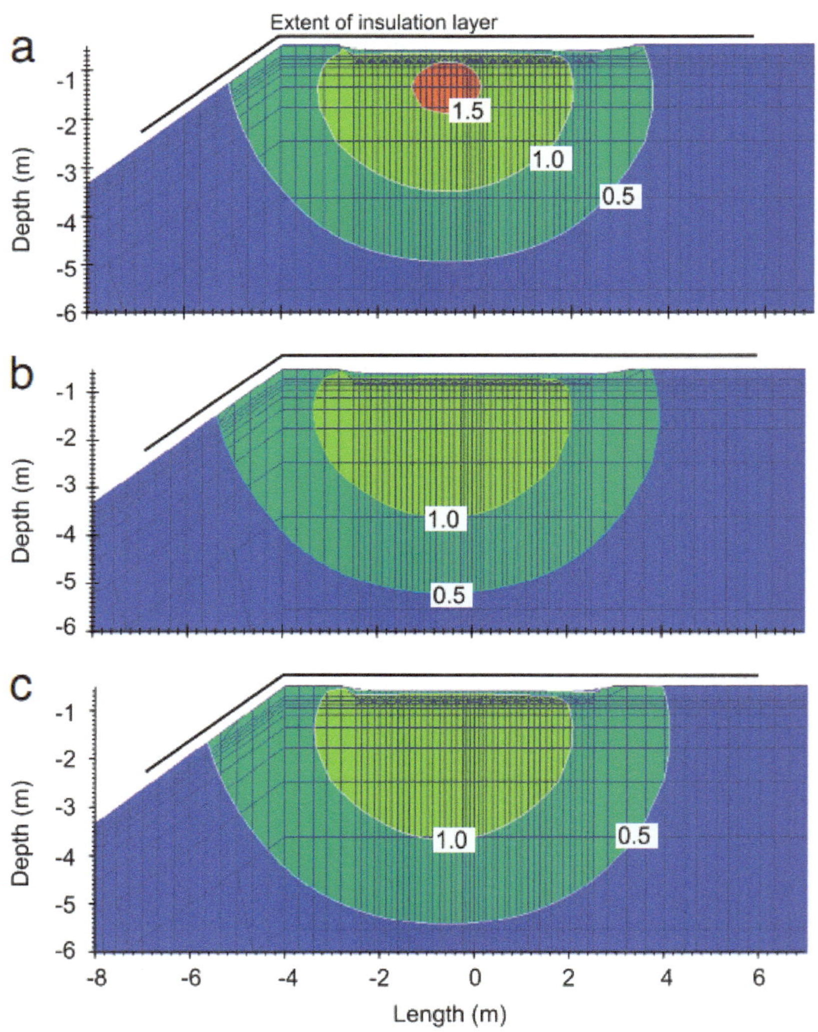

Figure 6. Temperature difference (°C) contours in region R2 at the end of simulation Period 3 for three thermal conductivities: (a) 0.96 W/m K, (b) 1.2 W/m K and (c) 1.44 W/m K.

Figure 7(a) shows the variation in time of thermal energy stored in the Storage Region calculated using Eq. (13). Figure 7(b) and (c) show the soil temperature variation at collector depth (0.1 m) and storage depth (0.875 m) respectively. Figure 7 also includes results from *Simulation* 4 using a near-ideal insulation layer for the base case. Thermal energy contained in the Storage Region starts to rise at the beginning of Period 3, reaches a maximum at day 42 and then decreases until the end of the simulation period is reached. In general, the accumulated thermal energy increases with thermal conductivity this is due to main two factors, firstly heat energy moves away from the pipes more easily to the wider storage mass and secondly this greater movement of heat results in lower soil temperatures at the soil–pipe interface and subsequently a higher thermal gradient across the interface and a higher heat flux. After day 42 the temperature difference between collector and storage no longer satisfies the activation criteria as climatic conditions cool and the temperature in the storage region increases. The subsequent decrease in thermal energy observed up until the end of the simulation period is due mainly to losses through the insulation layer. This is explored in *Simulation* 4 and it can be seen that about 20% more energy could be stored if these losses are minimized. Similar percentages are obtained for the 0.96 and 1.44 W/m K thermal conductivity cases. The temperature profiles at collector depth show negligible differences for the three analyzed cases. However, at storage depth it can be seen that higher temperatures are maintained throughout the duration of the numerical simulation for the case with the lowest value of thermal conductivity (0.96 W/m K) in the Storage Region. This behavior is due to the heightened concentration of thermal energy in the area close to the storage pipes (as shown by Figure 6) caused by the reduction in thermal energy transfer into the remaining storage region. The opposite is true for the case with higher thermal conductivity (1.44 W/m K). In Figure 7(c), the effect of a near-ideal insulation layer on the temperature at storage pipes depth can be observed with reduced thermal losses resulting in an increase in temperature towards the end of the simulation period.

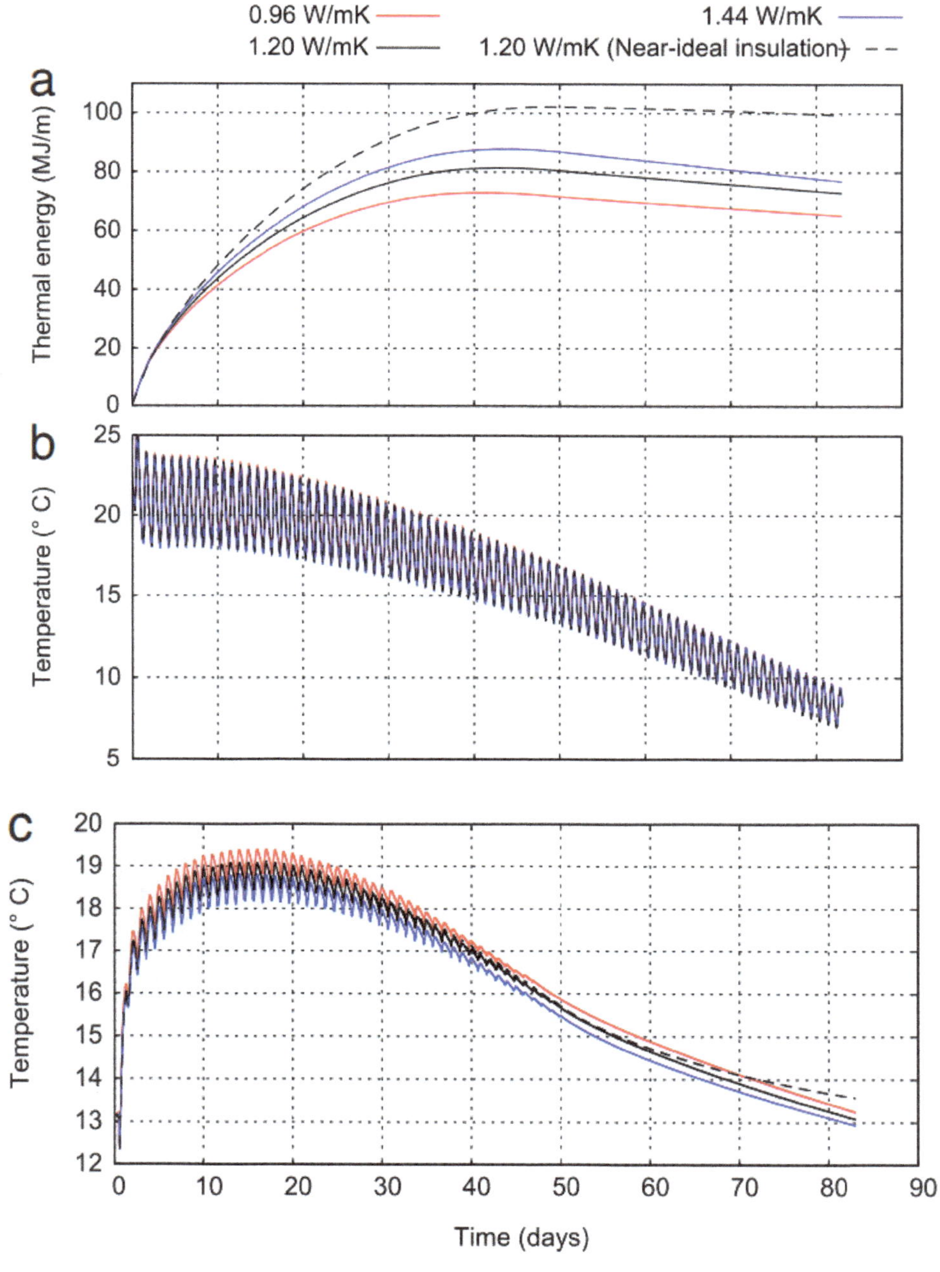

Figure 7. (a) Variation in time of thermal energy in region R2 for simulation Period 3 for three thermal conductivities and under near-ideal insulation conditions. (b) Corresponding temperature transients at collector depth (0.1 m) and (c) at storage pipes depth (0.875 m).

Varying Climatic Conditions

Simulations 5–9 investigate the effect of different climatic conditions on the thermal performance of an interseasonal heat storage facility. Figure 8 shows the thermal energy accumulated in the storage region during system activation in Period 3. It can be seen that the amount of energy collected in the storage region increases as the climatic conditions become warmer (cold to mild). However, it can also be seen that under hot climatic conditions the energy storage rate differs, being relatively slower at the beginning of the simulation period. This is believed to be related to the annual variation in air temperature as indicated by coefficients C and D (mid-summer daily average, mid-winter daily average) presented in Table 3. Under hot climatic conditions the variation of air temperature between seasons is ~5 °C while under cold and mild climatic conditions it is ~11 °C and ~12 °C respectively. This combined with the placement of insulation at the beginning of summer preventing the soil from naturally warming during the summer months allows a higher thermal gradient for the cold and mild conditions compared with hot conditions. However, this same seasonal variation prevents the system from collecting more thermal energy beyond day 42 as weather conditions cool with this change being more pronounced under cold and mild conditions. In a hot environment, although the weather conditions become cooler in winter, the change is smaller allowing the system to remain active and collect more energy during the whole simulation period. The results of *Simulations* 8 and 9 which explore the influence of the average air temperature amplitudes during summer and winter are shown in Figure 8. It can be seen that in general, higher summer air temperature amplitudes have a positive impact on the amount of energy that an interseasonal heat storage system is able to collect. However, the levels of thermal energy are still comparable.

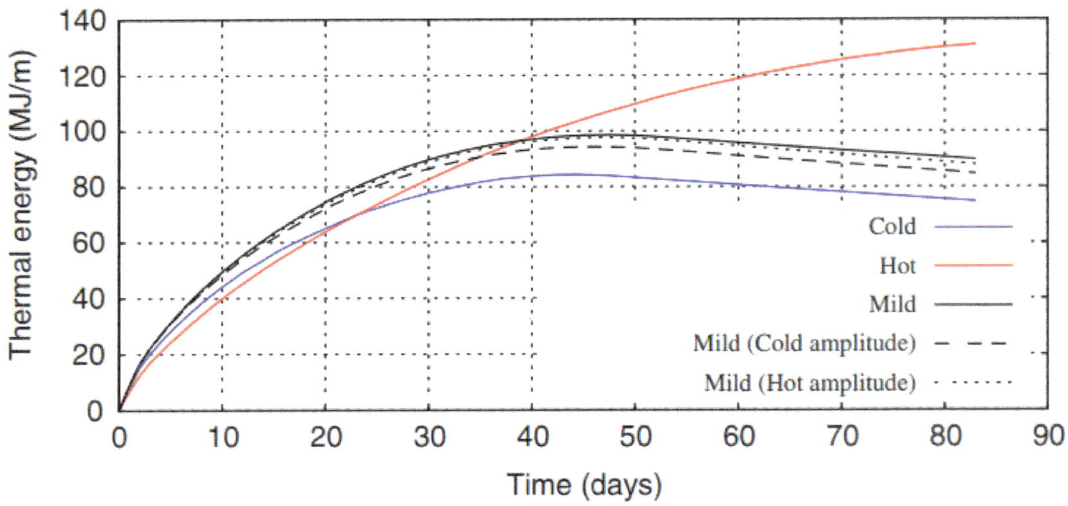

Figure 8. Variation in time of thermal energy accumulated in the storage region under different climatic conditions.

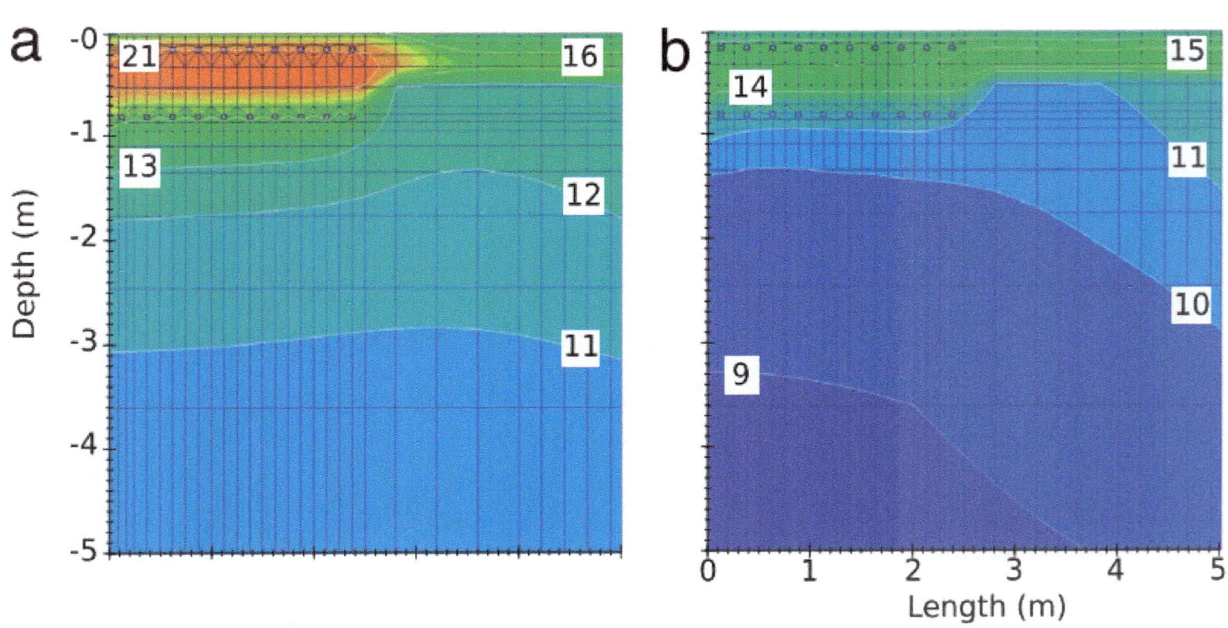

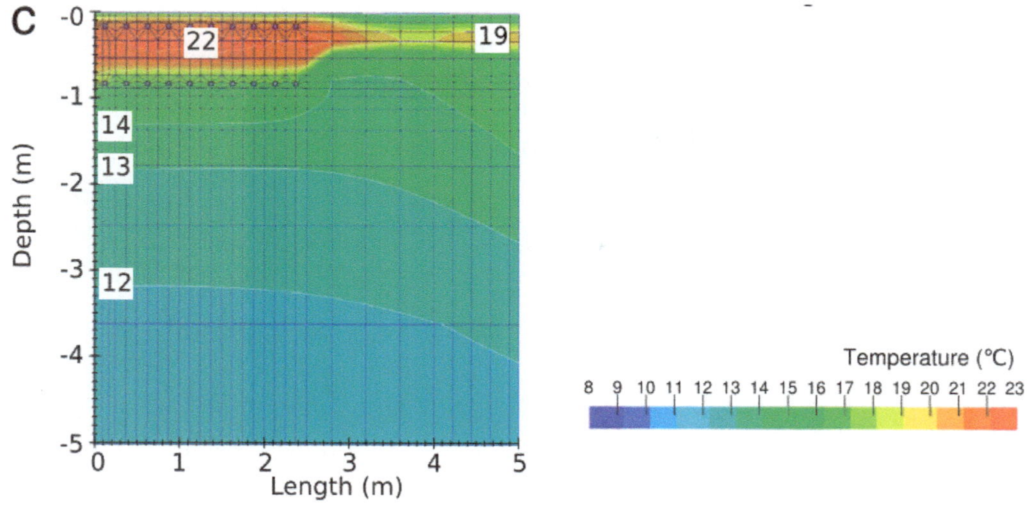

Figure 9. Temperature contours for region R1 at the beginning of simulation Period 3 using three different boundary conditions at the soil surface: (a) soil surface flux including vegetation cover model (*Simulation* 12), (b) fixed surface temperature (including pavement surface) defined by Eq. (17)(*Simulation* 10), (c) soil surface flux based on bare soil model (*Simulation* 11).

Different Soil Surface Conditions

The effect of different boundary conditions at the soil surface is considered in *Simulations*10–12. Figure 9 shows temperature contours in the region R1 (Figure 2(a)) at 0:00:00 h (end of Period 2/beginning of Period 3) on the day of system activation for the three different boundary conditions applied at the soil surface. The three temperature contours show significant differences, for example at the 5 m depth the vegetative cover case (Figure 9) simulates an almost homogeneous temperature of approximately 11 °C while the corresponding temperature under bare soil conditions (Figure 9(c)) is closer to 12 °C. Under fixed conditions, the temperature at this depth (Figure 9(b)) is not homogeneous varying between 9 °C and 10 °C. Surface temperature evolution at the center of the pavement (Point A) and at the far right hand side of the domain (Point B) during period 2 are shown in Figure 10. At the surface it can be seen that higher temperatures are obtained in the road section for both canopy covered and bare soil conditions (with the latter being higher), whereas for fixed conditions the temperatures are homogeneous (as expected). It is apparent that using suitable surface boundary conditions is of importance with significant differences in temperature fields obtained. It is worth noting that in the validation exercise, presented in Section 4, use of the more complex flux boundary condition to consider soil–atmosphere thermal interactions was necessary to correctly represent the observed behavior. A key point here is that, whilst air temperature data is readily available (from meteorological records) and surface type can be spatially defined, records of field measured soil–surface temperatures are much scarcer and require real-time measurement during a systems use. There is therefore a significant benefit in being able to predict soil surface temperatures rather than needing to define them *a priori* before any analysis.

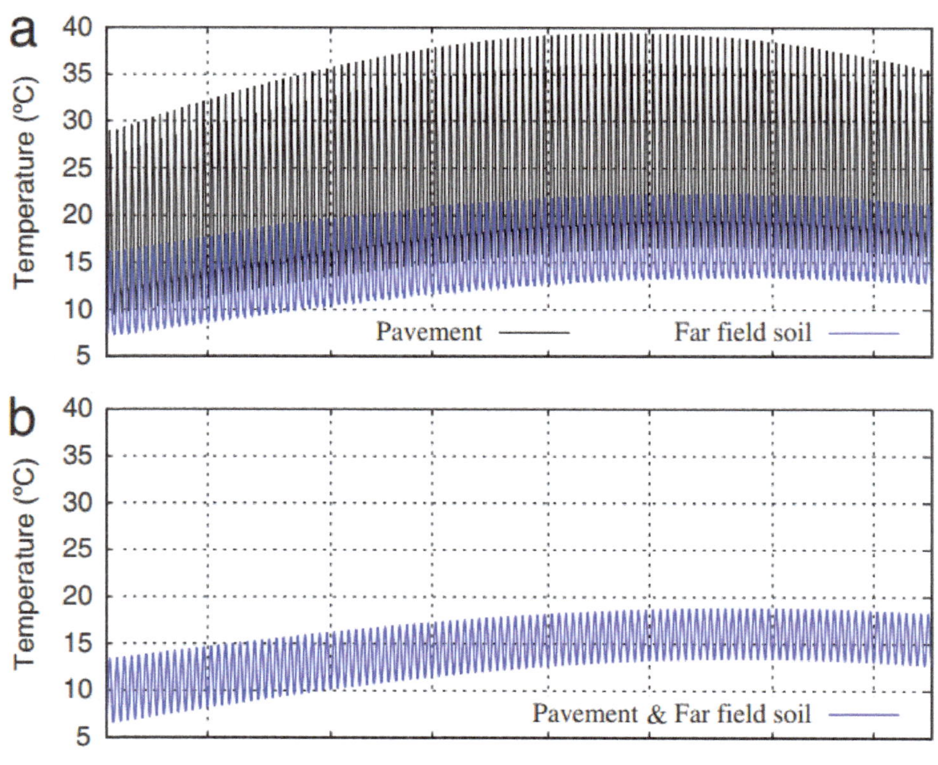

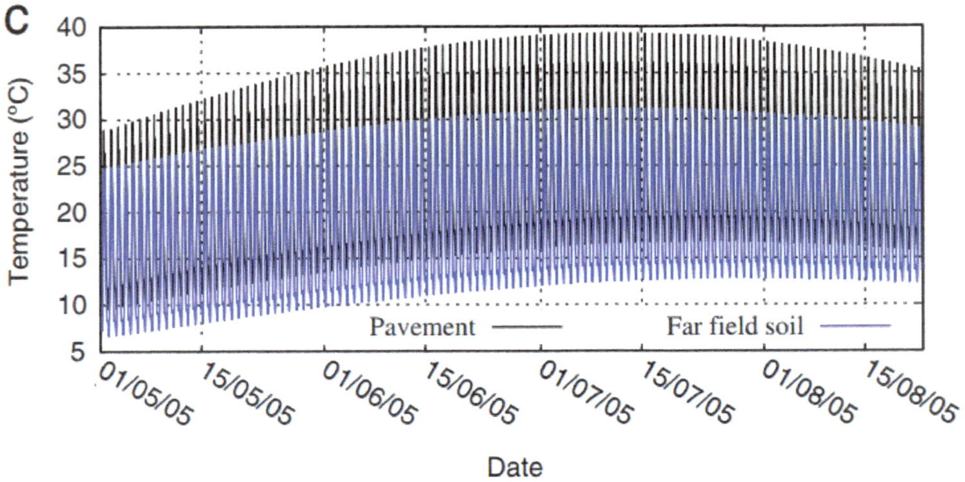

Figure 10. Surface temperatures at points A (Pavement) and B (Far field soil) for Period 2: (a) soil surface flux including vegetation cover model (*Simulation* 12), (b) fixed surface temperature (including pavement surface) defined by Eq. (17) (*Simulation* 10), (c) soil surface flux based on bare soil model (*Simulation*11).

CONCLUSIONS

In this paper, the influence of varying selected parameters (store thermal conductivity, climatic conditions and surface boundary conditions) on the performance of interseasonal heat storage systems has been investigated. The impact of different values of thermal conductivity in the storage region (affected by the installation process or prevention of natural wetting–drying process) is to reach higher temperatures close to the storage pipes (for low values of soil thermal conductivity) or an increased area of thermal influence (for higher values of soil thermal conductivity). In general higher values of thermal conductivity produce a higher amount of thermal energy being collected in the storage region. Despite this, however, the higher temperatures obtained close to the storage pipes for low thermal conductivities, might facilitate the extraction process or be more adequate for certain thermal applications. Subsequently, it can be concluded that the trade-off between total amount of thermal energy and temperature gradient needs to be optimized to the application under consideration.

Heat losses through a horizontal insulation layer have been found to be as high as 20% of the total thermal energy stored by the system. Higher thermal gradients near the insulation layer will increase these losses, while an increased storage area would require modifications to the insulation layer in order to decrease thermal energy lost through the edges. It is noted that use of innovative materials and systems where material properties can be controlled could lead to significant improvements in the system efficiency and these possibilities are currently under investigation.

The results under different climatic conditions have shown, as might be expected, that in general warmer weather conditions provide more thermal energy for potential storage. However, it was also observed that the seasonal variation of average air temperature has an effect on the rate of collection of thermal energy due to the temperature difference between the soil in the storage region and the air temperature. This effect is heightened if an insulation layer, which prevents natural warming of the soil underneath, is present keeping it close to winter temperatures.

The impact of using different boundary conditions applied on the soil surface has also been investigated. Two Robin-type boundary conditions have been compared, one including the presence of a layer of vegetation on the soil surface while the other assuming the soil as a bare surface. An additional boundary modelling technique has been considered by fixing the surface of the soil (including the paved surface) to the air temperature defined by a time varying mathematical expression. Using these approaches to obtain suitable initial conditions in the simulation periods before the activation of the interseasonal heat storage system have shown that the differences in temperature profiles (and thermal energy) generated in the soil domain are significant and that use of correct surface boundary conditions is critical in modelling the dynamics of the system under consideration.

Whilst it is recognized that the investigation undertaken here is limited to a particular system configuration, it is clear that it can be concluded that the performance of interseasonal heat storage systems is significantly affected by the meteorological conditions, surface flux boundary conditions and the thermal properties of the storage materials and that their correct representation is critical in both the assessment and design of such systems.

ACKNOWLEDGEMENTS

The authors gratefully acknowledge the support given to the first author whose Ph.D. studies were funded by CONACYT (the Mexican National Council of Science and Technology) and SEP (Mexican Secretariat of Public Education). Also the supply by TRL/HA of the source data published in (Carder et al. [5]), and the climatic data provided by Icelandic Meteorological Office (for the city of Reykjavík) and by the meteorological station staff from the School of Engineering of the Autonomous University of Yucatan (for the city of Merida) is gratefully acknowledged.

REFERENCES

1. Florides G, Kalogirou S. Ground heat exchangers—A review of systems, models and applications. Renew

Energy. 2007;32(15): 2461–2478. http://dx.doi.org/10.1016/j.renene.2006.12.014.

2. Wood CJ, Liu H, Riffat SB. An investigation of the heat pump performance and ground temperature of a piled foundation heat exchanger system for a residential building. In: The 3rd International Conference on Sustainable Energy and Environmental Protection, SEEP 2009. Energy. 2010;35(12):4932–4940. http://dx.doi.org/10.1016/j.energy.2010.08.032.
3. Rees SW, Adjali MH, Zhou Z, Davies M, Thomas HR. Ground heat transfer effects on the thermal performance of earth-contact structures. Renewable Sustainable Energy Rev. 2000;4(3):213–265. http://dx.doi.org/10.1016/S1364-0321(99)00018-0.
4. Zoras S. A review of building earth-contact heat transfer. Adv Build Energy Res. 2009;3(1):289–314. http://dx.doi.org/10.3763/ aber.2009.0312.
5. Carder DR, Barker KJ, Hewitt MG, Ritter D, Kiff A. Performance of an interseasonal heat transfer facility for collection, storage and re-use of solar heat from the road surface. Performance of an Interseasonal Heat Transfer Facility for Collection, Storage and Re-Use of Solar Heat from the Road Surface, Vol. 1, no. 1, 2007, pp. 1–114.
6. Ma X, Cheng B, Liu W, Li J. Simulation and analysis on the initial temperature profiles in soils. In: IEEE International Conference on Sustainable Energy Technologies, 2008. ICSET 2008, 2008, pp. 253–256. http://dx.doi.org/10.1109/ICSET.2008.4747012.
7. Qin Z, Berliner P, Karnieli A. Numerical solution of a complete surface energy balance model for simulation of heat fluxes and surface temperature under bare soil environment. Appl Math Comput. 2002; 130(1):171–200. http://dx.doi.org/10.1016/S0096-3003(01)00089-3.
8. Rajeev P, Chan D, Kodikara J. Ground–atmosphere interaction modelling for long-term prediction of soil moisture and temperature. Can Geotech J. 2012;49(9):1059–1073. http://dx.doi.org/10.1139/t2012- 068.
9. Liu C, Shi B, Tang C, Gao L. A numerical and field investigation of underground temperatures under Urban heat Island. Build Environ. 2011;46(5):1205–1210. http://dx.doi.org/10.1016/j.buildenv.2010. 12.015.
10. Yumrutaş R, Kanoğlu M, Bolatturk A, Bedir MŞ. Computational model for a ground coupled space cooling system with an underground energy storage tank. Energy Build. 2005;37(4):353–360. http://dx.doi.org/10.1016/j.enbuild.2004.07.004.
11. Shang Y, Li S, Li H. Analysis of geo-temperature recovery under intermittent operation of ground-source heat pump. Energy Build. 2011; 43(4):935–943. http://dx.doi.org/10.1016/j.enbuild.2010.12.017.
12. Wu Y, Gan G, Verhoef A, Vidale PL, Gonzalez RG. Experimental measurement and numerical simulation of horizontal-coupled slinky ground source heat exchangers. Appl Therm Eng. 2010;30(16): 2574–2583. http://dx.doi.org/10.1016/j.applthermaleng.2010.07.008.
13. Esen H, Inalli M, Esen M. Numerical and experimental analysis of a horizontal ground-coupled heat pump system. Build Environ. 2007;42(3):1126–1134. http://dx.doi.org/10.1016/j.buildenv.2005. 11.027.
14. Inallı M, Esen H. Experimental thermal performance evaluation of a horizontal ground-source heat pump system. Appl Therm Eng. 2004; 24(14–15):2219–2232. http://dx.doi.org/10.1016/j.applthermaleng.2004.01.005.
15. Congedo PM, Colangelo G, Starace G. CFD simulations of horizontal ground heat exchangers: A comparison among different configurations. Appl Therm Eng. 2012;33–34(February):24–32. http://dx.doi.org/10.1016/j.applthermaleng.2011.09.005.
16. Ramírez-Dávila L, Xamán J, Arce J, Álvarez G, Hernández-Pérez I. Numerical study of earth-to-air heat exchanger for three different climates. Energy Build. 2014;76(June):238–248. http://dx.doi.org/10. 1016/j.enbuild.2014.02.073.
17. Muñoz Criollo JJ. An investigation of inter-seasonal near-surface ground heat transfer and storage (Ph.D), Cardiff University; 2014:http://orca.cf.ac.uk/73226/.
18. Cengel YA. Heat Transfer: A Practical Approach. 2nd ed. McGraw-Hill; 2003.
19. Herb WR, Janke B, Mohseni O, Stefan HG. Ground surface temperature simulation for different land covers. J Hydrol. 2008;356(3–4): 327–343. http://dx.doi.org/10.1016/j.jhydrol.2008.04.020.
20. Jansson C, Almkvist E, Jansson P. Heat balance of an asphalt surface: Observations and physically-based simulations. Meteorol Appl. 2006; 13(2):203–212. http://dx.doi.org/10.1017/S1350482706002179.
21. North GR, Erukhimova TL. Atmospheric Thermodynamics: Elementary Physics and Chemistry. Cambridge University Press; 2009.
22. Deardorff J. Efficient prediction of ground surface-temperature and moisture, with inclusion of a layer of vegetation. J Geophys ResOceans Atmos. 1978;83(NC4):1889–1903. http://dx.doi.org/10.1029/JC083iC04p01889.
23. Best MJ. A model to predict surface temperatures. Bound-Layer Meteorol. 1998;88(2):279–306. http://dx.doi.org/10.1023/A: 1001151927113.
24. Cleall PJ, Muñoz-Criollo JJ, Rees SW. Assessment and representation of thermal surface fluxes in soils. In: Computer Methods Recent Advances Geomechanics. Kyoto, Japan: CRC Press; 2014:1321–1326, http://www.crcnetbase.com/doi/abs/10.1201/b17435-233.
25. Cleall PJ, Muñoz-Criollo JJ, Rees SW. Analytical solutions for ground temperature profiles and stored energy using meteorological data. Transp Porous Media. 2014;106(1):181–199. http://dx.doi.org/10.1007/s11242-014-0395-3.
26. Bangerth W, Hartmann R, Kanschat G. deal.II—a general-purpose object-oriented finite element library. ACM Trans Math Softw. 2007;33(4):http://dx.doi.org/10.1145/1268776.1268779. Article 24 (Preprint).

27. UK Meteorological Office. Met Office Integrated Data Archive System (MIDAS) Land and Marine Surface Stations Data (1853-Current). NCAS British Atmospheric Data Centre.
28. Online., 2012. Available at: http://badc.nerc.ac.uk/view/badc.nerc.ac.uk__ATOM__dataent_ukmomidas
29. Accessed: 03-Apr-2014..

CHAPTER 5 ROCKSLIDES ON LIMESTONE CLIFFS WITH SUBHORIZONTAL BEDDING IN THE SOUTHWESTERN CALCAREOUS AREA OF CHINA

Z. Feng[1], B. Li[1], Y. P. Yin[2], and K. He[3]

[1]Key Laboratory of Neotectonic Movement and Geohazard of MLR at Institute of Geomechanics, Chinese Academy of Geological Science, Beijing, 100081, China

[2]China Institute of Geo-Environment Monitoring, Beijing, 100081, China

[3]Chang'an University, Xi'an, 710054, China

ABSTRACT

Calcareous mountainous areas are highly prone to geohazards, and rockslides play an important role in cliff retreat. This study presents three examples of failures of limestone cliffs with subhorizontal bedding in the southwestern calcareous area of China. Field observations and numerical modeling of Yudong Escarpment, Zengzi Cliff, and Wangxia Cliff showed that pre-existing vertical joints passing through thick limestone and the alternation of competent and incompetent layers are the most significant features for rockslides. A "hard-on-soft" cliff made of hard rocks superimposed on soft rocks is prone to rock slump, characterized by shearing through the underlying weak strata along a curved surface and backward tilting. When a slope contains weak interlayers rather than a soft basal, a rock collapse could occur from the compression fracture and tensile split of the rock mass near the interfaces. A rockslide might shear through a hard rock mass if no discontinuities are exposed in the cliff slope, and sliding may occur along a moderately inclined rupture plane. The "toe breakout" mechanism mainly depends on the strength characteristics of the rock mass.

INTRODUCTION

The southwestern calcareous area of China covers a large area of 54.4×10^3 km^2, and it is highly prone to geohazards. Many multilayered carbonate rocks with sub-horizontal bedding have been deeply cut by rivers during crustal uplift and form significant topographic relief in steep slopes and cliffs. These sub-horizontal cliff slopes are usually dominated by two sets of sub-vertical conjugate joints and are characterized by slightly folded or faulted to massive rock masses. Rockslides from sub-horizontally bedded cliff failure and resulting catastrophes have occurred widely and frequently in the southwestern calcareous area of China.

Considerable research on cliff failure has been conducted in similar areas around the world (Abele, 1994; Von, 2002; Rohn, 2004; Embleton, 2007; Ruff et al., 2008; Palma et al., 2012), such as the North Calcareous Alps in Austria. It is believed that water, lithology, geological structure, and karstification are of primary importance in triggering rockslides from cliff failure (Kay et al., 2006). These factors dominate the tearing and shearing failure mechanism of the rock masses and joints, which are directly displayed in the consequent failure

Citation: Feng, Z., Li, B., Yin, Y. P., and He, K.: Rockslides on limestone cliffs with subhorizontal bedding in the southwestern calcareous area of China, Nat. Hazards Earth Syst. Sci., 14, 2627-2635, doi:10.5194/nhess-14-2627-2014, 2014.

behavior of the cliff slopes. Types of rockslides with different detachment mechanisms (slumps, plane slides, topples, and lateral spreading) in sub-horizontally bedded cliffs with particular geological settings and triggering mechanisms have been described by engineering geologists, e.g., the collapse of the Mt. Sandling and Mt. Raschberg limestone towers had a complex failure sequence from lateral spread to toppling followed by rock fall (Rohn et al., 2004). In addition to lateral spreading, a cliff slope with a geological formation of hard rock on a soft base may undergo the translational sliding or slumping of slab-shaped blocks (Poisel, 2005). In addition, the karst process is an inevitable factor in rockslide formation. Tectonic joints keep widening and extending to the deep of rock mass by constant dissolution and disgregation of underground water, forming boundaries of perilous rocks (Santo et al., 2007). It is believe that karst plays an important role in rockslides from carbonate mountains, especially for gently bedding-inclined slopes which are unfavorable to failure.

Active underground mining is widespread in the southwestern calcareous area of China. There is the potential for damage to cliff faces and overhangs when mining activity occurs beneath cliffs. This has been repeatedly shown to be true. Taking the Southern Coalfields of Australia as an example, several types of cliff failure have been witnessed in the goaf area, although no instabilities have been reported beyond the mining area (Kay, 2006). However, there is no available and widely accepted model that can predict the failure susceptibility of steep slopes close to mining. This is because of the complex interaction of factors influencing the stability of steep slopes, which include geometry, geology, geological structure, environmental factors, and technical mining parameters. An integrated method combining landslide science and mining subsidence science is promising as a direction for future research.

This study focuses on the failure mechanisms of cliffs with sub-horizontal bedding in the southwestern calcareous area of China and the recognition of these features in field investigations. The three cases presented occurred during the last 10 years and caused great damage to both human life and assets. Furthermore, post-failure behavior, such as rock avalanches and debris flows, is not included in descriptions of failure mechanisms of steep slopes. The term "failure mechanism" in this study particularly refers to the detachment mechanism of rock masses from the cliff slope.

Geological Background

The likelihood of rockslide instability in the southwestern calcareous China is high, mainly because of the local geological evolution. Tectonic movement during the Mesozoic was characterized by compression and formed the area into a fold belt, mainly striking NNE. Neotectonic uplift raised the thick carbonate rocks to high altitude and shaped the steep and folded landforms.

Many layers of carbonate rocks were deposited from Permian until Triassic time, and were interbedded with weak planes or soft strata. The carbonate rock masses possess great strength and integrity; thus, they usually form steep slopes hundreds of meters high, e.g., the Three Gorges, if there are no intercalated soft strata. Under these circumstances, slope movement is mainly controlled by discontinuities such as weak interlayers, karst, and fissures. The thick carbonate succession in the southwestern calcareous area of China contains multiple layers of weak shale planes, including carbonaceous shale and pyrobituminousshale. The strength of the shale planes is relatively low and significantly varies with the weathering process, from virgin rock to fractured planes to argillated layers. The strength of carbonaceous shales sandwiched in the Lower Permian limestone at Lianziya Cliff is low that the cohesion and internal friction angle are 0.08–0.39 MPa, 18–21° and 0.06–0.078 MPa, 18–19.8° for dry and saturated samples, respectively (Ding et al., 1990). The weak interlayers play a significant role in mass movement. When soft strata underlie hard rock, the yielding of the soft base may cause the uneven subsidence of the cap rock. Lateral spreading and slumps with backward tilting are caused by this type of destabilization mechanism, in which failure propagates uphill.

In the southwestern calcareous area of China, coal measures are common soft strata, accompanied by thick carbonate sequences. Except for coal seams, the coal measures are usually composed of associated interlayers of shale, carbonaceous shale, pyrobituminousshale, and bauxitic claystone. Mining is active in these coal measures using the techniques of room and pillar mining and longwall mining. The depth of cover ranges from dozens to hundreds of meters. It is believed that underground mining activities can change the engineering geological conditions and reduce the stability of cliffs (Tang, 2009; Altun et al., 2010; Marschalko et al., 2012; Lollino et al., 2013). Cliff failures and the resulting catastrophes, examples of which are discussed below, are related to underground mining.

Karstification is a significant factor when discussing the reasons for carbonate slope failure. It particularly affects steeply inclined faults and tectonic joints and slowly widens the discontinuities to create large open fractures, which foster rockslides (Santo el al., 2007; Parise, 2008). In addition, the karstification process can cause a significant reduction in the mechanical properties of the carbonate rock mass. This is very important for toe-constrained slides, which depend on the strength of the rock mass at the toe. In some cases, the underground karst voids play the same role as goaf, and cavern breakdown may also lead to landslides (White et al., 1969; Parise, 2010).

FAILURE MODE

Different types of rockslides have been observed in thickly and sub-horizontally bedded limestone escarpments. Gently inclined bedding is unfavorable for large translational slides. Slide, collapse, and slump slope failures are discussed below in detail to explain their mechanisms. Three examples are given: the Yudong Escarpment, Zengzi Cliff, and Wangxia Cliff (Figure 1).

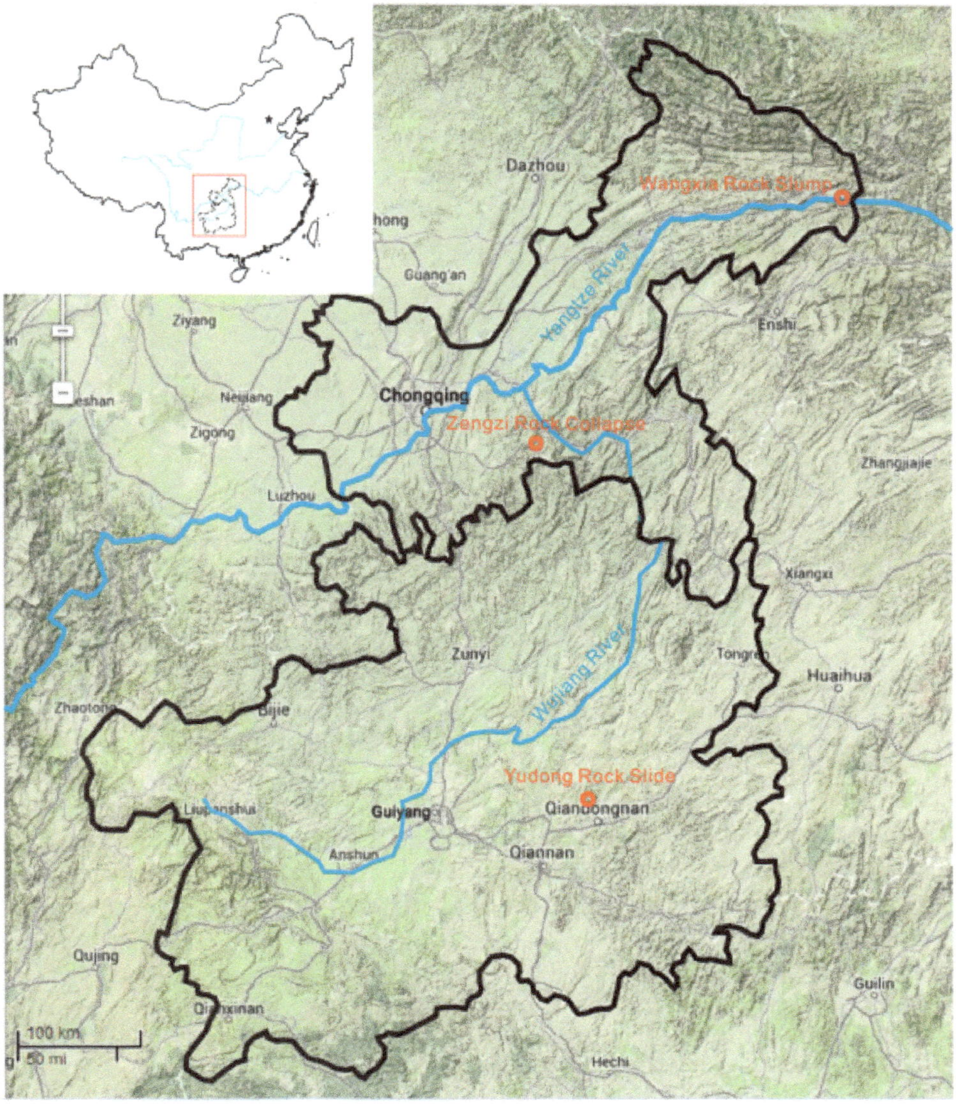

Figure 1. Locations of three rockslides: Wangxia Rock Slump, Zengzi Rock Collapse, and Yudong Rock Slide.

Rock Slide at the Yudong Escarpment

On 18 February 2013, a rock slide occurred in Longchang County, Guizhou Province. The mass movement is located in a high and steep bank of the Yudong River, close to an underground mining area (Figure 2). The S308 provincial road runs on the opposite bank of the river. The rock slide buried several houses and five people beneath the escarpment and blocked the river.

The Yudong Escarpment is about 220 m high and the slope angle is more than 80°. The steep slope faces in the direction SE100–130°, and the dip direction is NW325°, with slightly inclined bedding. The outcrop consists of thickly bedded P 2 1 limestone and underlying P 1 1 coal measures. The jointed limestone is moderately weathered, and the uniaxial compression strength of intact rock reaches 30 MPa. The coal measures are composed of thickly bedded argillaceous shale and coal seams. Coal extraction has been active in the soft basal unit, and the goaf is about 200 m deep behind the cliff face. The Yudong Escarpment lies in the gentle eastern flank of the Yudong Syncline, which strikes NNE. As a result, a set of NNE-trending joints and a conjugate set of NW-trending joints are present in the rock mass. The orthogonal 130° ∠70° and 80° ∠75° joints cut the 325° ∠9° rock beds into prism- and tower-shaped blocks on the edge of the cliff. The "2.18" Yudong rock slide originates from one of these unstable blocks with a volume of about 30×10^4 m ^{3}. Several types of karst are observed in the rock slide area, including sinkholes, karst tunnels, and dissolution fissures. A remote sensing image taken the day after the rock collapse shows a sinkhole at the crest with a diameter of 2.2 m immediately behind the fall. The pipe flow and induced groundwater dynamics acted on the unstable rock block and changed the instability. However, because of abundant vegetation, the sinkhole is not visible in a previous image that was taken in the summer of 2012 (Figure 3). The intense karstic erosion of a yellowish-orange color has been observed on the vertical scar. The selective karst widens and connects the pre-existing steep joints so that the area of the rock bridge covers less than 30 % of the main scar (Huang, 2013). The directed scratches indicate the brittle failure of the rock bridges and consequent fall of the rock mass. The rock slide left two scars originate from joints. The main scar is nearly parallel to the cliff face and about 80 m wide. The upper part of the main scar is a sub-vertical plane, and the lower parts are planar and dip out of the cliff. The conjugate side scar is about 24 m wide and perpendicular to the cliff face (Figure 2).

Figure 2. Photograph showing an overview of Yudong rock slide I.

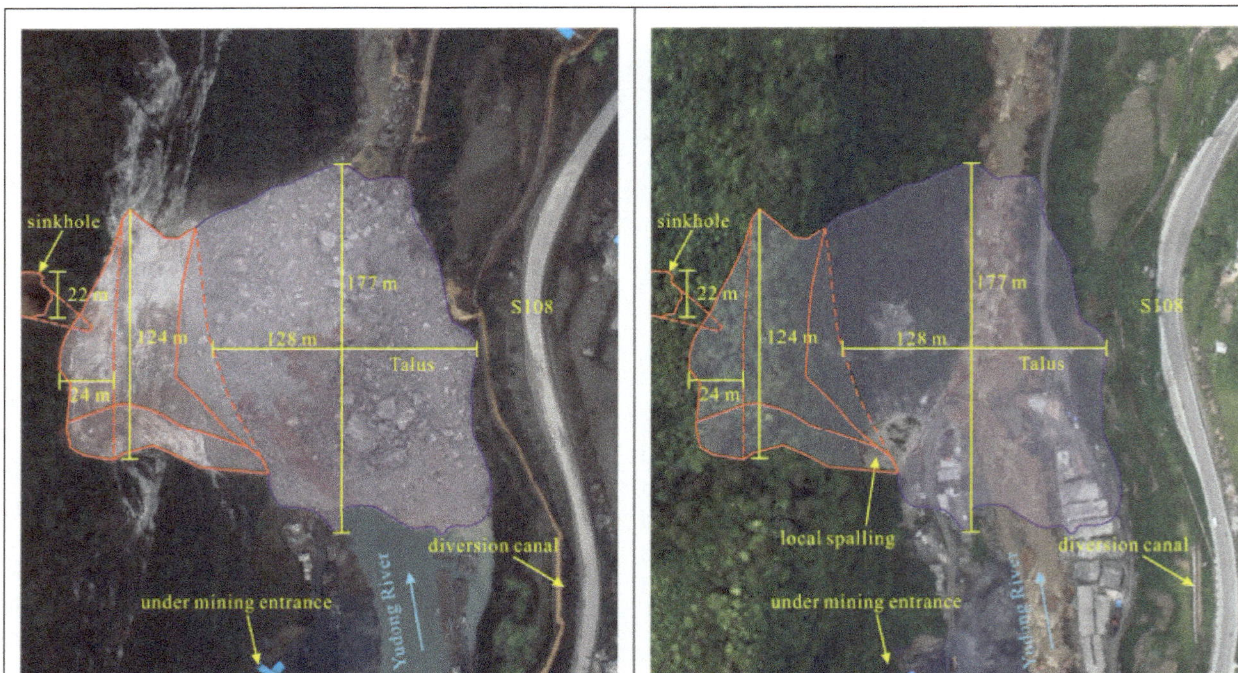

a	b

Figure 3. Aerial photos of Rock Slide Itaken in summer 2012 (a) and on 19 February 2013 (b). The zoning of the depleted mass and talus is projected in picture (a) prior to the catastrophe. The abundant vegetation and steep terrain make it difficult to distinguish the potential failure. However, there were signs of local spalling at the toe of the cliff face (a). A 22 m-diameter sinkhole was exposed in winter, which connected with the main back scar of the rock slide (b).

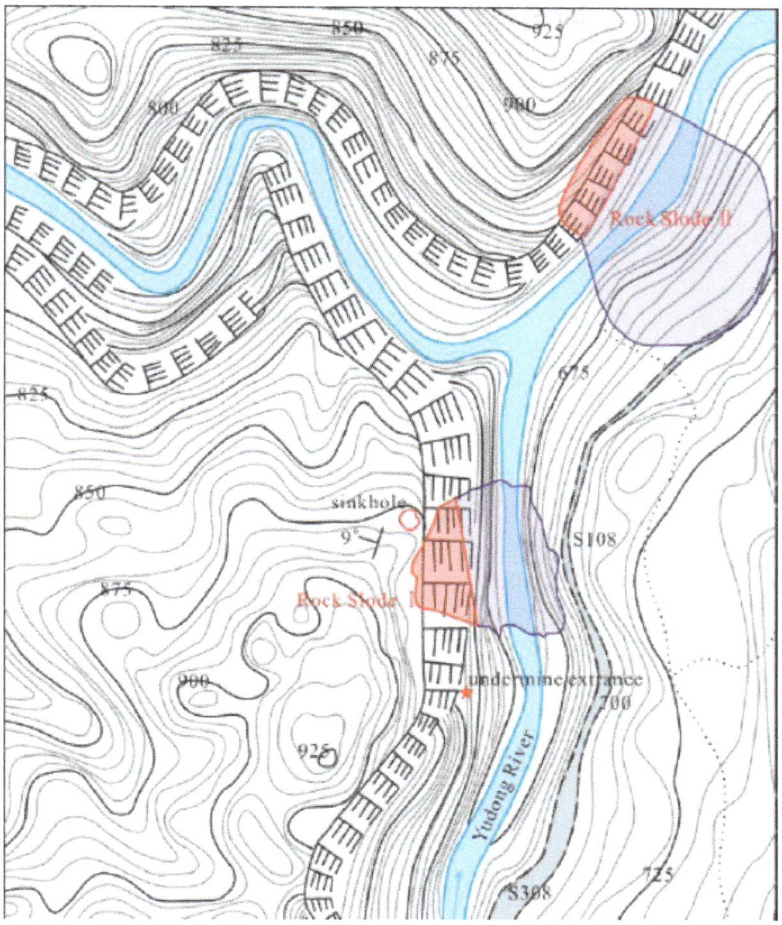

Figure 4. Rock slide II is located north of rock slide I and has the same geological conditions, except for undermining. The volume of the talus of rock slide II is 30×10^4 m^3. It blocked the Yudong River and covered 80 m of the S308 road.

Another massive rock slide (II) (Figure 4) occurred on 16 April 2013, about 100 m away from the 18 February rock slide (I). The morphology of the vertical scar of rock slide II is similar to that of rock slide I. The lower rupture surface forms a steep and irregular scar (Figure 5).

Figure 5. Photograph showing rock slide II at Yudong. The pre-existing vertical scars are coated with karst of a yellowish-brown color, while the rupture surfaces are white and gray, indicating brittle failure and planar sliding through the limestone.

These two rock slides have the same failure mechanism of a steep back scar separating the unstable block from the escarpment and the block breaking through the toe leading to free fall of the rock mass. The rupture surface implies a plane slide involving shear failure through the rock mass. There is little or no shear displacement along the rupture surface, and the velocity is very high. The brittle failure of the rock mass in the toe area can be explained by the brittle failure of the rock mass under uniaxial compression tests. In other words, it is largely dependent on the mechanical properties of the rock mass at the toe. The toe is sheared along random discontinuities of limited persistence traced by the rupture surface (Figure 6).

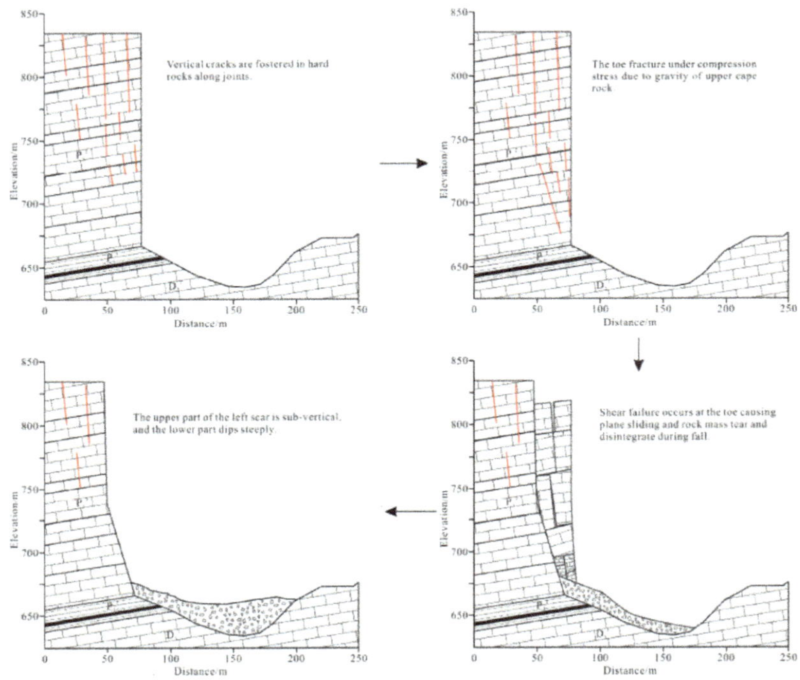

Figure 6. The evolutionary process of toe shear and planar sliding of the Yudong rockslides.

Rock Collapse at Zengzi Cliff

Zengzi Cliff lies in the gentle and wide core of the Jinfo Mountain Syncline, Nanchuan County of Chongqing, in a "hard on soft" landform. The vertical limestone consists of two platforms separated by softer rock (Figure 7). Two major tectonic joints in the hard rocks strike at N40–50° E and N30–50° W, with dip angles of 70–88°. These two sets of joints are approximately orthogonal to each other and perpendicular to bedding. The bedding trends at 300–305° with a very low inclination (4–7°). The upper platform is U in shape; hence, the steep slope varies from anaclinal to plagioclinal and cataclinal in different parts of the edge. Beneath the cliff, S1 silty shale forms a soft base and a gentle slope (20–30°). Talus occurs everywhere under the cliffs, indicating that cliff failure occurs as scarp retreat.

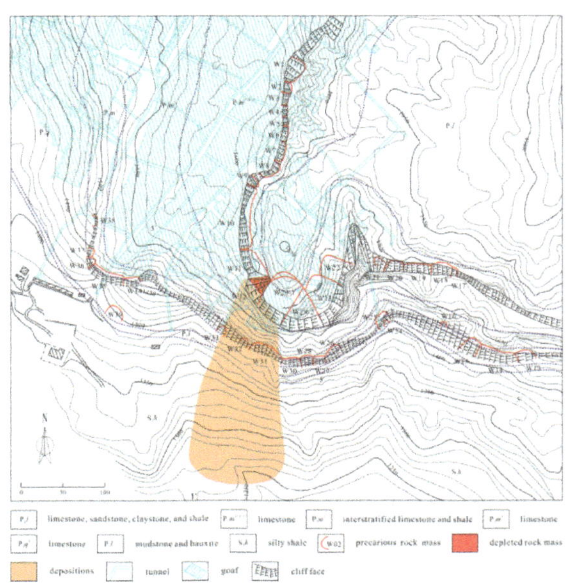

Figure 7. Topographic map of the Zengzi Cliff area and the distribution of precarious rock masses on its edge. Room and pillar mining has been conducted in the P1 l sequence since 1983, and the goaf extends from south to north. The depth of the goaf is about 100–120 m for the first step and 350–400 m for the second step.

Rocks frequently fall in the Zengzi Cliff area. On 12 August 2004, a massive rock collapse occurred in the upper platform, which is about 200 m high (Figure 8). The depleted mass is a prism block shaped by sub-vertical joints and bedding. Two vertical back scars and a groove surface in the underlain soft strata are exposed after the collapse. Large areas of the scars are coated with karst argillaceous fillings and yellowish-orange calcite, while the rock bridge failure scars are fresh. Field investigations showed that the head scars are wide open prior to the massive collapse. Seepage forces and frost from percolating water acted on the prism block over a long time and influenced the stability; however, only karst could alter the conditions in the rock adjoining the slope by increasing the connectivity of back scars. The underlying soft strata consist of alternating beds of thin-bedded shale and medium-bedded limestone, and as a result of weathering, form a 40 m-high slope.

a. Elevation view of Zengzi Cliff

b. Side view of the rock collapse

Figure 8. (a) Alternating strata of hard rocks and weak rocks; (b) Cracks (red lines) originated near the interface between the hard cap and weak base and propagated uphill.

Mining activity has been conducted for decades, but community monitoring started in 2001. The opening velocity of the head scar was very slow from 2001 to 2003 (Ren, 2005). It increased to 4–15 mm (10 d)$^{-1}$ in the period between April and July 2004 (Figure 9), and signs of pre-collapse, (rock falls) were repeatedly observed. At that time, the government issued warnings and evacuated residents and workers. The velocity abruptly increased from 10 August, and the increment reached 658 mm on the last day. The process of the Zengzi rock collapse was captured on video: the tower dropped vertically and disintegrated while falling. It is unusual that the failure initiated in the bottom of the hard block rather than the underlying soft strata. The splitting of the hard rock mass at the toe led to tower collapse. The curved surface in the soft strata will have been carved by collision.

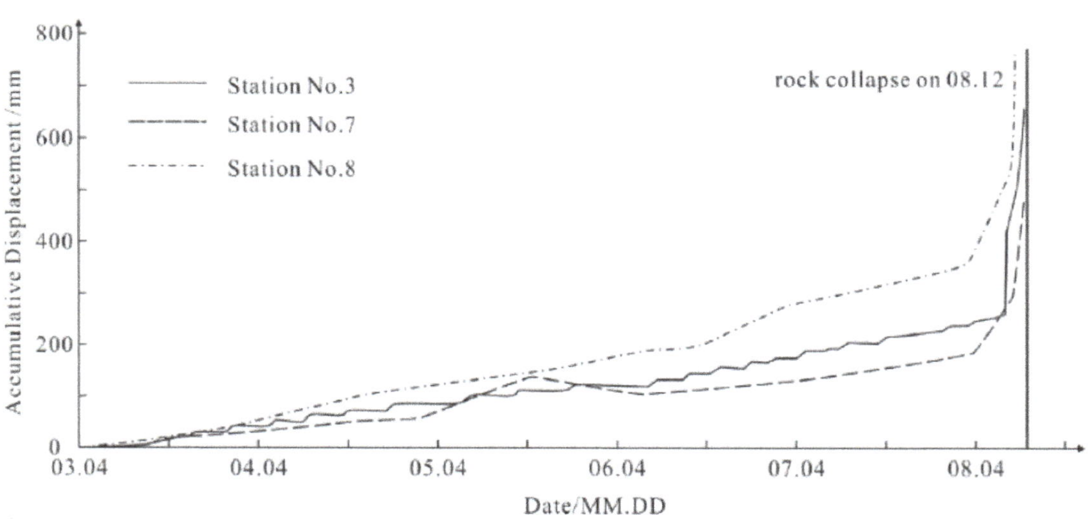

Figure 9. Cumulative displacement of monitoring stations at the crest of Zengzi Cliff in 2004 (Ren, 2005).

Tensile splitting has been observed in uniaxial compression tests of rock samples with high brittleness and strength, such as limestone. There are several reasons why the collapse did not start with shear failure in the underlying strata. The rock mass in the bottom of the tower block is in poor condition: fractured and weathered. Drill cores obtained strongly karstic rock containing dissolution pores, caves, tufa, and calcite. The drilling also revealed at least 11 fissures, some of which were filled with yellow clay (107 Geology Team, 1995). This is because the underlying strata are low-permeability soft rocks; hence, there are steady water flows immediately above. Furthermore, the soft strata are not sufficiently weak. These strata consist of interbedded thin layers of shale (0.1–0.2 m) and medium-thick limestone. The tower block could easily shear through the shale but not the intercalated limestone. Under these circumstances, tensile split failure of the tower bottom is a reasonable failure mechanism, involving compression fracturing and horizontal shearing in the shale beds (Figure 10). Field studies and deformability tests on rock masses all around the world have demonstrated that folded and flat-lying rock masses are prone to tensile splitting near thin weak planes. Compression fracturing and tensile splitting are important failure mechanisms for subhorizontally bedded slopes. The Zengzi cliff collapse exposed back scars in the nearby deformable rock mass, which propagated uphill (Figure 8b), and a fractured rock mass was observed near the shale interface.

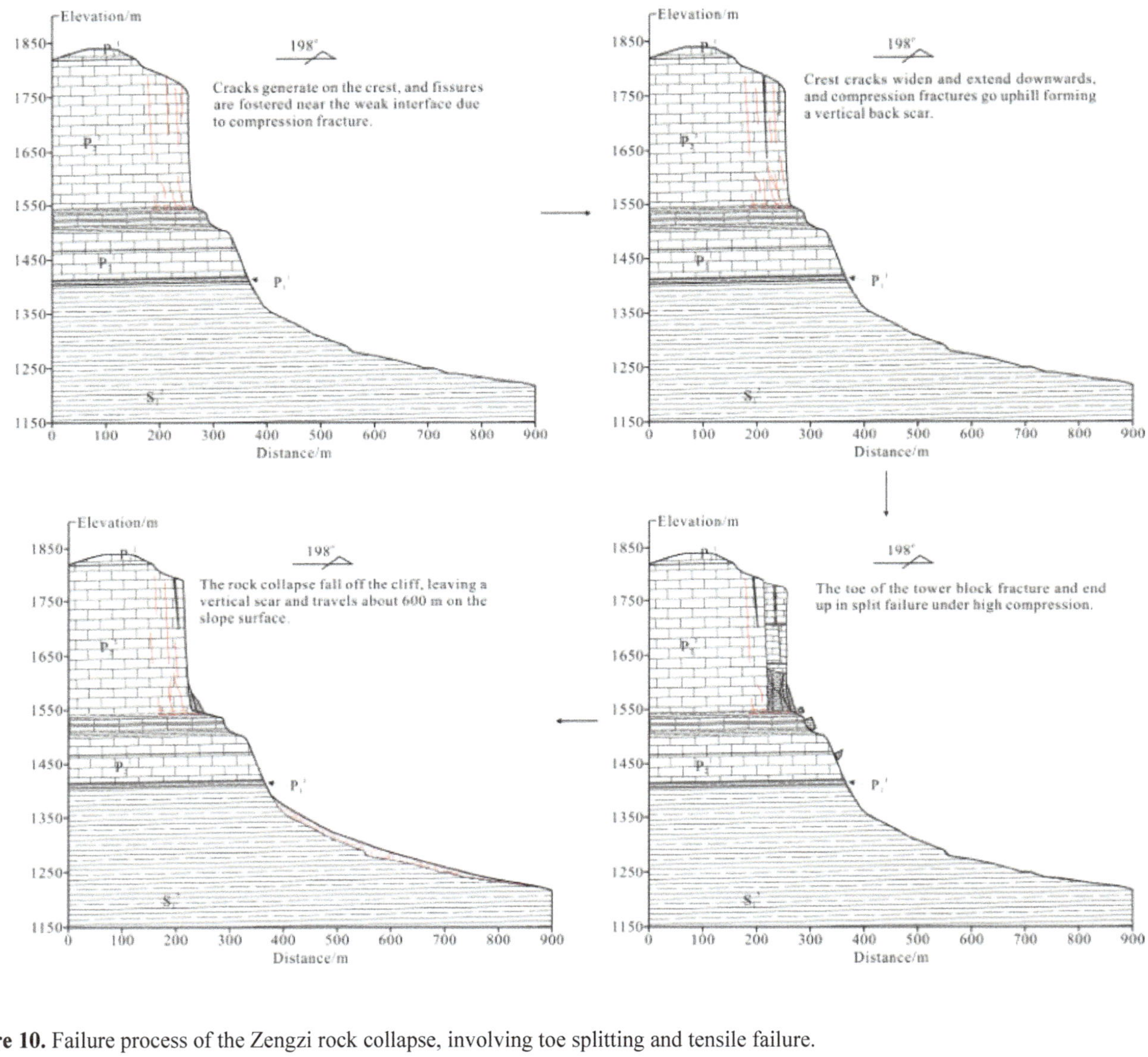

Figure 10. Failure process of the Zengzi rock collapse, involving toe splitting and tensile failure.

Rock Slump at Wangxia Cliff

When the base is sufficiently weak, and there is a steep fracture separating the column from the slope in the cap area, a rock slump with back-tilting mechanism is likely to occur. The Wangxia Cliff failure slump mechanism involved isolated limestone block breaking through shale at the toe, with rotational movement. The Wangxia Cliff is situated at the top of the right bank of the Yangtze River in Wushan County of Chongqing, about 1000 m above the water level (Figure 11). Located in the flat core of the Hengshixi Fold, the bedding is slightly inclined (3–8°) and strikes at 335–340°, opposite to the cliff face. Interbedded shales, mudstones, and coal seams form a gentle slope and separate the Upper Permian limestone into two cliff steps. A country road passes over the slope below the 70–75 m high limestone escarpment, which contains several isolated slab- and prism-shaped blocks. On 21 October 2010, a prism with a volume of 7×10^4 m^3 became a rock slump failure.

Figure 11. Overview of Wangxia Cliff prior to the catastrophic rock slump. G01 represents a GPS station, and Lw02–Lw07 indicate the displacement monitoring stations. The rock slump of 21 October 2010 was controlled by vertical scars T11 and T12. The red lines represent fissures in the rock mass.

The slope movement can be traced back to 18 June 1999. Four collapsecraters and nine cracks were observed in the crest area after 108.5 mm of rainfall on 15 and 16 June (Le et al., 2011). Intensive deformation began on 21 August 2010, after four days of concentrated rainfall. Repeated pre-collapse signs were observed. Crown cracks widened, and new cracks occurred on the crest. Rocks fell off from the cliff face and vertical scars in the rock mass. In addition, gravelly soils flowed out from steep cracks at the toe. Transverse ridges and cracks were observed on the country road beneath. The mass movement accelerated as a result of 70 mm of precipitation from 10 to 13 October. The fissures in the isolated blocks extended and widened. Rock falls became more obvious both in volume and frequency. Finally, a massive rock slump occurred on 21 October, in which the isolated prism slid downhill, breaking through the toe and leaning against the back scars. The underlying weak rocks were squeezed out and scattered over the slope surface. Displacement monitoring showed that the crest was dominated by subsidence, while the horizontal and vertical displacements were about the same at the base. The incompetent base showed horizontal movement and little subsidence. These phenomena were caused by ductile yielding and rotational shearing in the incompetent base (Figure 13). The squeezing out contributed to the uphill propagation of cracks and disintegration of the mountain into slab- and prism-like blocks.

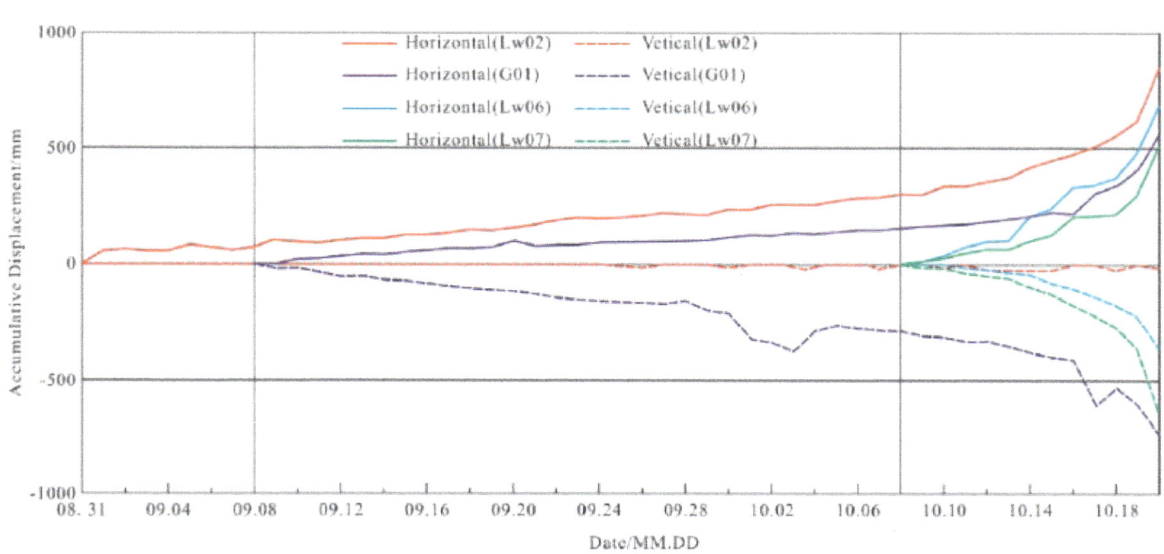

Figure 12. Cumulative displacement at monitoring stations at the Wangxia rock slump (Le et al., 2011).

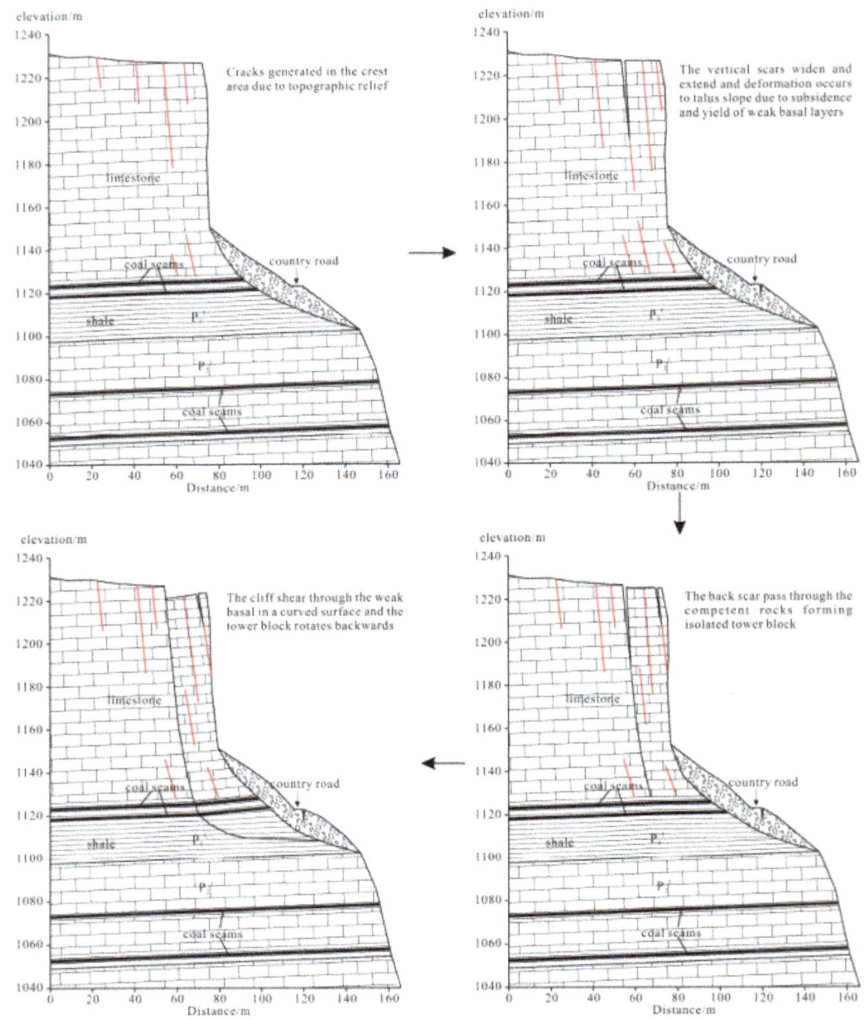

Figure 13. Deformation and failure processes of the Wangxia rock slump.

It is worth mentioning that there is a good correlation between the acceleration of displacement and concentrated rainfall. Because of cracks and fissures in the limestone, water could easily and rapidly percolate into the weak strata. The unconfined compressive strength (UCS) of the shale shows a substantial decrease when saturated, and the softening factor is about 0.62. However, there is a 2–4 day time lag between concentrated rainfall and acceleration of displacement, because the infiltration of water into poorly permeable shale takes time.

Numerical Back-Analysis

Numerical analysis is a sophisticated method for assessing the potential failure modes of rock slopes. We use it herein to validate the cases discussed above, which show different failure mechanisms for sub-horizontally bedded cliffs (compound slide, rock collapse, and rock slump) and to explain the backgrounds to the slides. The computation was performed using the computer program UDEC; the parameters used in the simulation are given in Table 1.

Table 1. Parameters used in the numerical simulation

Parameter	Limestone	Shale		
		Rock slide	Rock collapse	Rock slump
Thickness (m)	170	20	2	20
Density(kg m^{-3})	2700	2640		
Young's modulus, E(GPa)	65	5		
Poisson's ratio	0.13	0.2		
Friction angle of rock mass (°)	32	28	20	25
Cohesion of rock mass (MPa)	1.3	0.8	0.4	0.5
Tensile strength of rock mass (MPa)	0.3	0.1	0.04	0.04
Friction angle of joints (°)	30	25		
Cohesion of joints (MPa)	1.0	1		
FOS	–	1.01	1.21	1.07

The natural stress field is characterized by a tensile zone on the crest and stress concentration at the toe. When a thick incompetent basal layers is present, the jointed slope tends to fail as a slump. A yielding curved surface gradually emerges in the basal layers under long-term gravitational compression of the upper cap rock. A large plastic zone is present in the weak basal layers and nearby hard rock mass (Figure 14c). The horizontal movement of the weak stratum is prominent. Pre-existing joints in the hard rock mass widen upward and slip with the ductile flow of the basal layers. The movement of the massive unstable rock mass is dominated by subsidence in the crest area and back tilting at the toe (Figure 15c).

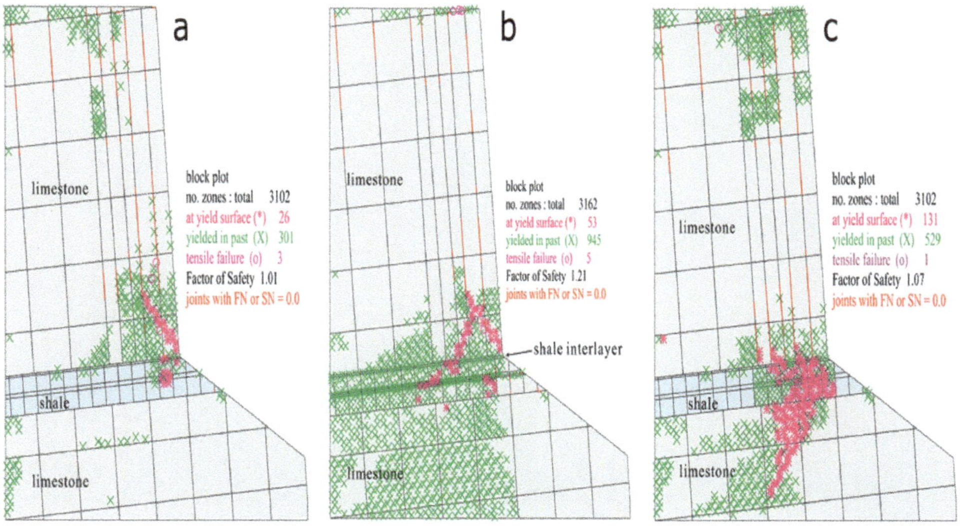

Figure 14. Plastic state and joint opening for rockslide initiation in cli ff slopes.

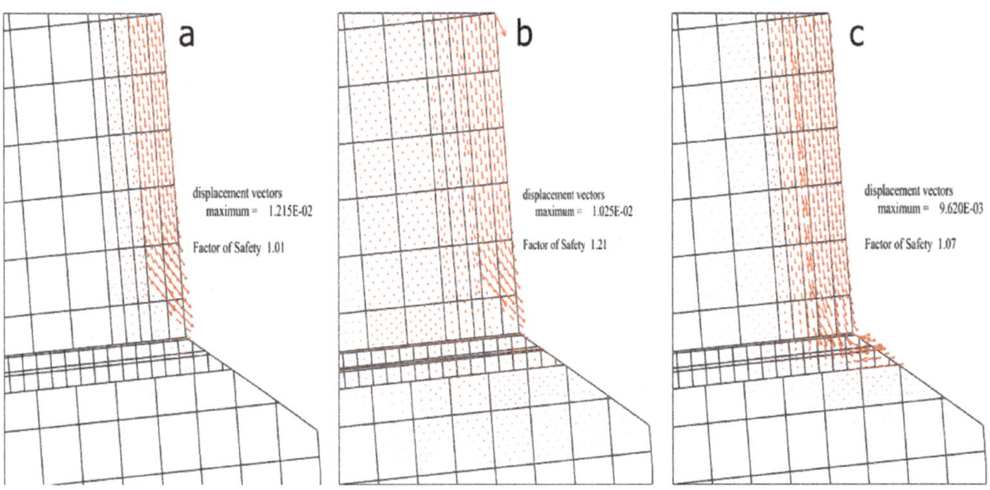

Figure 15. Displacement vectors of the initial failures of cliff slopes.

When the weak basal is a thin interlayer, rotational slide through the toe is unlikely to occur. The horizontal shear stress determines the possibility of shear failure in the beds. For a sub-horizontal bedded slope with gentle tectonic disturbance during geologic evolution, the horizontal to vertical stress ratio is generally less than 1 (Zhang et al., 1994). This indicates that it is possible for a weak interlayer with a frictional angle less than 45° to shear horizontally. Under these circumstances, the plastic zone ranges over the interlayer as well as the overlying competent rocks. The rock mass at the toe is in a plastic state as a result of compression fractures. Two yield surfaces dipping in opposite directions are formed (Figure 14b), similar to tensile failure behavior for some brittle rock samples. The displacement prior to collapse is limited, and remarkable squeezing out is unlikely to be observed in the toe area (Figure 15b).

A rock slide might burst out in hard rock where back scars cut through, and no discontinuities or weak strata are exposed at the toe. Irregular scars caused by brittle and shear failure through rock mass dip out of the cliff faces. The yield zone is mainly located at the block toe but not in the underlying rocks (Figure 14a). The fracture of the toe rock mass gives rise to joints opening immediately above. Numerical computation gives a plane yield surface at the bottom of the separated block, which is called a potential failure surface (Figure 15a). A small displacement appears before the outbreak of a compound slide. However, ground fractures on the crest and spalling in the toe area might occur.

UNDERGROUND MINING

The failures at the three locations described above are related to large areas of mining out. Pells (2008) used a continuum 2-D model to assess macro-scale movements of cliff faces caused by total extraction and proved that the steep slope tends to tilt outwards when mining occurs beneath the cliff, and extraction well behind the cliff face causes back tilt. Our similar simulations are shown in Figure 16. The rock slide model in Figure 14a was adopted. The roof tends to collapse, and the surrounding rocks gradually fracture. The undermining-induced subsidence of the crest causes the dilation and tensile failure of the rock mass (Figure 16a). The FOS decreases from 1.01 to 0.98 when the underground mining is located behind the cliff face. When the extraction is directly beneath the cliff, the rock mass is subjected to a small constraint; hence, cliff failure break-out through the fractured rock mass between the goaf and open face is feasible (Figure 16b). The maximum principal stress on both sides of the goaf increases while that of the overlying rocks decreases (Figure 16c).

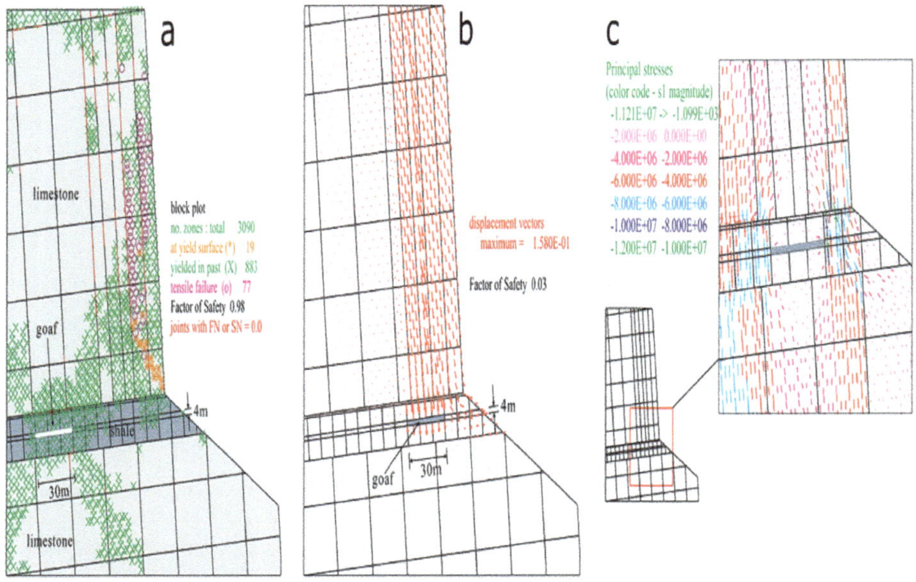

Figure 16. Numerical simulations of mining in cli ff slopes.

CONCLUSIONS

In this study, three different examples of failure in sub-horizontally bedded limestone cliffs are discussed. The failures are characterized by pre-existed vertical joints passing through thick limestone. The Yudong rock slide originated in a limestone cliff edge and left a moderately inclined rupture plane, implying shear failure through the hard rock. Rock collapse caused by compression fracture and tensile splitting of the rock mass near the interface between the hard cap and weak stratum occurred at Zengzi Cliff.

The Wangxia cliff failure showed a slow rock slump sheared through the underlying incompetent rock mass along a curved surface. The mechanism of toe breaking mainly depends on the strength characteristics of the rock mass.

Considering that the rock mass near the cliff face is not laterally constrained, it is reasonable to assume that the there is a massive UCS failure at the toe of the vertical tower and slab blocks. Some failure characteristics of jointed rock mass in in situ UCS tests have been observed all around the world, for both shear and tensile failure. Of special concern is tensile failure in horizontally bedded rock masses with interlayers. Compression fractures emerge near the interfaces and form vertical slabs, which eventually split. Another possibility is squeezing out of weak interlayers as a result of compression fracture, causing the hard rock nearby to yield to tensile stress and disintegrate. It is worth noting that the UCS of the rock mass in compression fracturing is much lower than that of intact rock. A criterion based on horizontal shear failure along weak interlayers and the UCS ratio of the cap and underlying rock masses could be used to assess the failure mode of tower and slab blocks on cliff edges.

ACKNOWLEDGEMENTS

This work was carried out with financial support from the China Geological Survey (No. 12120114079101), the Ministry of Science and Technology of the People's Republic of China (No. 2012BAK10B01) and the National Natural Science Foundation of China (No. 41302246).

REFERENCES

1. 107 Geology Team of Sichuan Geology and Mineral Bureau: Geology Survey Report on Bauxite Deposit of Loujiashan in Nanchuan, Sichuan, 1995 (in Chinese). Abele, G.: Large rockslides: their causes and movement on internal sliding planes, Mt. Res. Dev., 14, 315–320, 1994.
2. Altun, A. O., Yilmaz, I., and Yildirim, M.: A short review on the surficial impacts of underground mining, Sci. Res. Essays, 5, 3206–3212, 2010.
3. Ding, W. W., Yu, Y. Z., Deng, W. L. et al.: Stability analysis of reservoir bank in The Three Gorges of The Yangtze River, Center of Hydrogeology and Engineering Geology, Ministry of Geology and Mineral Resource, Beijing, Open File Rep. 81291, 127 pp., 1990.
4. Embleton-Hamann, C.: Geomorphological Hazards in Austria, Geomorphology for the Future, Innsbruck University Press, Innsbruck, 33–56, 2007.

5. Huang, R. Q.: Mechanisms of large-scale landslides and pre-recognition. Key note on the Six National Academic Conference of Geological Hazard, 11–12 April 2013, Beijing, China, 2013.

6. Hungr, O. and Evans, S. G.: The occurrence and classification of massive rock slope failure, Felsbau, 22, 16–23, 2004.

7. Kay, D., Barbato, J., Brassington, G., and De Sommer, B.: Impacts of longwall mining to rivers and cliffs in the Southern Coalfield, in: Coal 2006: Coal Operators' Conference, edited by: Aziz, N., University of Wollongong & the Australasian Institute of Mining and Metallurgy, 2006, 327–336, 2006.

8. Le, Q. L., Wang, H. D., Xue, X. Q., Gao, Y. L., Jin, X. H., Zhang, J. Y., and Pan, S. T.: Deformation monitoring and failure mechanism of Wangxia Dangerous Rock Mass in Wushan County, J. Eng. Geol., 19, 823–830, 2011.

9. Lollino, P., Martimucci, V., and Parise, M.: Geological survey and numerical modeling of the potential failure mechanisms of underground caves, Geosys. Engin., 16, 100–112, 2013.

10. Marschalko, M., Yilmaz, I., Bednárik M., and Kubečka, K.: Deformation of slopes as a cause of underground mining activities: three case studies from Ostrava–Karviná coal field (Czech Republic), Environ. Monit. Assess., 184, 6709–6733, 2012.

11. Palma, B., Parise, M., Reichenbach, P., and Guzzetti, F.: Rock-fall hazard assessment along a road in the Sorrento Peninsula, Campania, southern Italy, Nat. Hazards, 61, 187–201, 25 2012.

12. Parise, M.: Rock failures in karst, in: Landslides and Engineered Slopes, Proc. 10th Int. Symp. on Landslides, Xi'an (China), edited by: Chen, Z., Zhang, J., Li, Z., Wu, F., and Ho, K., 30 June–4 July 2008, 1, 275–280, 2008.

13. Parise, M.: Hazards in karst, in: Proceedings International Interdisciplinary Scientific Confer- 30 ence "Sustainability of the Karst Environment. Dinaric Karst and Other Karst Regions": Series on Groundwater, IHP-UNESCO, Plitvice Lakes, Croatia, 155–162, 2010.

14. Pells, P. J. N.: Assessing parameters for computations in rock mechanics, in: Proceedings First Southern Hemisphere International Rock Mechanics Symposium (SHIRMS), edited by: Potvin, Y., Carter, J., Dyskin, A., and Jeffrey, R., 1, 39–54, 2008.

15. Rainer, P., Martin, B., Rudolf, H., and Pavel, L.: Geomechanics of hazardous landslides, J. Mt. Sci., 2, 211–217, 2005.

16. Ren, Y. R., Chen, P., Zhang, J., and Chen, P.: Early-warning analysis on the rockfall for Zengziyan W12# dangerous rock mass in Nanchuan City of Chongqing, The Chinese Journal of Geological Hazard and Control, 16, 28–31, 2005.

17. Rohn, J., Resch, M., Schneider, H., Fernandez-Steeger, T. M., and Czurda, K.: Large-scale lateral spreading and related mass movements in the Northern Calcareous Alps, B. Eng. Geol. Environ., 63, 71–75, 2004.

18. Ruff, M. and Rohn, J.: Susceptibility analysis for slides and rockfall: an example from the Northern Calcareous Alps (Vorarlberg, Austria), Environ. Geol., 55, 441–452, 2008.

19. Santo, A., Del Prete, S., Di Crescenzo, G., and Rotella, M.: Karst processes and slopeinstability: some investigations in the carbonate Apennine of Campania (southern Italy), in: Natural and Anthropogenic Hazards in Karst Areas: Recognition, Analysis and Mitigation, edited by: Parise, M. and Gunn, J., Geol. Soc. London, sp. publ. 279, 59–72, 2007.

20. Tang, F. Q.: Research on mechanism of mountain landslide due to underground mining, Journal of Coal Science and Engineering (China), 15, 351–354, 2009.

21. Von Poschinger, A.: Large rockslides in the Alps: a commentary on the contribution of G. Abele (1937–1994) and a review of some recent developments, in: Catastrophic Landslides: Effects, Occurrence, and Mechanisms, edited by: Evans, S. G., 237–255, 2002.

22. White, E. and White, W.: Processes of cavern breakdown, Bull. Natl. Speleol. Soc., 31, 83–96, 1969.

23. Zhang, Z. Y., Wang, S. T., and Wang, L. S.: Principle of Engineering Geology Analysis, China Geology Press, Beijing, 66 pp., 1994.

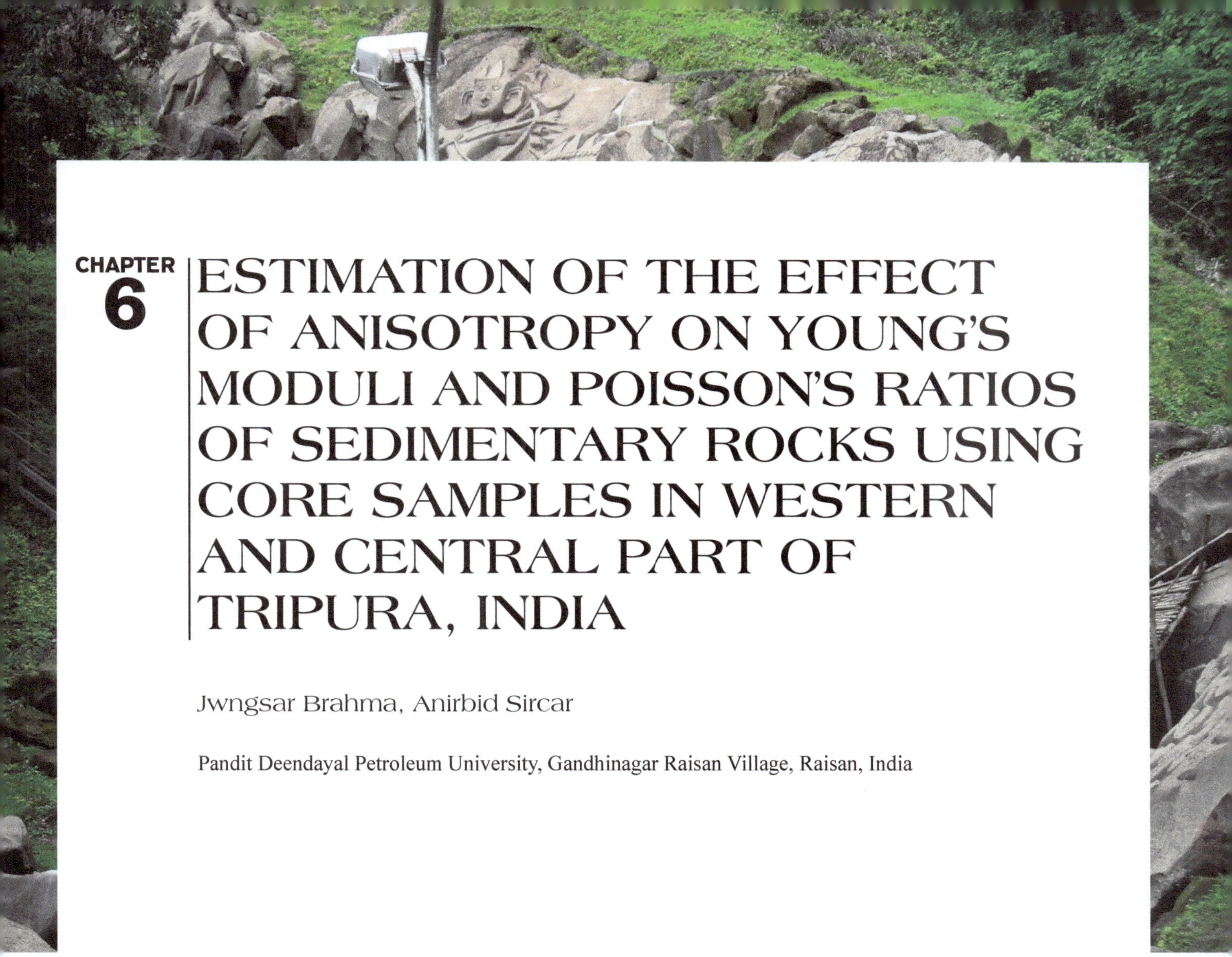

CHAPTER 6

ESTIMATION OF THE EFFECT OF ANISOTROPY ON YOUNG'S MODULI AND POISSON'S RATIOS OF SEDIMENTARY ROCKS USING CORE SAMPLES IN WESTERN AND CENTRAL PART OF TRIPURA, INDIA

Jwngsar Brahma, Anirbid Sircar

Pandit Deendayal Petroleum University, Gandhinagar Raisan Village, Raisan, India

ABSTRACT

The velocity anisotropy parameters and elastic constants play a very important role to estimate Young's modulus and Poisson's ratios accurately. For geomechanics applications such as hydraulic fracturing design, analysis of wellbore stability and rock failure, determination of in situ stress and assessment of the response of reservoirs and surrounding rocks to changes in pore pressure and stress, Young's modulus and Poisson's ratios play a very important role. Four rock samples were collected from four different wells situated in study area. The ultrasonic transmission method has been used to measure P-wave, Sh-wave and Sv-wave travel times as a function of orientation and confining pressure. The five independent stiffnesses constants, Young's moduli, Poisson's ratios and Bulk moduli of the samples were estimated. The Poisson's ratios ($\vartheta_{12}, \vartheta_{13}$ and ϑ_{31}) are varying as the confining pressure is changed. The axial strain is larger than the lateral strain, resulting $\vartheta_{31} < \vartheta_{12}$ $_w$ang##Bracket#. For shales, the Young's modulus measured parallel to bedding E1 i_s usually greater than the Young's modulus measured perpendicular to bedding E3. Through this study it has been observed that, there is a strong effect of anisotropy parameters on Young's modulus and Poisson's ratio.

Keywords

Elastic Stiffness; Shales; Sandstone; Sedimentary Rock; Anisotropy

---

Citation: Brahma, J. and Sircar, A. (2014) Estimation of the Effect of Anisotropy on Young's Moduli and Poisson's Ratios of Sedimentary Rocks Using Core Samples in Western and Central Part of Tripura, India. International Journal of Geosciences, 5, 184-195. doi: 10.4236/ijg.2014.52020.

INTRODUCTION

In hydrocarbon exploration and production theory, one of the common assumptions in conventional seismic exploration is the subsurface consisting of a series of elastically homogenous isotropic layers. That assumption is not true; numerous investigations during the previous decades have demonstrated the presence of seismic anisotropy in different geological settings and on various scales [1]. Ignoring these anisotropy effects can adversely influence the result of most basic seismic data processing and interpretation steps, such as NMO correction, velocity analysis, stacking, migration, DMO correction, time to depth conversion and AVO analysis [1-4].

The earth material is said to be transversely isotropic with vertical axis of symmetry if the elastic properties vary vertically but not horizontally in the simplest horizontal or layered case. Detecting and quantifying this type of anisotropy are very useful for correlation of sonic logs in vertical and deviated wells and for borehole surface seismic imaging and studies of amplitude variation with offset (AVO). The yield information about rock stress and fracture density and orientation can be estimated by identifying and quantifying the elastic anisotropy and these parameters are useful for designing hydraulic fracture jobs and for understanding horizontal and vertical permeability anisotropy. To understand the elastic anisotropy for more complex cases, such as dipping layers, fractured layered rocks or rocks with multiple fracture sets, superposition of the effects of the individual anisotropies is necessary.

In spite of the fact that shale formations exhibit high anisotropies, the study of fracture related anisotropy has been more intensive, both theoretically [5-12] and experimentally [13-15]. This is most likely explained because when anisotropy is observed it can be associated with fractures, which have a strong impact on permeability. Highly permeable rocks can be a good oil reservoir. However, in order to associate anisotropy with fracture zone, a correction for the anisotropic effect of the upper layers of the sedimentary column that is traversed by the wave field is necessary. More recently, the influence of pore fluid on elastic properties of shale has been investigated. Hornby measured compressional and shear wave velocities up to 80 MPa on two fluid saturated shale samples under drained conditions [16]. One sample from Jurassic outcrop shale was recovered from under sea and stored in its natural fluid, and the other is Kimmeridge clay taken from a North Sea borehole. Measurements were made on cores parallel, perpendicular, and at 45° to bedding. Values of anisotropy were up to 26% for compressional and 48% for shear wave velocity, and were found to decrease with increasing pressure. The effect of reduced porosity was, therefore, concluded to be more influential on anisotropy than increased alignment of minerals at higher pressure. The elastic constants, velocities and anisotropies in shales can be obtained from traditionally measured on multiple adjacent core plugs with different orientations [17]. To derive the five independent constants for transversely isotropic (TI) rock, Wang measured three plugs separately (one parallel, one perpendicular, and one $\pm 45^{\circ}$ to the symmetric axis). The advantage of this three plugs method is redundancy for calculation of the five independent elastic constants, since each core plug measurement yields three velocities.

Sandstones are rarely clean; they often contain minerals other than quartz, such as clay minerals, which can affect their reservoir qualities as well as their elastic properties. The presence of clay mineral and clastic sheet silicates strongly influences the physical and chemical properties of both sandstones and shales [18]. Clay can be located between the grain contacts as structural clay, in the pore space as dispersed clay, or as laminations [19]. The distribution of the clay will depend on the conditions at deposition, on compaction, bioturbation and diagenesis. While most reservoirs are composed of relatively isotropic sandstones or carbonates, their properties may be modified by stress. Non-uniform compressive stress will have a major effect on randomly distributed microcracks in a reservoir. When the rock is unstressed all of the cracks may be open, however, compressional stresses will close cracks oriented perpendicular to the direction of maximum compressive stress, while cracks parallel to the stress direction will remain open. Elastic waves passing through the stressed rock will travel faster across the closed cracks (parallel to maximum stress) than across the open ones. The sands and sandstone are intrinsically isotropic unless they are fractured, finely layered or clay bearing. Wang shows that the brine-saturated Africa reservoir sands, which are essentially clay free, have very little anisotropy (average P-anisotropy and Sanisotropy are probably within the measurement uncertainties) [17]. For the brine saturated tight sands, the anisotropy is 5.0% for the P-wave and 3.3% for the S-wave when averaged over all samples at all pressures. At the net reservoir pressure of 7500 psi (51.7 MPa), anisotropy is slightly lower, averaging 4.6% and 3.2% for P-and S-waves, respectively. Gas-saturated tight sands and shaly sands show some degrees of anisotropy, ranging from 0% to 36% for P-waves and 0.3% to 19.5% for S-waves. When averaged at all pressures, the anisotropy is 9.9% for the P-wave and 5.5% for the S-wave.

There are various methods to estimate anisotropy parameters. Wang presented a method for measuring seismic velocities and transverse isotropy in rocks using a sing core plug [17]. Wang pointed out that shale cores must be preserved. Otherwise, once a shale core is dehydrated, fractures will develop along the bedding plane, and the measured seismic velocities and anisotropy will no longer be accurate [20].

Leslie [21] determined the Thomsen anisotropic parameters ε and δ by measurement of head wave velocities along the seismic lines. This method obtained more realistic anisotropic parameters than those measured in the laboratory, but needs to ensure that the spread is of sufficient length so that the refractor velocities measured are from rocks below the weathered layer.

In this paper four kinds of core samples; Dry Sandstone, Shale, Sandy Shale and Saturated Sandstone were collected from Tripura oil field. The purpose of this paper is to estimate effects of elastic anisotropic on Young's moduli and Poison's ratios. We used the ultrasonic transmission method to measure V_p, V_{sh} and V_{sv} wave velocities, as function of confining pressure from 1 MPa to 40.3 MPa. Using these velocities five independent stiffnesses have been calculated and hence Thomsen's parameters as well as geo-mechanical parameters.

METHODOLOGY

Sample Preparation

The four kinds of core samples namely dry sandstone, shale, sandy shale and saturated sandstone were collected from the

Tripura field in the zone between 1910 - 2200 m and were analyzed. Each specimen had diameter 25.4 mm and length 50 mm. The samples had significant lattice preferred orientation. Before performing the test the specimens were cored using diamond drill and the ends of the specimens were ground flat and parallel to ± 0.001 mm/mm To get more effective result the samples were also checked for flaws and defects. In a vacuum oven at 170°C for 20 hours the specimens were dried. Next, using digital balance the dry mass of samples was measured. The dry bulk density was computed by dividing the volume by the mass of the samples. After completing, the sandstone was water saturated and the bulk density was computed.

Experiment Procedure

For each specimens three velocities were measured, one compressional wave and two orthogonal shear wave as a function of pore pressure and confining pressure. Each specimen was jacketed and secured between a matched set of ultrasonic transducers having resonant frequency 1.0 MH. Specimen reference system has the x-axis parallel to the lamination in the foliation, the y-axis normal to the lineation in the foliation and z-axis perpendicular to the foliation plane. Main frequency of ultrasound transducer is 1 MHz. Volume change can obtained from linear strains in three directions which is determined from the piston displacement through necessary correction. Each polarization is sequentially propagated through the rock and each waveform is determined from the data using the appropriate correction for the travel time through the transducer assembly. The experiments, on dry and saturated specimens, were conducted at confining pressures from 1 Mpa to 40.3 Mpa.

It is necessary to core specimens in three directions: vertical, horizontal and at an intermediate angle between vertical and the plane of isotropy (Figure 1), to measure traverse isotropy completely. Three velocities were measured in each direction, leading total nine velocities for each sample. The particle polarization and the direction of propagation of the three modes of propagation, P S_h and S_v in the stratigraphic column within the zone of interest are shown in Figure 2.

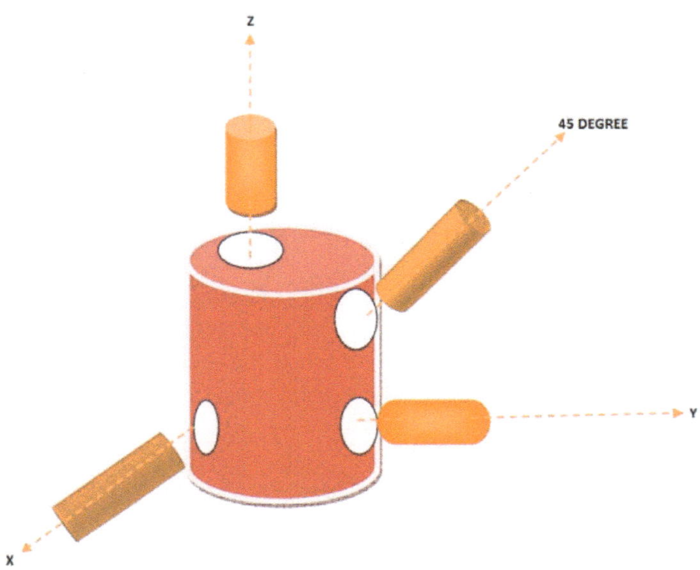

Figure 1. Traditional three-plug method for measuring transverse isotropy in laboratory core samples. Three adjacent core plugs (one parallel, one perpendicular, and one ±45 to symmetry axis) must be cut from whole cores and velocities measured to derive the five elastic constants (Wang, 2002).

Both in room dry and full water saturation conditions, sandstone specimens were tested. But for shale and sandy shale it is difficult in saturating these rocks. So, only in room dry conditions the shale and sandy shale specimens were tested.

It was assumed that, the effective hydrostatic pressure is the difference between confining pressure and pore pressure. The pore pressure was kept constant while the confining pressure varied (drained regime). The confining pressure was varied from atmospheric pressure up to the reservoir effective pressure condition in dry specimen. But the effective pressure never exceeded the reservoir pressure in saturated specimens. However the pore pressure was varied from atmospheric pressure up to a pressure approximately 32% higher than the reservoir pressure condition.

Mathematical Analysis

In this paper only the standard three plug method will be discussed since it is commonly used in oil exploration. Standard three plug method is by far the most common measurement approach. The method requires the extraction of three core plug along prescribed orientation relative to the assumed symmetric axes. These orientations are parallel, perpendicular and typically 45° to the vertical symmetry axis. Either static or dynamic ultrasonic laboratory measurements can be performed on these plugs to provide the magnitude of tensor elements C_{11}, C_{33}, C_{44}, C_{66} and C_{13} (Mavko, 1986). These five elastic constants can be recast into more geophysical meaningful parameter [1].

THE RELATION BETWEEN STRESS (S_{IJ}) AND STRAIN (E_{KL}) IN VERTICAL TRANSVERSE ISOTROPY (VTI) IS GIVEN BY THE FOLLOWING RELATION AS:

$$\sigma_{ij} = C_{ijkl} e_{kl} \tag{1}$$

where C_{ijkl} is the Voigt matrix [22].

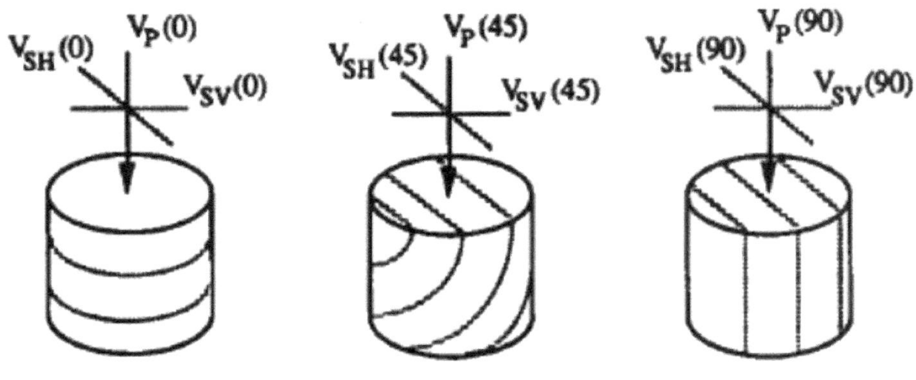

Figure 2. Scheme of sample preparation and velocity measurements in shales. Wave propagation and polarization with respect to bedding-parallel lamination is shown. Numbers in parentheses indicate the phase velocity angle with respect to the bedding-normal symmetry axis (Vernik and Nur, 1992).

From Figure 1 it is seen that if the axis is denoted by Z, the other two principal axes (X and Y) are parallel to the transversely isotropic plane. In this coordinate system, the stiffness matrix C is expressed as:

$$C = \begin{bmatrix} C_{11} & C_{12} & C_{13} & 0 & 0 & 0 \\ C_{12} & C_{22} & C_{13} & 0 & 0 & 0 \\ C_{13} & C_{13} & C_{33} & 0 & 0 & 0 \\ 0 & 0 & 0 & C_{44} & 0 & 0 \\ 0 & 0 & 0 & 0 & C_{44} & 0 \\ 0 & 0 & 0 & 0 & 0 & C_{66} \end{bmatrix} \tag{2}$$

From the above Equation (2), Voigt matrix C, there are five non-zero independent elastic constants: C_{11}, C_{33}, C_{44}, C_{66} and C_{13}. The sixth elastic constant is $C_{12} = C_{11} - 2C_{66}$. Here, C_{11} is the in-plane compressional modulus, C_{33} is the out-of-plane compressional modulus, C_{44} is the out-of-plane shear modulus, and C66 is the in-plane shear modulus, C_{13} is an important constant that controls the shape of the wave surfaces.

Using five elastic stiffness C_{ij}, a bulk modulus, one vertical Young's moduli, E_3 and one horizontal Young's modulus E_1 and three dynamic Poisson's ratios can be determines for a hexagonal material as follows [23]:

$$K = \frac{C_{33}\left(C_{11} + C_{12}\right) - 2C_{13}^2}{2C_{33} + C_{11} + C_{12} - 4C_{13}} \tag{3}$$

$$E_1 = \frac{\left[C_{33}\left(C_{11} + C_{12}\right) - 2C_{13}^2\right]\left(C_{11} - C_{12}\right)}{C_{33}C_{11} - C_{13}} \tag{4}$$

$$E_3 = \frac{\left[C_{33}\left(C_{11} + C_{12}\right) - 2C_{13}^2\right]}{C_{11} + C_{12}} \tag{5}$$

$$\vartheta_{31} = \frac{C_{13}}{C_{11} + C_{12}} \tag{6}$$

$$\vartheta_{12} = \frac{C_{33}C_{12} - C_{13}^2}{C_{11}C_{33} - C_{13}^2} \tag{7}$$

$$\vartheta_{13} = \frac{C_{13}\left(C_{11} - C_{12}\right)}{C_{11}C_{33} - C_{13}^2} \tag{8}$$

These dynamic Poison's ratios Equations (7) and (8), ϑ_{ij}, are indirect measure of the ratio of the lateral to axial strains when the uniaxial stress is applied in the same direction of axial strain.

There are three types of velocity anisotropies (ε ,g and d) known as Thomsen's parameters and can be estimated by the following mathematical equations suggested by Thomsen [1]:

$$\varepsilon = \frac{C_{11} - C_{33}}{2C_{33}} = \frac{V_p^2\left(90^\circ\right) - V_p^2\left(0^\circ\right)}{2V_p^2\left(0^\circ\right)} \tag{9}$$

$$\gamma = \frac{C_{66} - C_{44}}{2C_{44}} = \frac{V_{sh}^2\left(90^\circ\right) - V_{sh}^2\left(0^\circ\right)}{2V_{sh}^2\left(0^\circ\right)} \tag{10}$$

$$\delta = \frac{\left(C_{13} + C_{44}\right)^2 - \left(C_{33} - C_{44}\right)^2}{2C_{33}\left(C_{33} - C_{44}\right)} \tag{11}$$

RESULT AND DISCUSSION

Four types of rock samples that represent elastic and anisotropy behavior in the sedimentary column: dry sandstone, saturated sandstone, shale and sandy shale have been collected from Tripura oil filed. Each of these rocks types exhibit varying sensitivities to confining pressure, and as a consequence the anisotropies are affected differently as the confining pressure varies. The behavior of the elastic anisotropy for the different rock types as the confining pressure is increased (Figures 3-6).

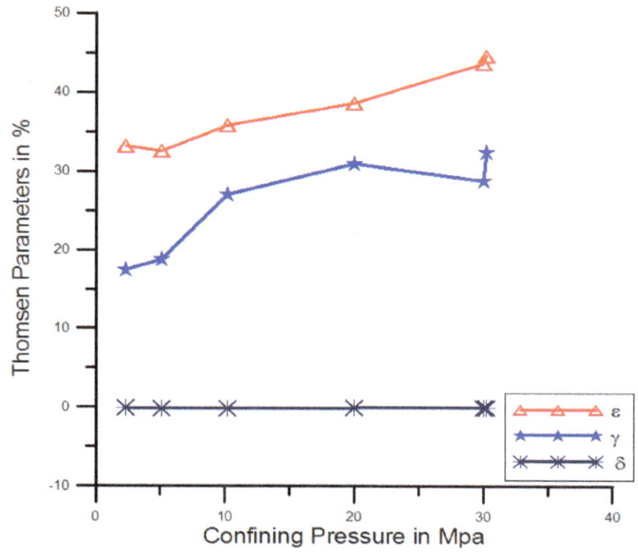

Figure 3. Comparison of Thomsen's parameters as a function of confining pressure for dry sandstone.

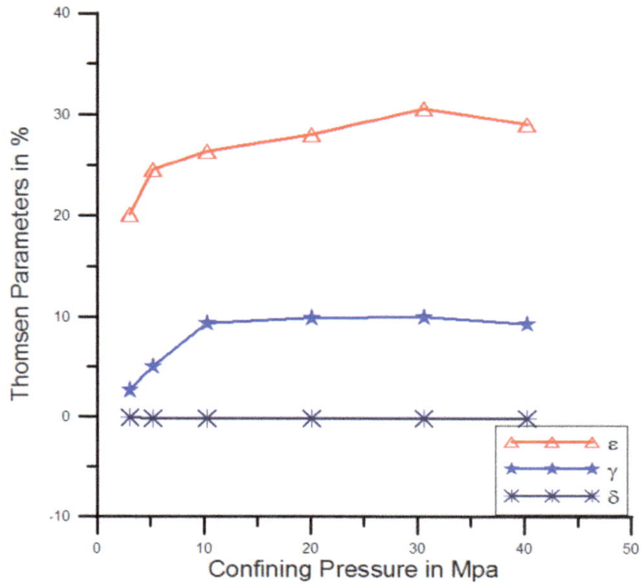

Figure 4. Comparison of Thomsen's parameters as a function of confining pressure for shale.

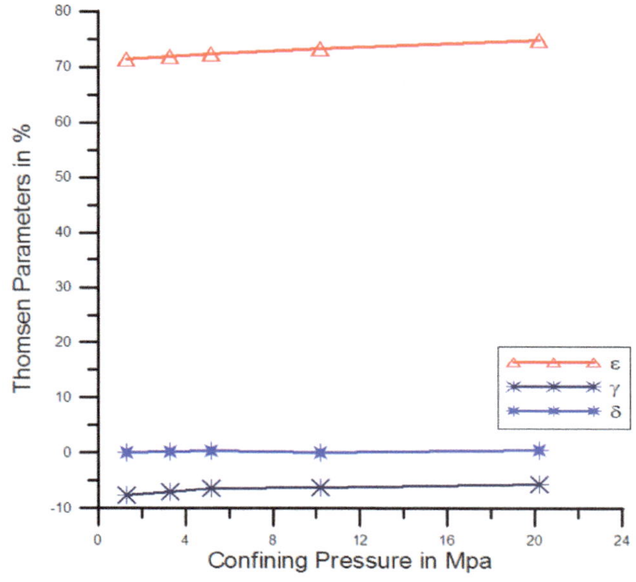

Figure 5. Comparison of Thomsen's parameters as a function of confining pressure for sandy shale.

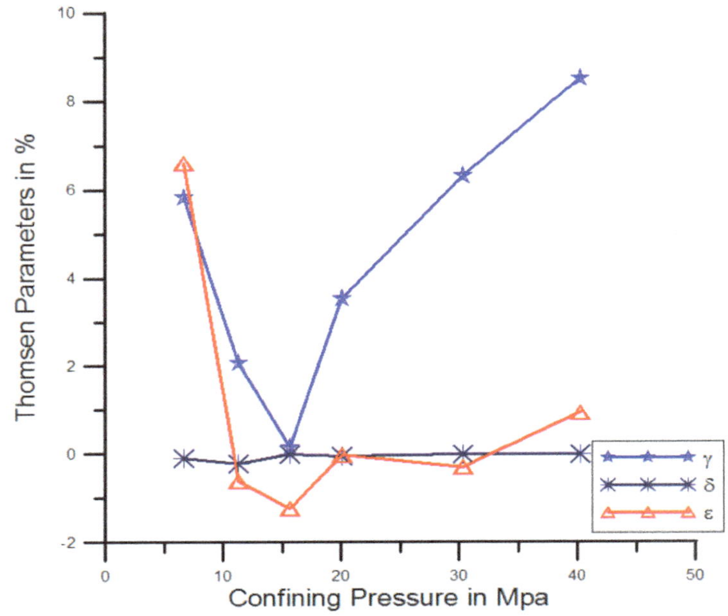

Figure 6. Comparison of Thomsen's parameters as a function of confining pressure for saturated sandstone.

The analysis shows that the anisotropy parameters are increased as the confining pressures are increased for all the rock samples except saturated sandstone. Figure 6, shows that the anisotropy parameters are decreased when confining pressure are increase up to 15 Mpa, but again anisotropy parameters are increased with increasing confining pressure beyond 15 Mpa up to 40 Mpa for saturated sandstone.

Shale

Figure 7 shows that the Poisson's ratio $\vartheta_{12}, \vartheta_{13}$ and ϑ_{31} varies as the confining pressure is raised. It has been observed that ϑ_{13} and ϑ_{31} are slightly decreased as confining pressure is increased. The difference between ϑ_{13} and ϑ_{31} remains approximately constant over the whole range of frequency, while ϑ_{12} remains approximately constant as the confining pressure is raised. ϑ_{12} is an indirect measure of the ratio of lateral to axial strain when the specimen is under uniaxial stress in the direction X. The high and almost constant value of ϑ_{12} indicated that the rock is very stiff and linearly elastic in the plane XY as consequence of the absence of voids and discontinuities in this plane.

ϑ_{31} is an indirect measurement of the ratio of lateral strain to axial strain when the sample is subjected to uniaxial stress in the direction normal to the plane of isotropy Z ϑ_{31} is smaller than ϑ_{12} because the specimen is softer in the vertical direction than in the horizontal direction. The vertical direction is perpendicular to the foliation plane. Between the layers, there are discontinuities or void space filled with gas, liquid, or viscous organic material that increase the compliance of the rock in this direction. Thus the axial strain is larger than the lateral strain, resulting $\vartheta_{31} < \vartheta_{12}$. Consequently, the axial strain rate is also less than the lateral strain rate, resulting in a decrease in ϑ_{31} as the pressure increased. The analysis of the sample shows that ϑ_{13} has similar behavior to ϑ_{31}, however, the coefficients larger because the strain in the X direction is smaller than the strain in the Z direction, when the rock is compressed in the Z direction.

Figure 4 shows that the variation of velocity anisotropy as a function of confining pressure. Our analysis found that δ remains approximately constant while confining pressure is increased, but ε and g are increased slightly as the pressure is increased.

Sandy Shale

Figure 5 shows that, the V_p anisotropy, ε, is bigger that the Sh-anisotropy, g within the confining pressure range. Oriented fractures affect the V_p-anisotropy more than the V_{sh}-anisotropy. When an Sh-wave propagates in the vertical and horizontal directions, the particle is polarized parallel to the plane of isotropy, and it always encounters more rigid material. However, when a P-wave propagates in the vertical and horizontal directions, the polarization of the particle and the direction of propagation are perpendicular and parallel to the plane of isotropy, respectively. The P-wave polarized in the vertical direction encounter softer material than the P-wave polarized in the horizontal direction. But our analysis shows that the V_p-anisotropy and V_{sh}-anisotropy are not approaching same value within our confining pressure range. That implies that the fracture of the specimen is not close within our pressure range. However, the anisotropic caused by preferred orientation of minerals, by fracture, and by layering are always due to the fact that the material is softer in the direction perpendicular to the direction of preferential orientation than in the direction parallel to it.

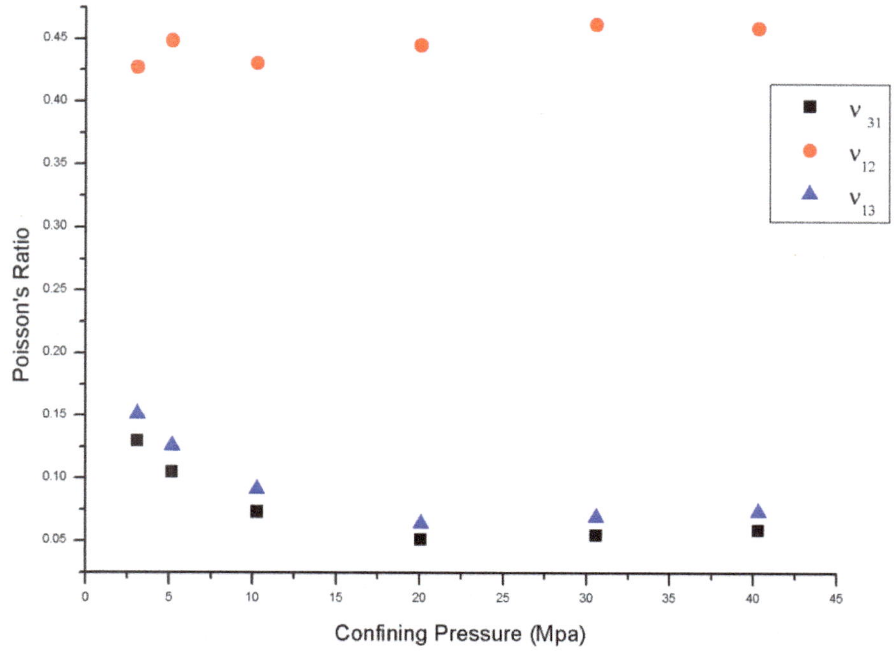

Figure 7. Relationship between Poisson's ratio and confining pressure for shale.

Dry and Saturated Sandstone

At low pressure, the V_p anisotropy ε, is higher than the V_{sh} anisotropy g, and they trend to approach the same value at higher confining pressure. However, d shows almost constant at all pressure. At higher pressure, the anisotropy is still controlled by the incomplete closures at the cracks.

After saturation $V_p(0°)$ and $V_p(45°)$ increase , while $V_{sv}(0°)$ decreases. $V_p(0°)$ and $V_p(45°)$ increase because the material is less compressible. A pore filled with fluid resists compression in a similar way when it is filled with solid material. The difference between $V_p(0°)$ for dry and for saturated sandstone decreases as the confining pressure is increased However, the difference between $V_s(0°)$ for dry and for saturated sandstones remain approximately constant The decrease in $V_s(0°)$ is caused by a significant increase in the bulk density, while C_{44} remains approximately constant (Figure 8). This explains why the difference between $V_s(0°)$ for dry and $V_s(0°)$ for saturated does not change as the confining pressure is increased. $V_p(90°)$ shows a significant increase after saturation. The significant increase is due to the polarization of the particle and the directions of propagation are perpendicular to the plane of isotropy, and fluid have a strong effect in this direction. The stiffness C_{33} shows a pronounced increase reaching the same value as C_{11}. C_{44}remains approximately constant and C_{66} shows a slight decrease (Figure 8) after saturation.

After saturation, the P-wave anisotropy significantly decrease. However, g shows only a slight increase and remains increasing as the confining pressure is increased.

The Effect of Anisotropy on the Young's Moduli and Poison's Ratios

The strong effects of anisotropy parameters ε,g and d on vertical and horizontal Young's moduli and the Poisson's ratios are shown in the Figures 9-16. The vertical Poisson's ratio significantly decreases with ε for all four rock samples, while the horizontal Poisson's ratio increases. On the other hand, both of the Young's moduli increase with ε. The anisotropy parameter g has a completely different type of effect, the horizontal and vertical moduli show slightly constant at low g but it increases as g increased. The vertical and horizontal Poisson's ratio show opposite effects with g.

We showed that anisotropy has significant effect on the two key parameters; the Young's modulus and the Poisson's ratio. In fact there are two Young's moduli and at least two Poisson's ratios and they vary significantly with the magnitudes of the anisotropy parameters ε, g and d.

CONCLUSIONS

The study found that, the elastic anisotropy parameters are key factors which affect the estimation of hydrocarbon reservoir characterization parameters. To estimate more accurate hydrocarbon reservoir characterization parameters, we need vertical P-wave and S-wave velocities and three anisotropy parameters. With high quality and high resolution surface seismic data, we can reason ably estimate ε, g and δ, and for the remaining parameters, we must depend on the downhole data, wire line measurements including sonic scanner and borehole seismic such as walk away vertical seismic profiling. Our laboratory experiments on core samples will only be beneficial to obtain valid empirical relationships among some of the parameters to develop a starting model.

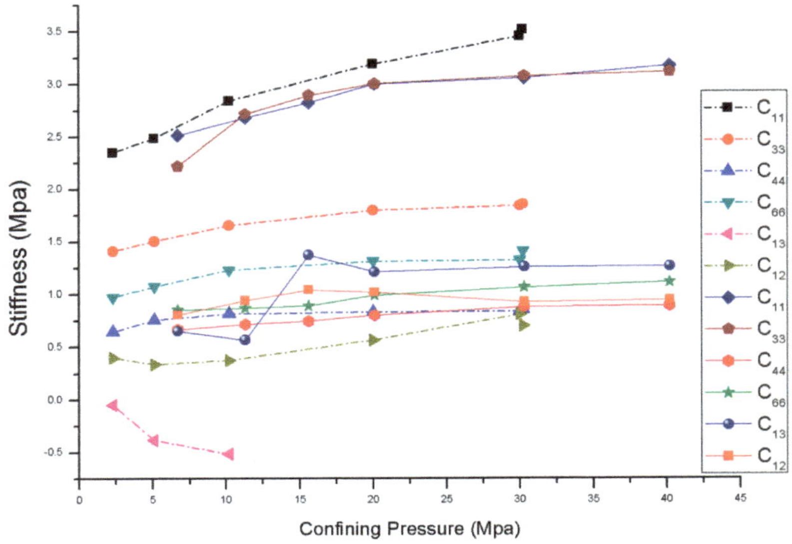

Figure 8. Stiffness for dry and saturated sandstone specimens. The dashed curves show the stiffness in dry condition.

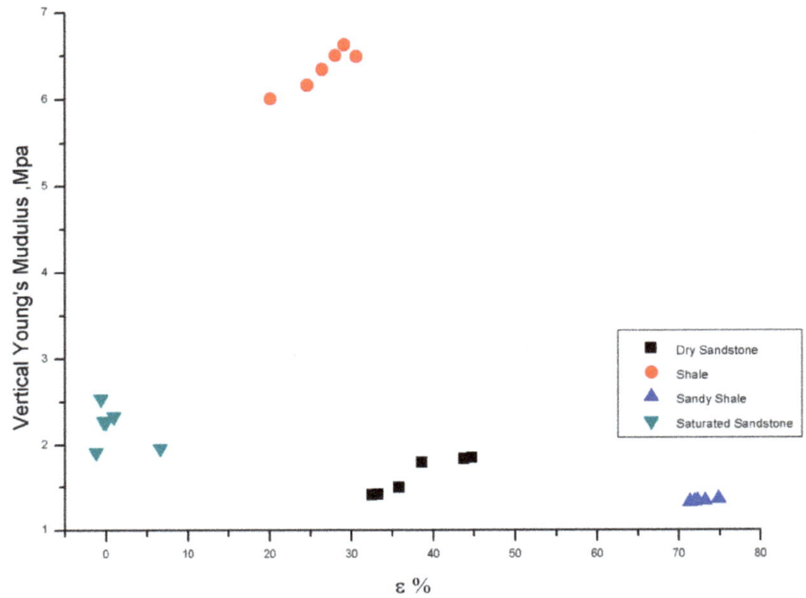

Figure 9. Relationship between vertical Young's modulus and Epsilon (ε).

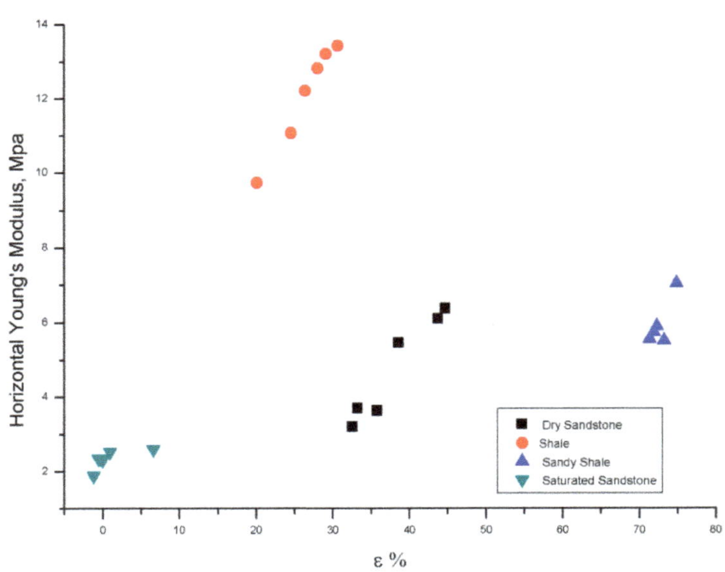

Figure 10. Relationship between horizontal Young's modulus and Epsilo (ε).

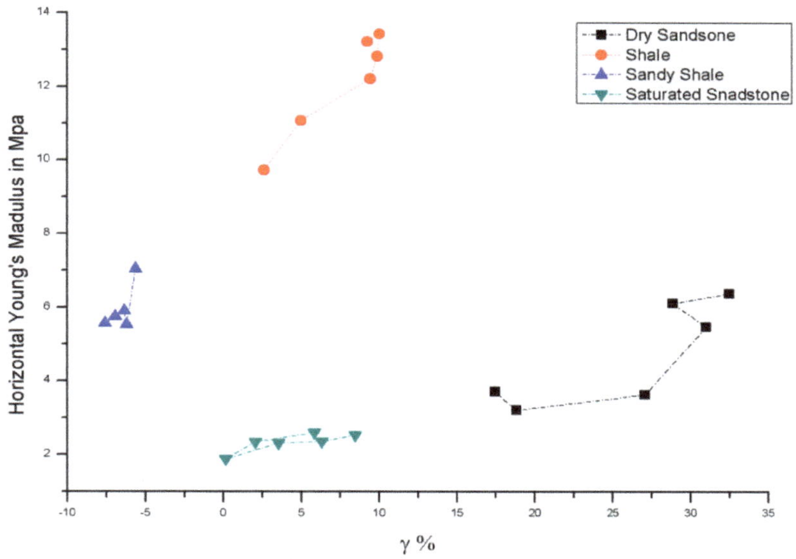

Figure 11. Relationship between horizontal Young's modulus and Gamma (l)

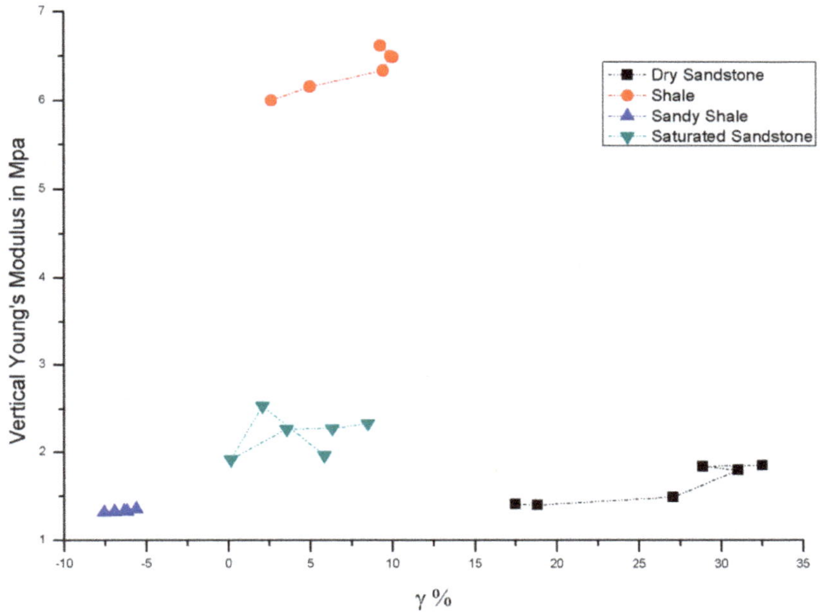

Figure 12. Relationship between vertical Young's modulus and Gamma (l).

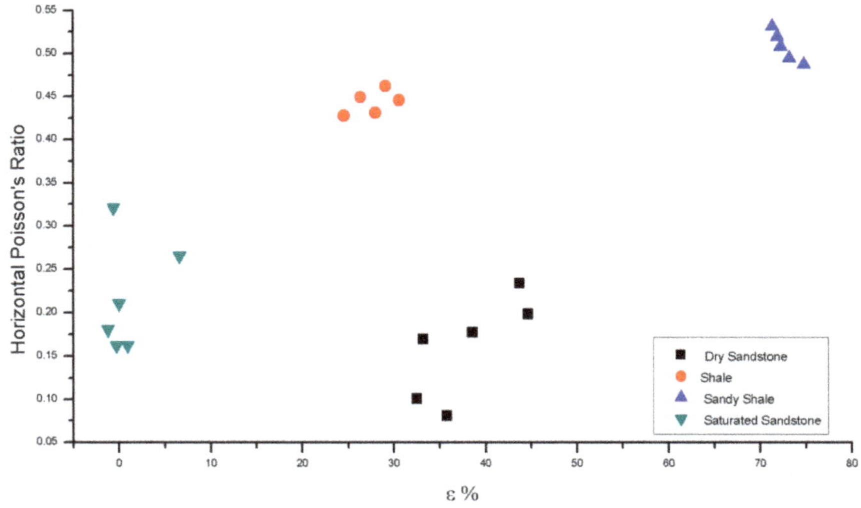

Figure 13. Relationship between horizontal Poisson's ratio and Epsilon (ε).

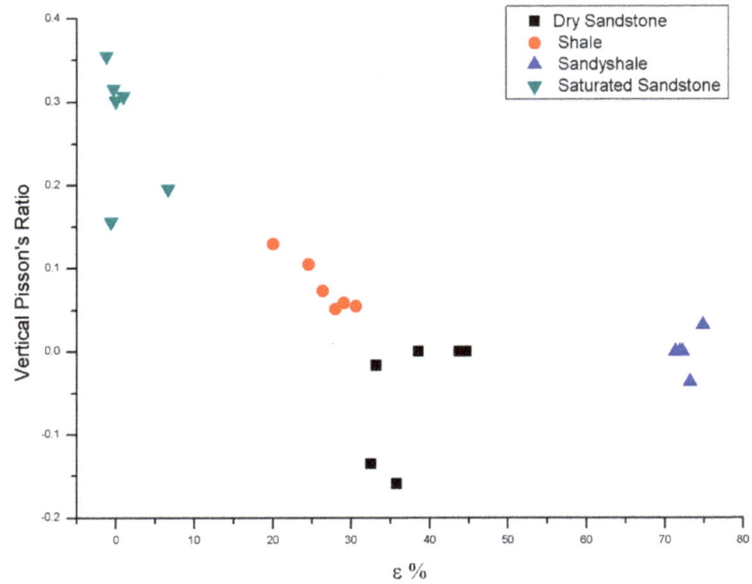

Figure 14. Relationship between vertical Poisson's ratio and Epsilon (ε).

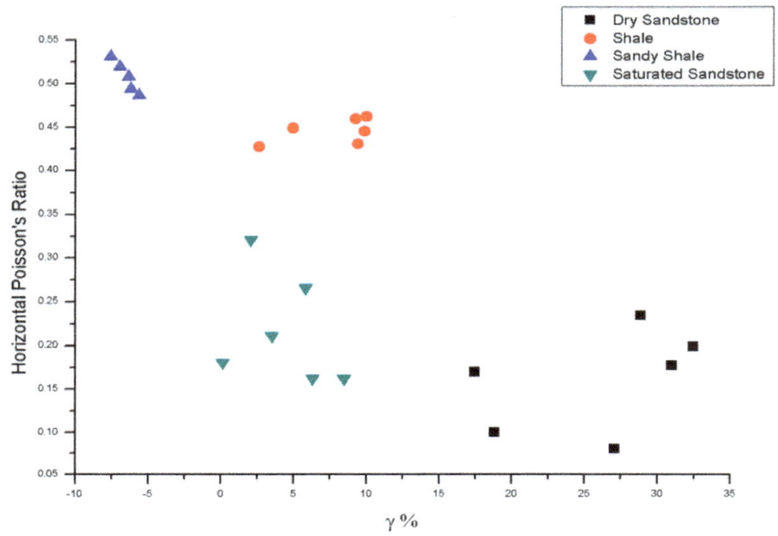

Figure 15. Relationship between horizontal Poisson's ratio and Gamma (λ).

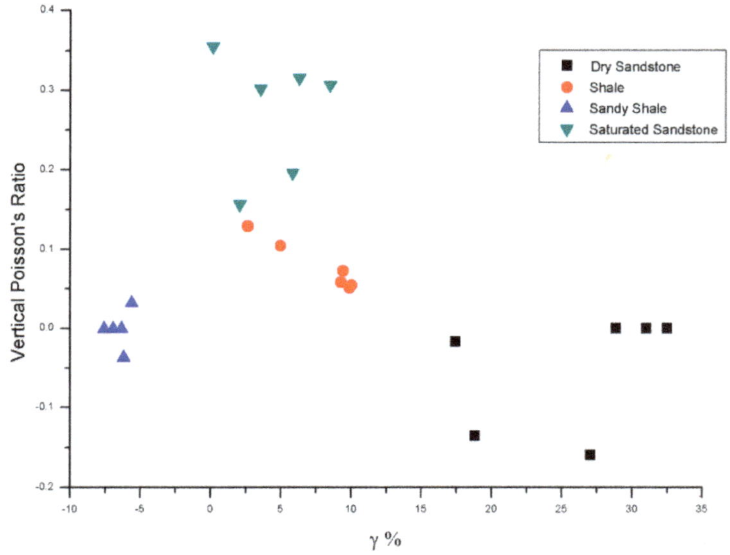

Figure 16. Relationship between vertical Poisson's ratio and Gamma (λ).

Thus through this study we proposed a prestack full waveform AVO inversion in a VTI medium constraining elastic parameters based on depth velocity analysis of P-wave surface seismic data and other information available.

This study also suggested that the Young's moduli and Poisson's ratios that the driller must have are the static moduli, not the dynamic moduli estimated through seismic experiment. The relationship established themselves in this study is laboratory based. The static moduli is larger than dynamic Young's moduli by the factor 2 and it may vary basin to basin.

The initial model can be integrated by adding drilling and micro seismic monitoring, initial production, and estimated ultimate recovery data for updation.

ACKNOWLEDGEMENTS

We thank M/S Pandit Deendayal Petroleum University for supporting in preparation of manuscripts.

REFERENCES

1. L. Thomsen, "Weak Elastic Anisotropy," Geophysics, Vol. 51, No. 10, 1986, pp. 1954-1966.http://dx.doi.org/10.1190/1.1442051
2. T. Alkhalifah and I. Tsvankin, "Velocity Analysis for Transversely Isotropic Media," Geophysics, Vol. 60, No. 5, 1995, pp. 1550-1566. http://dx.doi.org/10.1190/1.1443888
3. N. C. Banik, "Velocity Anisotropy of Shales and Depth Estimation in the North Sea Basin," Geophysics, Vol. 49, No. 9, 1984, pp. 1411-1419. http://dx.doi.org/10.1190/1.1441770
4. D. F. Winterstein, "Velocity Anisotropy Terminology for Geophysicists," Geophysics, Vol. 55, No. 8, 1990, pp. 1070-1088. http://dx.doi.org/10.1190/1.1442919
5. R. Brown and J. Korrigan, "On the Dependence of the Elastic Properties of a Porous Rock on the Compressibility of the Pore Fluid," Geophysics, Vol. 40, No. 4, 1975, pp. 608-616.http://dx.doi.org/10.1190/1.1440551
6. J. A. Hudson, "Overall Properties of a Cracked Solid," Mathematical Proceedings of the Cambridge Philosophical Society, Vol. 88, No. 2, 1980, pp. 371-384. http://dx.doi.org/10.1017/S0305004100057674
7. J. A. Hudson, "Wave Speeds and Attenuation of Elastic Waves in Material Containing Cracks," Geophysical Journal International, Vol. 64, No. 1, 1981, pp. 133-150. http://dx.doi.org/10.1111/j.1365-246X.1981.tb02662.x
8. J. A. Hudson, "Overall Elastic Properties of Isotropic Materials with Arbitrary Distribution of Circular Cracks," Geophysical Journal International, Vol. 102, No. 2, 1990, pp. 465-469.http://dx.doi.org/10.1111/j.1365-246X.1990.tb04478.x
9. J. A. Hudson, "Crack Distribution Which Accounts for a Given Seismic Anisotropy," Geophysical Journal International, Vol. 104, No. 3, 1991, pp. 517-521.http://dx.doi.org/10.1111/j.1365-246X.1991.tb05698.x
10. J. A. Hudson, E. Liu and Crampin, "The Mechanical Properties of Materials Which Interjected Cracks and Pores," Geophysical Journal International, Vol. 124, No. 1, 1996, pp. 105-112.http://dx.doi.org/10.1111/j.1365-246X.1996.tb06355.x
11. T. Mukerji and G. Mavo, "Pore Fluid Effects on Seismic Velocity in Anisotropic Rocks," Geophysics, Vol. 59, No. 2, 1994, pp. 233-244. http://dx.doi.org/10.1190/1.1443585
12. L. Thomsen, "Elastic Anisotropy Due to Aligned Cracks in Porous Rock," Geophysical Prospecting, Vol. 43, No. 6, 1995, pp. 805-829. http://dx.doi.org/10.1111/j.1365-2478.1995.tb00282.x
13. T. D. Jones, "Wave Propagation in Porous Rock and Models for Crustal Structures," Ph.D. Thesis, Standford University, 1983.
14. N. M. Lucet and P. A. Tarif, "Shear—Wave Birefringence and Ultrasonic Shear Wave Attenuation Measurements," SEG Annual Meeting, Anaheim, 30 October-3 November 1988, pp. 922-924.
15. M. Zamora and J. P. Porier, "Experimental Study of Acoustic Anisotropy and Birefringence in Dry and Saturated Fontainebleau Sandstone," Geophysics, Vol. 55, No. 11, 1990, p. 1455.http://dx.doi.org/10.1190/1.1442793
16. B. E. Hornby, "Experimental Laboratory Determination of the Dynamic Elastic Properties of Wet, Drained Shales," Journal of Geophysical Research: Solid Earth, Vol. 103, No. B12, 1998, pp. 29945-29964.
17. Z. Wang, "Seismic Anisotropy in Sedimentary Rocks," Part 1: A Single-Plug Laboratory Method," Geophysics, Vol. 67, No. 5, 2002, pp. 1415-1422. http://dx.doi.org/10.1190/1.1512787
18. K. Bjørlykke, "Clay Mineral Diagenesis in Sedimentary Basins—A Key to the Prediction of Rock Properties. Examples from the North Sea Basin," Clay Minerals, Vol. 33, 1998, pp. 15-34.
19. M. S. Sam and A. Andrea, "The Effect of Clay Distribution on the Elastic Properties of Sandstones," Geophysical Prospecting, Vol. 49, No. 1, 2001, pp. 128-150.http://dx.doi.org/10.1046/j.1365-2478.2001.00230.x
20. Z. Wang, "Seismic Anisotropy in Sedimentary Rocks, Part 2: Laboratory Data," Geophysics, Vol. 67, No. 5, 2002, pp. 1423-1440. http://dx.doi.org/10.1190/1.1512743
21. J. M. Leslie and D. C. Lawton, "A Refraction Seismic Field Study to Determine the Anisotropic Parameters of Shales," The Leading Edge, Vol. 17, No. 8, 1998, pp. 1127-1129.http://dx.doi.org/10.1190/1.1438105
22. G. Mavko, T. Mukerji and J. Dvorkin, "Rock Physics Handbook: Tools for Seismic Analysis in Porous Media," Cambridge University Press, 1998.
23. F. R. Pena, "Elastic Properties of Sedimentary Anisotropic Rocks," Dissertation, Massachusetts Institute of Technology, 1998, pp. 19-26.

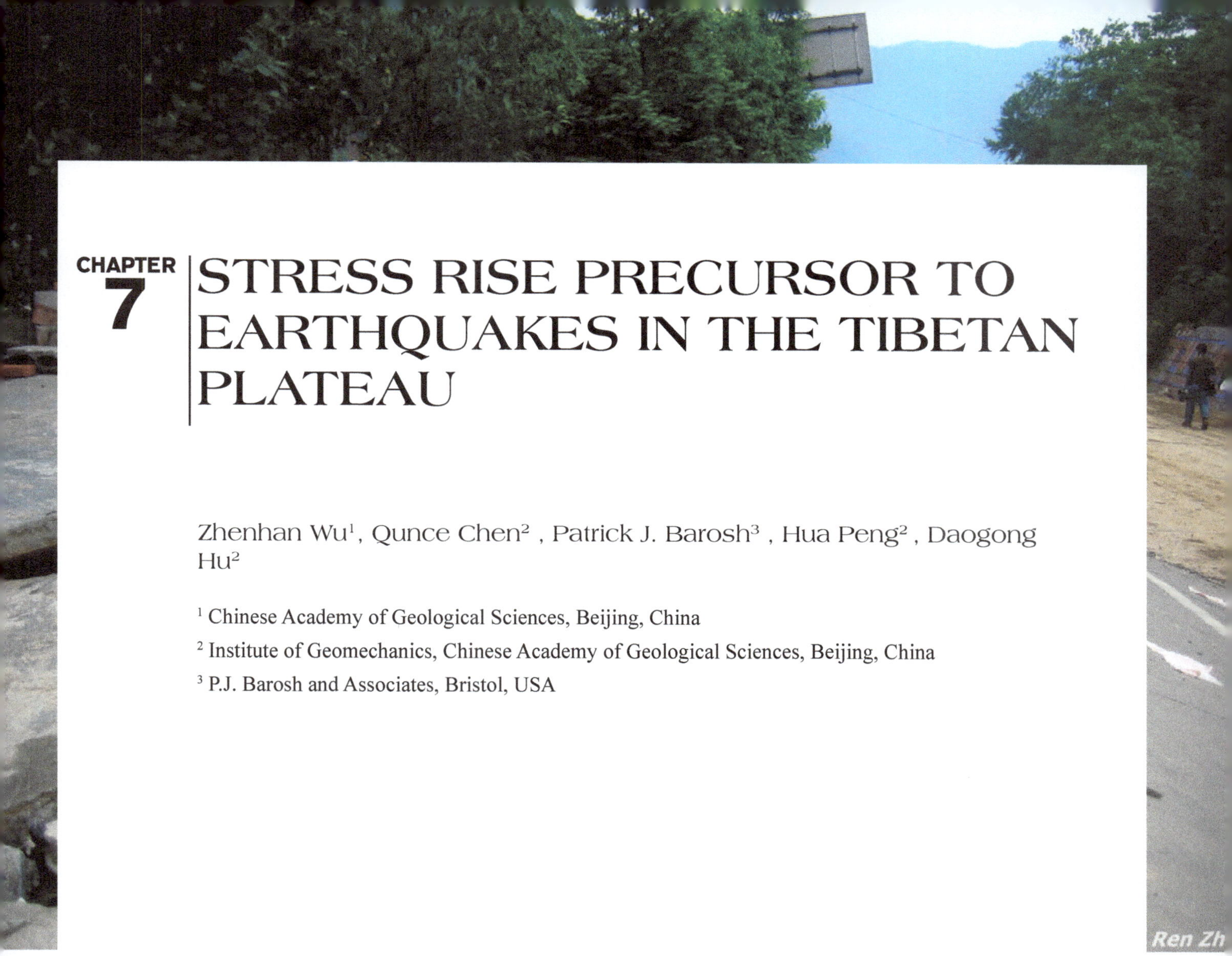

CHAPTER 7

STRESS RISE PRECURSOR TO EARTHQUAKES IN THE TIBETAN PLATEAU

Zhenhan Wu[1], Qunce Chen[2] , Patrick J. Barosh[3] , Hua Peng[2] , Daogong Hu[2]

[1] Chinese Academy of Geological Sciences, Beijing, China

[2] Institute of Geomechanics, Chinese Academy of Geological Sciences, Beijing, China

[3] P.J. Barosh and Associates, Bristol, USA

ABSTRACT

Earthquake prediction thus far has proven to be a very difficult task, but changes in situ stress appear to offer a viable approach for forecasting large earthquakes in Tibet and perhaps other continental regions. High stress anomalies formed along active faults before large earthquakes and disappeared soon after the earthquakes occurred in the Tibetan Plateau. Principle stress increased up to ~2 - 5 times higher than background stress to form high stress anomalies along causative faults before the Ms 8.1 West Kunlun Pass earthquake in November 2001, Ms 8.0 Wenchuan earthquake in May 2008, Ms 6.6 Nimu earthquake in October 2009, Ms 7.1 Yushu earthquake in April 2010 and the Ms 7.0 Lushan earthquake in April 2013. Stress near the epicenters rapidly increased 0.10 - 0.12 MPa over 45 days, ~8 months before the Ms 6.6 Nimu earthquake occurred. The high principle stress anomalies decreased quickly to the normal stress state in ~8 - 12 months after the Ms 8.1 West Kunlun Pass and the Ms 8.0 Wenchuan earthquakes. These high stress anomalies and their demise appear directly related to the immediate stress rise along a fault prior to the earthquakes and the release during the event. Thus, the stress rise appears to be a viable precursor in prediction of large continental earthquakes as in the Tibetan Plateau.

Keywords

Earthquake Prediction; High Stress Anomalies; *In-Situ* Stress Measurement; Large Earthquakes; Seismic Fault; Tibetan Plateau

INTRODUCTION

Large continental earthquakes are a major geologic hazard and commonly result in great disasters as recently occurred in China. The 2008 Ms 8.0 Wenchuan earthquake destroyed towns and villages along and near the seismic faults, and killed 69,227 people with 17,923 still missing [1]. The 2010 Ms 7.1 Yushu earthquake killed 2192 people with an additional 78 missing, and the 2008 Ms 6.6 Nimu earthquake resulted in significant economic loss. The Ms 7.0 Lushan earthquake occurred on April 20, 2013 killed 196 people with 21 still missing in southwestern Longmenshan Mts. Another event, the 2001 Ms 8.1 West Kunlun Pass earthquake, might have caused further devastation, but for it stroke a sparsely populated

Citation: Wu, Z. , Chen, Q. , Barosh, P. , Peng, H. and Hu, D. (2013) Stress rise precursor to earthquakes in the Tibetan Plateau. Natural Science, 5, 46-55. doi: 10.4236/ns.2013.58A1006.

region. The Chinese government has encouraged scientists to develop effective ways for tracing and predicting large earthquakes, aiming at decreasing seismic hazards. However, earthquake prediction has thus far proven a difficult task for the world's scientists even though seismic monitoring systems are established in many countries. Here we use stress to investigate these four large earthquakes in the Tibetan Plateau in hopes of finding a viable precursor for future earthquake prediction.

Ever since Reid [2] formulated the elastic rebound theory following the 1903 San Francisco earthquake, researchers have considered the relation of stress change and earthquakes. Li and his colleagues advocated the prediction of earthquakes by in-situ stress measurement in the 1960s [3,4] and considerable work has been accomplished ever since on measuring stress and delineating stress regions in China [5-7]. The lithosphere stress state now has been widely probed by in-situ stress measurements [8,9], but care must be used in relating these to earthquakes, because the measured stress may reflect the stress unrelated to the present tectonic movement [6,10]. Furthermore, although the delineation of general levels of stress can relate to general levels of seismic activity [11] this does not relate to activity along specific fault zones. Neither did estimates of deep stress arrive from epicentral distribution and focal plane solutions [12].

Direct measured changes in the stress fields, however, reflect the present day movement in the crust [4,11] and where these changes are adjacent to active faults they may relate to the seismic activity along the fault. Even small changes in stress, such as due to reservoir filling, are enough to trigger earthquakes sometimes [3]. Investigations into the crustal stress present and possible changes have been underway for the past decade for both engineering purposes and earthquake studies in the Tibetan Plateau now have revealed several changes in measured stress.

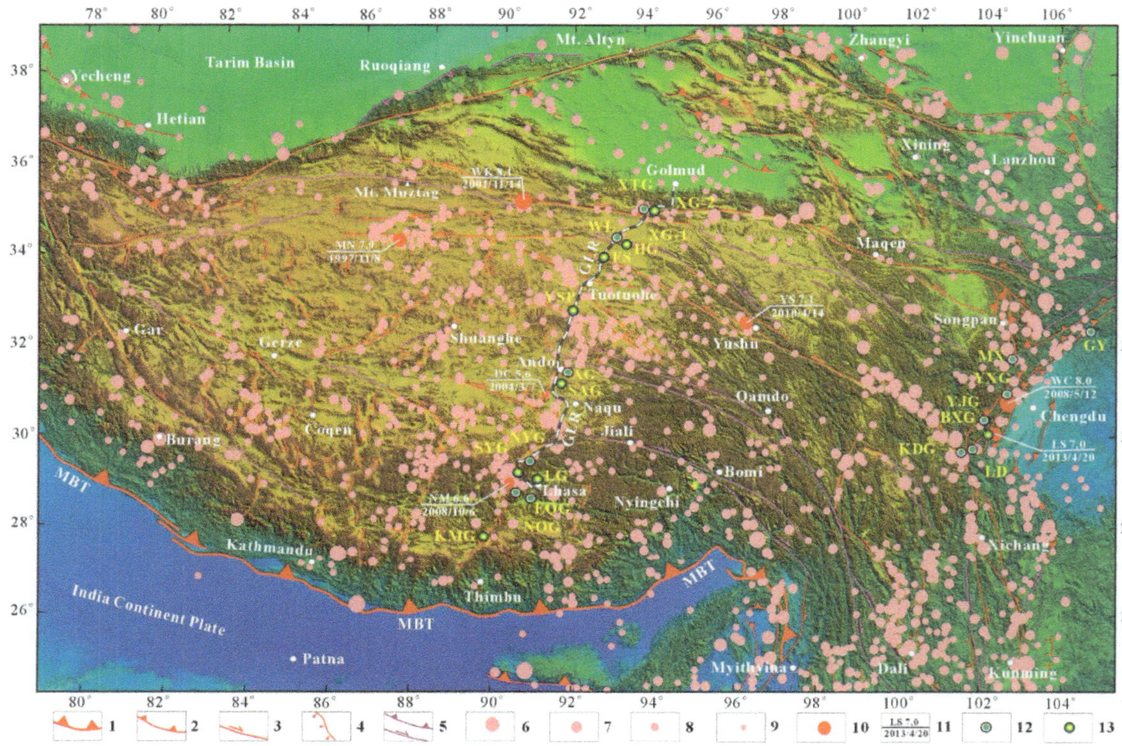

Figure 1. Tectonic map of active faults and seismicity in the Tibetan Plateau. Explanation: 1, Main Boundary Thrust (MBT); 2, Holocene thrust fault; 3, Holocene strike-slip fault; 4, Holocene normal fault; 5, Pleistocene thrust and strike-slip faults; 6, Ms 8.0 - 8.5 earthquake; 7, Ms 7.0 - 7.9 earthquake; 8, Ms 6.0 - 6.9 earthquake; 9, Ms 5.0 - 5.9 earthquake; 10, epicenter of earthquake in recent years; 11, location, magnitude and date of earthquake; 12, hydraufracturing stress measurement; 13, piezomagnetic stress measurement. The GLR refers to the Golmud-Lhasa Railway. The MN, WK, DC, YS, NM, WC and LS refer to the Mani, West of Kunlun Pass, Dong Co, Yushu, Nimu, Wenchuan and Lushan earthquakes, respectively. The XTG, XG-1, XG-2, WL, SYG, KMG, KDG, YJG, YXG, MX, BXG and others marked by yellow refer to the sites for stress measurement.

Intensive seismicity in the Tibetan Plateau is caused by the continued northward subduction of Indian Continental Plate that has created a variety of seismically active faults [13-15], which have produced large earthquakes over the past ten years in western China (Figure 1). The Ms 8.1 West Kunlun Pass earthquake occurred on November 14, 2001 and formed a fracture zone 350 - 426 km long on the Kusai Lake fault with a co-seismic sinistral slip of 7.6 - 8.0 m near Kusai Lake [16] and 3.5 - 4.7 m near Kunlun Pass [17]. The Ms 8.0 Wenchuan earthquake of May 12, 2008 formed a seismic rupture zone with a length of 240 - 270 km along the BeichuanYingxiu fault [18,19] and triggered thousands of largescale landslides, huge rock falls and debris flows in the Longmenshan Mountains [1], at the eastern margin of the Tibetan Plateau. Co-seismic offsets of 4.6 m dip-slip and 6.1 m dextral-slip were measured in Yingxiu and Beichuan, respectively [1,20]. The Ms 7.1 Yushu earthquake, which occurred on April 14, 2010, formed a seismic fracture zone with a length up to 23 km along the Garze-Yushu fault and co-seismic sinistral slips of 1.2 - 1.8 m happened near the epicenter northwest of Yushu. And the Ms 7.0 Lushan earthquake, which occurred on April 20, 2013, formed seismic hazards, landslides and rock falls without seismic rupture zone along the eastern marginal thrust in southwestern Longmenshan Mountains. These large earthquakes, together with the Ms 7.9 Mani earthquake that occurred on November 8, 1997, constitute the most recent activity on the faults bounding the Hohxil-Bayanhar-Garze structural block of the northern Tibetan Plateau (Figure 1). Earthquakes that occurred in the central Tibetan Plateau, such as the Ms 5.6 Dongco earthquake on March 7, 2004 and the Ms 6.6 Nimu earthquake on October 6, 2008, resulted from activity along the boundary normal faults of the YadongYangbajain-Gulu graben system.

STRESS CHANGES BEFORE AND AFTER LARGE EARTHQUAKES

Stress measurements by both the hydraulic fracturing and piezomagnetic methods have been taken in boreholes about the Tibetan Plateau since 2001 to provide data to characterize the stress field in which the large earthquakes occur and for engineering work. The great majority of the sites are in granite and gabbro. Hydraulic fracturing stress measurements were made in Xidatan granite (XTG), Wudaoliang limestone (WL), Ando gabbro (AG), Northeastern Yangbajain granite (NYG), Northwestern Quxi granite (NQG), Eastern Quxi granite (EQG) [21] along Golmud-Lhasa Railway across the Tibetan Plateau (Table 1). Hydraulic fracturing stress measurements in the Yaoji granite (YJG) [22], Yingxiu granite, Kangding granite (KDG) [19], Maoxian (MX) and Guanyuan (GY) [23] were conducted before and after the Ms 8.0 Wenchuan earthquake in central Longmenshan Mountains at the eastern margin of the Tibetan Plateau (Figure 1).

Piezomagnetic stress measurements were undertaken in the Xidatan granite (XG-1) and Xidatan gabbro (XG-2) [24], Hohxil granite (HG), Fenghuoshan siltstone (FS), Yanshiping sandstone (YSP), South Ando granite (SAG) [7], Southeastern Yangbajain granite (SYG), Lhasa granite (LG) and Kangma granite (KG) [25] (Figure 1). These contributed valuable data (Table 2) for construction of the Golmud-Lhasa railway and revealed an obvious stress change before and after the Ms 8.1 West Kunlun Pass earthquake [24]. And Piezomagnetic stress measurement in the Baoxing Granite (BXG) and Hydraulic fracturing stress measurement in the Luding (LD) were conducted before the Ms 7.0 Lushan earthquake in the southwestern Longmenshan Mountains.

Stress is measured at different depths in order to formulate the stress increase due to the additional rock mass with depth. Stress was measured by hydraulic fracturing on December 6, 2005 in a 301 m deep borehole in the Northeastern Yangbajain granite (NYG), east of Yangbajain graben. The maximum principle stress (S_H) and minimal principle stress (S_h) at NYG increase from 4.62 MPa and 3.82 MPa, respectively, at a depth 60 m to 11.34 MPa, and 10.20 MPa at 286 m (MPa = 145.0377 psi) (Table 1). The stress change with depth is found to be $S_H = 2.56 + 0.0340$ D and $S_h = 1.34 + 0.0325$ D with D referring to depth in meters. The NYG is located in central Lhasa block, ~9 km away from the active fault along the western boundary of the Yangbajain graben, and seismicity surrounding NYG is very weak (Figure 1). The stress recorded in NYG is therefore taken as representative for relatively stable areas in the Tibetan Plateau and is used as the background stress in analyzing stress anomalies (Figures 2 and 3).

Several other sites, AG, WL, LG and GY, also are located in relative stable areas away from seismic faults and seismic zones (Figure 1), and are characterized by relatively lower principle stress similar to NYG (Figures 2 and 3). Measurements along the Golmud-Lhasa railway found S_H and S_h in AG changed from 4.50 MPa and 3.82 MPa, respectively, at a depth of 50 m to 5.56 MPa and 5.00 MPa, respectively, at 135 m; the S_H and S_h in WL changed from 3.33 MPa and 2.30 MPa, respectively, at a depth of 28 m to 7.78 MPa and 5.10 MPa at 116 m; and the S_H and S_h in LG was 4.6 MPa and 2.6 MPa, respectively, at a depth of 18 m (Figures 2 and 3). At the northwest margin of the South China Block, S_H increased from 6 MPa at 100 m depth to 20 MPa at 347 m and S_h increased from 5 MPa at a depth of 100 m to 10 MPa at 347 m in Guanyuan (GY), northeast of Chengdu (Figures 2 and 3).

In contrast, the maximum principle stress S_H found along the active faults is abnormally high; always 2 - 5 times higher than in NYG; and the minimal principle stress S_h along the fault zones also increased before large earthquakes in the northern, eastern and central Tibetan Plateau. These formed high stress zones in S_H-D (Figure 2) and S_h-D diagrams (Figure 3).

West Kunlun Pass Earthquake (WK)

The principle stress S_H and S_h increased to 12.9 MPa and 12.1 MPa, respectively, at XG-1 at a depth of 18 m and 6.8 MPa and 4.4 MPa at XG-2 at a depth of 14 m in August 2001 (Table 2), ~3 months before the Ms 8.1 West Kunlun Pass earthquake [24]. The S_H and S_h then decreased to 3.5 MPa and 3.2 MPa, respectively at XG-1 at a depth of 18 m and 2.2 MPa and 1.2 MPa, respectively, in XG-2 at 14 m depth in July 2002 (Table 2), ~8 months after the Ms 8.1 West Kunlun Pass earthquake [24]. The principle stress S_H and S_h at XTG decreased to 2.32 - 7.88 MPa and 2.20 - 5.70 MPa, respectively, between depths of 44 and 152 m and 11.07 - 10.74 MPa and 7.0 - 6.5 MPa, at depths of 168 - 176 m (Figures 2 and 3).

Wenchuan Earthquake (WC)

The Ms 8.0 Wenchuan earthquake on May 12, 2008 resulted from an eastward thrust of the Songpan-Garze terrain over the South China Block [18,20]. The maximum principle stress (S_H) in the eastward thrust sheet was measured at 14.44-25.53 MPa from depths of 180.3 m to 280.5 m at YJG in 1999, ~9 years before the earthquake. This was in contrast with general values of 7.8 MPa at depths of 170 m found to the northwest [6]. The stress then had decreased to 3.48 - 8.34 MPa between depths of 90 m to 142 m at YXG and 1.74 - 6.41 MPa from depths of 55 m to 186 m at KDG in 2009 (Figure 2), one year after the Wenchuan earthquake. The minimal principle stress (S_h) also changed from high anomaly at YJG before the Wenchuan earthquake to normal in YXG at depths of 90 - 142 m and in KDG at depths of 55 - 186 m after the earthquake (Figure 3). The principle stress still remained at higher values at greater depths. The S_H and S_h was 13.11 - 15.62 MPa and 11.81 - 8.35 MPa, respectively, at depths of 171 - 185 m in Yingxiu (YXG) along the seismic Beichuan-Yingxiu fault one year after the Wenchuan earthquake (Tables 1 and 2). High principle stress also occurred in Maoxian (MX) along the seismic Beichuan-Yingxiu fault, and the S_H and S_h were 14 MPa and 8 MPa, respectively, in MX at a depth of 229 m (Table 1). However, the stress had remained low with S_H, 6 - 7.5 MPa, and S_h, 5.0 - 7.0 MPa in GY at depths of 100-200 m in the stable South China Block [23], ~4 years before the Wenchuan earthquake.

Table 1. Stress measured by hydraulic fracturing in the Tibetan Plateau

Site and rock/date measured	Depth (m)	Hydraulic fracturing parameters (MPa)					Stress (MPa)			Orientation of S_H
		Pb	Pr	Ps	Po	T	S_H	S_h	S_v	
Xidatan granite (XTG)*/August, 2005	44.0	5.92	4.25	2.20	0.03	1.67	2.32	2.20	1.15	N21°E
	64.0	7.66	4.03	2.30	0.23	3.63	2.64	2.30	1.67	
	84.0	8.92	6.40	3.50	0.44	2.52	3.66	3.50	2.20	N29°E
	126.0	7.40	7.03	5.20	0.86	0.37	7.71	5.20	3.29	
	152.0	10.29	8.10	5.70	1.12	2.18	7.88	5.70	3.98	N43°E
	168.0	9.36	8.66	7.00	1.27	0.70	11.07	7.00	4.38	N34°E
	176.0	8.70	7.40	6.50	1.36	1.30	10.74	6.50	4.59	N42°E
Wudaoliang limestone (WL)*/August, 2006	50.0	7.97	6.24	3.82	0.50	1.73	4.72	3.82	1.50	
	65.0	8.03	6.30	3.82	0.66	1.73	4.50	3.82	1.98	N10°W
	93.0	8.67	7.01	4.82	0.93	1.66	6.52	4.82	2.79	
	129.0	9.45	8.10	5.14	1.29	1.35	6.03	5.14	3.88	N9°W
	135.0	10.01	8.08	5.00	1.36	1.93	5.56	5.00	4.07	
Ando gabbro (AG)**/October, 2005	28.0	4.74	3.40	2.30	0.16	1.33	3.33	2.30	0.79	N11°W
	55.0	6.13	3.22	2.20	0.43	2.90	2.94	2.20	1.55	
	75.0	7.13	5.12	3.60	0.63	2.01	5.05	3.60	2.11	N24°W
	91.0	5.92	5.62	4.00	0.79	0.30	5.58	4.00	2.56	
	116.0	8.23	6.48	5.10	1.04	1.75	7.78	5.10	3.26	N13°W
Northeastern Yangbajain granite (NYG)***/December, 2005	60.0	7.97	6.24	3.82	0.60	1.73	4.62	3.82	1.81	
	85.0	8.13	6.30	3.82	0.86	1.83	4.30	3.82	2.57	N55°E
	113.0	8.87	7.31	4.82	1.14	1.56	6.02	4.82	3.41	
	139.0	10.25	9.10	6.14	1.40	1.15	7.92	6.14	4.19	
	165.0	11.31	10.08	6.16	1.66	1.23	6.75	6.16	4.97	N46°E
	209.0	14.20	12.41	8.39	2.09	1.79	10.67	8.39	6.27	
	222.0	14.40	12.30	8.10	2.23	2.10	9.77	8.10	6.68	
	243.0	16.60	14.10	9.20	2.44	2.50	11.06	9.20	7.31	
	261.0	18.30	16.90	10.8	2.62	1.40	12.88	10.80	7.85	N65°E
	286.0	18.73	16.40	10.2	2.86	2.33	11.34	10.20	8.59	N53°E
Northwestern Quxi granite (NQG)*/July, 2007	73.8	16.84	13.34	8.54	0.54	3.50	11.73	8.54	1.96	
	115.8	23.58	11.76	9.03	0.96	11.82	14.36	9.03	3.07	N46.5°W
	122.3	16.18	10.72	8.63	1.03	5.46	14.15	8.63	3.24	N39.6°W
	134.8	18.52	9.72	7.65	1.15	8.80	12.07	7.65	3.57	
Eastern Quxi granite (EQG)**/After Cao *et al.*, 2003 [21]	133.0	25.43	6.84	5.75	0.90	18.60	9.51	5.75	3.60	
	170.0	-	6.12	6.80	1.27	-	13.01	6.80	4.60	
	241.0	24.10	12.60	10.90	1.98	11.50	18.12	10.90	6.50	N3.0°E
	283.0	16.43	11.00	10.31	2.40	5.43	17.53	10.31	7.60	
	296.0	24.80	11.12	9.08	2.53	13.60	13.59	9.08	8.00	
Yaoji granite (YJG)/after Mao *et al.*, 1999 [22]	180.3	9.04	8.34	8.28	1.08	0.7	15.42	8.28	4.77	
	187.6	12.72	8.95	8.19	1.15	3.77	14.44	8.19	4.96	N57°E
	224.4	14.23	12.46	11.61	1.51	1.77	20.86	11.61	5.49	N63°E
	233.3	16.69	13.66	11.11	1.60	3.03	18.07	11.11	6.18	N1°W
	250.3	12.29	10.89	10.80	1.77	1.40	19.74	10.80	6.63	N55°E
	259.1	13.69	11.81	11.68	1.85	1.88	21.38	11.68	6.86	
	280.5	15.60	13.00	13.53	2.06	2.60	25.53	13.53	7.42	

	90.0	4.30	3.50	2.60	0.82	0.80	3.48	2.60	2.39	
	128.0	13.62	12.78	7.44	1.20	0.84	8.34	7.44	3.39	
Yingxiu granite (YXG)/after Wu *et al.*, 2009 [19]	142.0	9.95	8.70	5.95	1.34	1.25	7.81	5.95	3.76	N56°W
	171.0	18.21	18.18	11.81	1.63	0.03	15.62	11.81	4.53	
	185.0	12.73	10.17	8.35	1.77	2.56	13.11	8.35	4.90	
	55.0	4.47	3.27	1.70	0.31	1.20	1.74	1.70	1.46	N5°E
	81.0	4.66	3.91	2.44	0.57	0.75	2.84	2.44	2.15	N81°E
	94.0	4.10	3.61	2.52	0.70	0.49	3.25	2.52	2.49	
Kangding granite (KDG)/after Wu *et al.*, 2009 [19]	117.0	5.03	3.93	2.31	0.93	1.10	2.46	2.42	3.10	
	136.0	9.81	6.12	3.55	1.12	3.69	3.67	3.55	3.60	
	172.0	17.32	11.47	5.92	1.48	5.85	6.41	6.12	4.56	
	186.0	10.91	6.89	4.94	1.62	4.02	6.31	4.94	4.93	
Luding (LD)	150.0						16.00	10.00	4.00	N60°W
Maoxian (MX)	229.0						14.00	8.00	6.00	N55°W
	100.0						6.00	5.00	2.65	N30°W
Guanyuan (GY)/after An *et al.*, 2004 [23]	200.0						7.50	7.00	5.30	N35°W
	347.0						20.00	10.00	9.20	

Explanation: P_b, break down pressure; P_r, re-opening pressure; P_s, shut-in pressure; P_o: pore pressure; T, tensor strength; S_H, horizontal maximum principle stress; S_h, horizontal minimal principle stress; S_v, vertical stress caused by overloading rocks with density 2.63 $(g/cm^3)^*$, 2.70 $(g/cm^3)^{**}$ or 3.02 $(g/cm^3)^{***}$. Stress of LD, MX and GY came from An *et al.*, 2004 [23]. Sites of measured stress are marked by yellow in **Figure 1**.

Dongco Earthquake (DC)

The Ms 5.6 Dongco earthquake, which was due to normal fault offset in the Yadong-Yangbajain-Gulu graben system, was preceded by a high stress anomaly in the South Ando granite (SAG) (Figures 2 and 3). Principle stress S_H and S_h reached 8.1 MPa and 4.8 MPa, respectively, in SAG at a depth of 14 m in July 2003; this is more than 2 times higher than background stress (Figures 2 and 3), and ~8 months later the Dongco earthquake occurred southwest of SAG in the Ando-Dongco graben on March 7, 2004 (Figure 1).

Nimu Earthquake (NM)

A high stress anomaly also was found adjacent to the Yadong-Yangbajain graben system before the Ms 6.6 Nimu earthquake occurred on October 6, 2008. Principle stress S_H and S_h was as high as 5.7 - 10.4 MPa and 4.6 - 8.4 MPa, respectively, in SYG at a depth of 12 - 13 m in October-November 2001, and reached 11.73-14.36 MPa and 7.65 - 9.03 MPa at a depth of 73.8 - 134.8m in NQG in July 2007 (Figures 2 and 3), one year before the Nimu earthquake. Stress in NQG at a depth of 31 - 35 m increased quickly ~8 months before the earthquake; S1, oriented N40°W, changed from 7.855 MPa to 7.958 MPa; S_2, oriented S80°W changed from 9.627 MPa to 9.751 MPa; and S_3, oriented N20°E, changed from 7.312 MPa to 7.426 MPa. Thus S_1, S_2 and S_3 increased 0.103 MPa, 0.124 MPa and 0.114 MPa, respectively, in 45 days from January 13 to February 26, 2008 (Figure 4). However, the principle stress in relatively stable areas away from Yadong-Yangbajain graben system remained lower (Figure 2) even though the measurement sites are not located too far from the epicenter of the Nimu earthquake. The S_H and S_h in NYG ranged from 4.62 MPa and 3.82 MPa, respectively, at 60 m depth to 11.34 MPa and 10.20 MPa at a depth of 286 m in December 2005, and S_H and S_h in EQG ranged between 9.51 - 18.12 MPa and 5.75 - 10.90 MPa, respectively, at depths of 133 - 283 m in 2003 (Table 1).

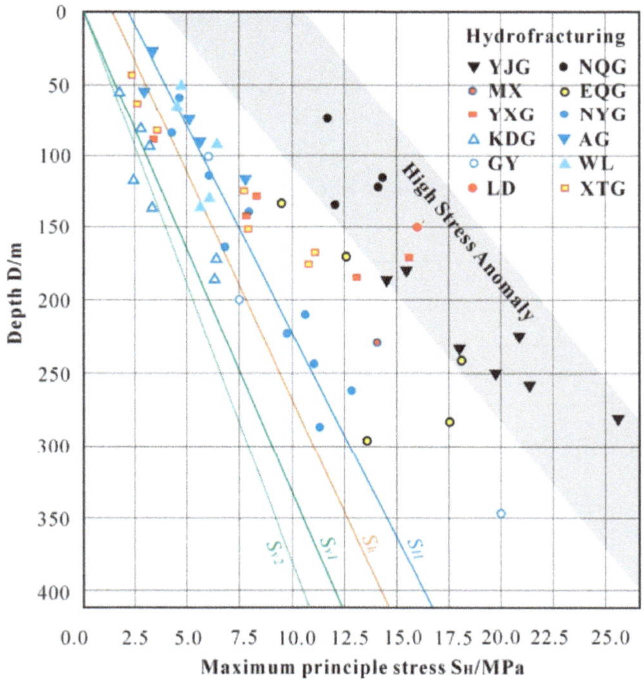

Figure 2. Vertical change of maximum principle stress in the Tibetan Plateau. Explanation: S_H, maximum principle stress of Yangbajain granite at NYG; S_h, minimal principle stress of Yangbajain granite at NYG; Sv1 and Sv2, vertical stress caused by overlying rocks with average densities of 3.02 g/cm3 and 2.7 g/cm3 respectively. The shaded area refers to high stress zone for large earthquakes.

Yushu Earthquake (YS)

The Ms 7.1 Yushu earthquake followed high stress in the Hohxil granite (HG) and Fenghuoshan siltstone (FS). Principle stress S_H and S_h were 6.8 MPa and 3.2 MPa, respectively, in HG at a depth of 20 m in May 2002; 4.6 - 5.5 MPa and 2.8 - 2.9 MPa, respectively, in FS at depths of 12 - 16 m in August 2002; being 1.8 - 2.7 times higher than background stress (Figure 5). The high stress in HG and FS formed along the Wudaoliang and Fenghuoshan sinistral-slip faults, which are en echelon or branch faults of Garze-Yushu fault which broke during the earthquake. The Yushu earthquake occurred on April 14, 2010 along southern boundary faults of Hohxil-Bayanhar-Garze block (Figure 1).

Lushan Earthquake (LS)

The Ms 7.0 Lushan earthquake occurred along the Eastern Marginal Thrust, southwestern Longmenshan Mts. on April 20, 2013, ~5 years after the Ms 8.0 Wenchuan earthquake along the Beichuan-Yingxiu fault, central Longmenshan Mts. High stress anomalies were found in the Luding (LD) (Figures 2 and 3) and Baoxing granite (BXG) (Figure 5) before the Ms 7.0 Lushan earthquake. The S_H and S_h were 16 MPa and 10 MPa, respectively, in LD at 150 m depth (An et al., 2004), and the principle stress S_H and S_h increased to 9.8 MPa and 7.9 MPa, respectively, in Baoxing Granite (BXG) at depth of 18 - 22 m in September 2009, ~3.5 years before the Ms 7.0 Lushan earthquake (Table 2).

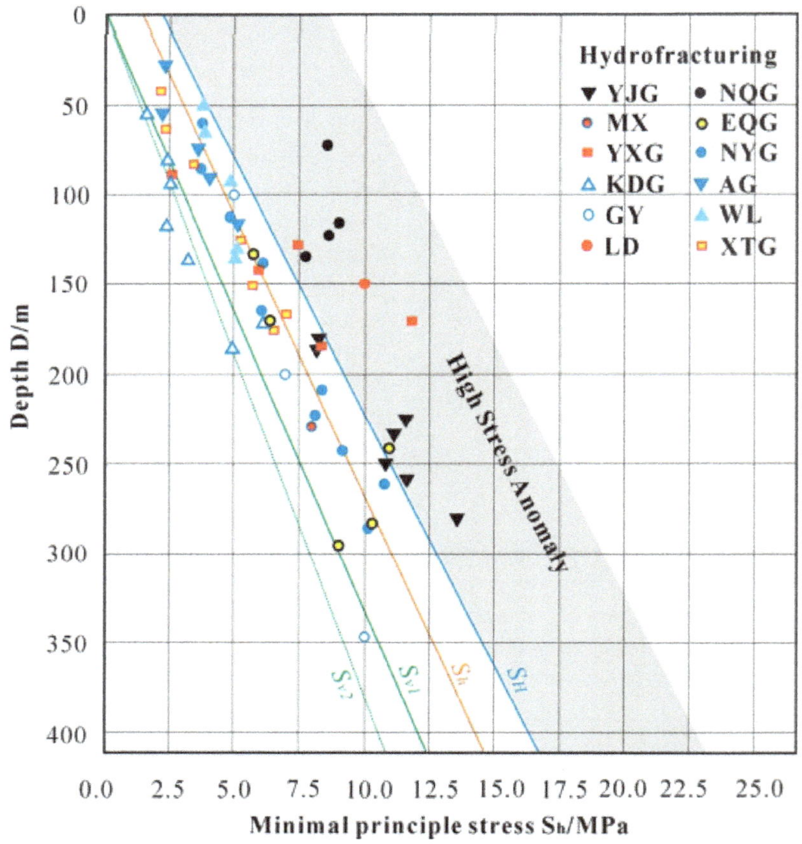

Figure 3. Vertical change of minimal principle stress in the Tibetan Plateau. Explanation: same as Figure 2.

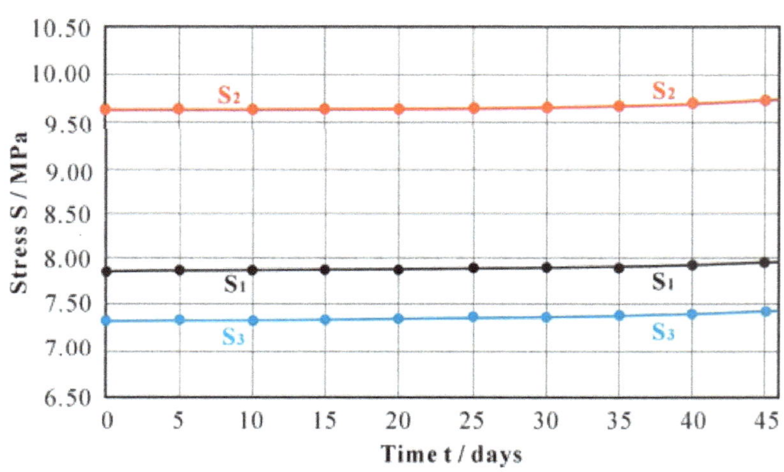

Figure 4. Diagram of stress changes in Northwestern Quxi granite. Explanation: S_1, S_2 and S_3 represents piezomagnetic stress measured in bore hole at depth of 31 - 35 m at orientations of N40°W, S80°W and N20°, respectively, from January 13 to February 26, 2008.

CONCLUSIONS AND DISCUSSION

The principle stress along the seismically active faults generally rose to ~2 - 5 times higher than background stress before large earthquakes occurred in the Tibetan Plateau and high stress anomalies near seismic faults were precursors of large earthquakes (Figures 2, 3 and 5). Maximum principle stress (S_H) and minimal principle stress (S_h) along the seismic fault increased to 6.8 - 12.9 MPa and 4.4 - 12.1 MPa, respectively, at a depth of 14 - 18 m, being 2.5 - 5.2 times higher than background stress ~3 months before the Ms 8.1 West Kunlun Pass earthquake in the northern plateau, and decreased to normal stress state ~8 months after the earthquake (Table 2). The S_H and S_h in Yaoji granite on eastward thrust sheet increased to 14.44 - 25.53 MPa at depths of 180.3 - 280.5 m before the Ms 8.0 Wenchuan earthquake, being ~2 times higher than background stress, and recovered to a normal stress state one year after the earthquake in Longmenshan Mountains at the eastern margin of the Tibetan Plateau (Figures 2 and 3). The S_H and S_h increased to 9.8 MPa and 7.9 MPa, respectively, in BXG at depth of 18 - 22 m, and 16 MPa and 10 MPa, respectively, in LD at 150 m depth before the Ms 7.0 Lushan earthquake. The principle stress S_H and S_h increased to 2.2 - 4.2 times higher than background stress before the Ms 6.6 Nimu earthquake; stress in 3 orientations rapidly increased 0.10 - 0.12 MPa within 45 days near seismic epicenter ~8 months before the Nimu earthquake in the southern Tibetan Plateau (Figure 4). A high stress anomaly also formed in Fenghuoshan (FS) and Hohxil (HG) along seismic faults before the Ms 7.1 Yushu earthquake occurred in April 2010.

Table 2. Stress measured by piezomagnetic method in Tibetan Plateau

Site and rock	Date	Depth (m)	S_H (Mpa)	S_h (Mpa)	S_H orientation	Reference
Xidatan granite (XG-1)	August 2001	18	12.9	12.1	N45°E	[24]
Xidatan granite (XG-1)	July 2002	18	3.5	3.2	N66°E	[24]
Xidatan gabbro (XG-2)	August 2001	14	6.8	4.4	N58°E	[24]
Xidatan gabbro (XG-2)	July 2002	14	2.2	1.2	N5°W	[24]
Hohxil granite (HG)	May 2002	20	6.8	3.2	N26°E	[7]
Fenghuoshan siltstone (FS)	August 2002	16	5.5	2.9	N84°E	[7]
	August 2002	12	4.6	2.8	N61°E	[7]
Yangshiping sandstone (YSP)	June 2003	13	5.6	4.4	N47°E	[7]
South Ando granite (SAG)	July 2003	14	8.1	4.8	N64°W	[7]
Southeastern Yangbajain granite (SYG)	October 2001	13	10.4	8.4	N70°E	[25]
	October 2001	12	5.7	2.7	N81°E	[25]
	November 2001	12	6.6	4.6	N45°E	[25]
Lhasa granite (LG)	October 2003	18	4.0	2.6	N38°W	[25]
Kangma granite (KMG)	September 2003	13	5.2	4.4	N49°W	[25]
Baoxing granite (BXG)	September 2009	18 - 22	9.8	7.9	N51°W	[19]
Yingxiu Granite (YXG)	September 2009	20	4.3	2.7	N19°E	[19]
Kangding Granite (KDG)	September 2009	20	2.6	1.8	N39°E	[19]

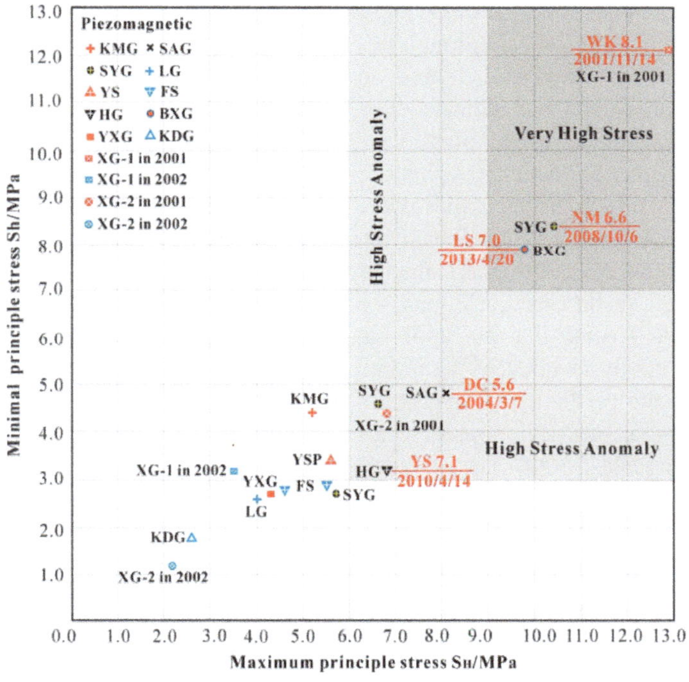

Figure 5. Diagram of piezomagnetic principle stress change in the Tibetan Plateau. Explanation: the SYG, KMG, FS and BXG et al. refer to the sites for piezomagnetic stress measurement, and the YS 7.1/2010/4/14 et al. refer to the related earthquakes occurred in recent years.

Additional high stresses were also found in the Yanshiping sandstone (YSP) where the S_H and S_h were 5.6 MPa and 4.4 MPa, respectively, at a depth of 13 m in June 2003, and in the Kangma Granite (KMG) where the S_H and S_h were 5.2 MPa and 4.4 MPa, respectively, at a depth 13m (Table 2). If the maximum stress S_H continues to increase as much as 0.5 - 1.0 MPa in the related areas, then the YSP and KMG will be plotted in the high stress anomaly in S_H-S_h diagram (Figure 5). This high stress anomaly could be connected with some Ms 6.0 - 7.0 earthquake in the near future.

Significantly different types of fault zones show the stress increase even though they might be compressional or extensional structures. The West Kunlun Pass and Yushu earthquakes were due to strike-slip offset, the Wenchuan earthquake was due to oblique-slip thrust offset, the Lushan earthquake was due to thrust offset and the Nimu and Dongco earthquakes were due to normal fault offset. Thus, the fault type makes no obvious difference in the development of stress anomalies prior to an earthquake. Stress variations relative to rock mechanic changes [26] and tectonic divisions [6], local stress from non-tectonic processes as permafrost freezing [27] and ancient residual stress [10] might cause local stress anomalies, but these factors rarely lead to regional stress anomalies along active faults and result in regular stress changes before and after large earthquakes.

Apparently deep barriers acting as friction decrease slip rates or even stop slip along active faults and increase stress along the blocked segments of active faults, leading to high stress anomalies before large earthquakes (Figures 2, 3 and 5). This is a reflection of the strain build up near faults as proposed by Reid [2], but the resulting stress seems more restricted and the buildup is more rapid than what might be expected. These characteristics aid in using high stress anomalies in the upper crust surface as viable precursors for prediction of large earthquakes within continents, such as the Tibetan Plateau.

Other indices as Coulomb failure stress on faults [28,29], tectonic stress and stress variations inferred from numerical modeling [12,30-32], and hydrochemistry anomalies [33] also have been suggested in forecasting earthquakes, but none are as sensitive to earthquakes as the in-situ measured stress in the Tibetan Plateau. The GPS measurements indicated low slip rates of active faults and low velocity of terrains in the Longmenshan Mountains [15], where the Wenchuan earthquake occurred, and no high velocity or strain anomalies were discovered by GPS measurements before the Ms 8.0 Wenchuan earthquake, the Ms 8.1 West Kunlun Pass earthquake, the Ms 6.6 Nimu earthquake, the Ms 7.1 Yushu earthquake and the Ms 7.0 Lushan earthquake in the Tibetan Plateau, implying that GPS measurements are not nearly as effective as in-situ stress measurements for earthquake prediction. Using stress to monitor earthquakes appears a practical approach for forecasting large continental earthquakes.

ACKNOWLEDGEMENTS

This research was supported by the Ministry of Land and Resources of China under grant No. 201211095 and China Geological Survey under grant Nos. 1212011120185 and 1212011221111. Thanks to the Tibetan Bureau of Geology and Mineral Resources for their help in field drilling, and thanks to the expert who reviewed the manuscript.

REFERENCES

1. Wu, Z.H., Barosh, Patrick, J., Zhang, Z.C. and Liao, H.J. (2012) Effects from the Wenchuan Earthquake and seismic hazard in the Longmenshan Mountains at the eastern margin of the Tibetan Plateau. Engineering Geology, 143-144, 28-36. doi:10.1016/j.enggeo.2012.06.006
2. Reid, H.F. (1910) Mechanics of the earthquake, the California Earthquake of April 18, 1906. Report of the State Investigation Commission, Carnegie Institution of Washington, Washington DC.
3. Li, S.-G. (1974) Earthquake geology. Science Press, Beijing.
4. Li, F.Q. (2010) In-situ stress measurement is an important approach to realize earthquake prediction—Developing J.S. Lee's scientific ideas on earthquake prediction. In: Xue and Furen, Eds., Rock Stress and Earthquakes, CRC Press, Taylor and Francis Group, London, 757-759.
5. Chen, Q.X. (1998) Analysis of rock mechanics and tectonic stress field. Geology Publishing House, Beijing.
6. Yao, R., Yang, S.X., Lu, Y.Z., Cui, X.F., Chen, Q. and Mi, Q. (2010) Characteristics of tectonic stress in the east of Tibetan Plateau and its neighboring region inferred from in-situ stress measurements. In: Xue and Furen, Eds., Rock Stress and Earthquakes, CRC Press, Taylor and Francis Group, London, 687-693.
7. Wu, M.L., Zhang, C.S., Liao, C.T., Ma, Y.S. and Ou, M.Y. (2005) The recent state of stress in the central Qinghai-Tibet Plateau according to in-situ stress measurements. Chinese Journal of Geophysics, 48, 327-332.
8. Zoback, M.L. (1992) First and second order patterns of stress in the lithosphere: The world stress map project. Journal of Geophysical Research, 97, 11703-11728. doi:10.1029/92JB00132
9. Heidbach, O., Tingay, M., Barth, A., Reinecker, J., Kurfe, D. and Miller, B. (2008) The release 2008 of the world stress map. www.world-stress-map.org
10. Barosh, P.J. (1986) Neotectonic movement, earthquakes and stress state in the eastern United States. Tectonophysics, 132, 117-152. doi:10.1016/0040-1951(86)90029-6
11. Xie, F.R., Zhang, H.Y. and Du, Y. (2010) The recent tectonic stress districts and strong earthquakes in China. In: Xue and Furen, Eds., Rock Stress and Earthquakes, CRC Press, Taylor and Francis Group, London, 35-40.
12. Zhang, Y.-Q. and Xie, F.-R. (2010) Background stress state estimated from 2008 Wenchuan earthquake sequence. In: Xue and Furen, Eds., Rock Stress and Earthquakes, CRC Press, Taylor and Francis Group, London, 707-712.
13. Molnar, P. and Tapponnier, P. (1978) Active tectonics of Tibet. Journal of Geophysical Research, 83, 5361-5375. doi:10.1029/JB083iB11p05361
14. Tapponnier, P., Peltzer, A. Y., Le, Dain, Amijo, R. and Cobbold, P. (1982) Propagating extrusion tectonics in Asia: New insights from simple experiments with plasticine. Geology, 10, 611-616. doi:10.1130/0091-7613(1982)102.0.CO;2

15. Zhang, P.Z., Shen, Z.K., Wang, M., Gan, W.J., Burgmann, R., Molnar, P., Wang Q., Niu, Z.J., Sun, J.Z., Wu, J.C., Sun, H.R. and You, X.Z. (2004) Continuous deformation of the Tibetan Plateau from global positioning system data. Geology, 32, 809-812. doi:10.1130/G20554.1

16. China Seismological Bureau (2003) Album of the Ms 8.1 earthquake occurred in west of Kunlun Pass, China. Seismological Press, Beijing, 1-105.

17. Wu, Z.H., Wu, Z.H., Hu, D.G., Wang, W., Zhang Z.C. and Lei, W.Z. (2005) Album of active faults and geological hazards along the Golmud-Lhasa Railway across the Tibetan Plateau. Seismological Press, Beijing, 1-150.

18. Xu, X.W., Yu, G.H., Chen, G.H., Ran, Y.K., Li, C.X., Chen, Y.G. and Chang, C.P. (2009) Parameters of coseismic reverse- and oblique-slip surface ruptures of the 2008 Wenchuan earthquake, eastern Tibetan Plateau. Acta Geologica Sinica, 83, 673-684. doi:10.1111/j.1755-6724.2009.00091.x

19. Wu, M.L., Zhang, Y.Q., Liao, C.T., Chen, Q., Ma, Y.S., Wu, J.S., Yan, J.F. and Ou, M.Y. (2009) Preliminary results of in-situ stress measurements along the Longmenshan fault zone after the Wenchuan Ms 8.0 earthquake. Acta Geologica Sinica, 83, 746-753. doi:10.1111/j.1755-6724.2009.00098.x

20. Wu, Z.H., Dong, S.W., Barosh, P.J., Zhang Z.C. and Liao, H.J. (2009) Dextral-slip thrust faulting and seismic events of the Ms 8.0 Wenchuan earthquake Longmenshan Mountains, eastern margin of the Tibetan Plateau. Acta Geologica Sinica, 83, 685-693. doi:10.1111/j.1755-6724.2009.00092.x

21. Cao, Z.Q., Xie, P., Jin, H., Chen, Q. and Mao, J.Z. (2003) Variety characteristics of the crustal stress field near the Brahmaputra fault zone. Progress in Geophysics, 18, 167- 172.

22. Mao, J.Z., Li, F.Q., Zhang, Z.G., Chen, Q., Ceng, J.Q. and Yang, X. (1999) Report on stress measurement by hydraulic fracturing in Yaoji reservoir dam. Institute of Crust Dynamics, China Seismological Bureau, Beijing.

23. An, Q.M., Ding, L.F., Wang, H.Z. and Zhao, S.G. (2004) Research of property and activity of Longmenshan Mountain fault zone. Journal of Geodesy and Geodynamics, 24, 115-119.

24. Liao, C.T., Zhang, C.S., Wu, M.L., Ma, Y.S. and Ou, M.Y. (2003) Stress change near the Kunlun Fault before and after the Ms 8.1 Kunlun Earthquake. Geophysical Research Letter, 30, 2027-2030. doi:10.1029/2003GL018106

25. Zhang, C.S., Wu, M.L., Liao, C.T., Ma, Y.S. and Ou, M.Y. (2007) The result of current stress measurements and stress state analysis in the region of Yangbajain-Kangmar in Tibet. Chinese Journal of Geophysics, 50, 517-522.

26. Hudson, J.A. and Feng, X.T. (2010) Variability of in situ rock stress. In: Xue and Furen, Eds., Rock Stress and Earthquakes, CRC Press, Taylor land Francis Group, London, 3-10.

27. Wu, Z.H., Barosh, P.J., Wang, L.J., Hu, D.G. and Wang, W. (2008) Numerical modeling of stress and strain associated with bending of an oil pipeline by a migrating pingo in the permafrost region of the northern Tibetan Plateau. Engineering Geology, 96, 62-77. doi:10.1016/j.enggeo.2007.10.001

28. Shan, B., Xiong, X., Zheng, Y. and Diao, F.Q. (2009) Stress changes on major faults caused by Mw 7.9 Wenchuan earthquake, May 12, 2008. Science in China Series D: Earth Sciences, 52, 593-601. doi:10.1007/s11430-009-0060-9

29. Hori, T. and Kaneda, Y. (2004) Physical criterion to evaluate seismic activity associated with the seismic cycle of great intraplate earthquakes. Journal of Seismology, 8, 225-233. doi:10.1023/B:JOSE.0000021364.93301.2a

30. Spudich, P., Guatteri, M., Otsuki, K. and Minagawa, J. (1998) Use of fault striations and dislocation models to infer tectonic stress during the 1995 Hyogo-Ken Nanbu (Kobe) earthquake. Bulletin of Seismological Society of America, 88, 413-427.

31. Kato, N. and Hirasawa, T. (1999) The variation of stresses due to aseismic sliding and its effect on seismic activity. Pure and Applied Geophysics, 155, 425-442. doi:10.1007/s000240050273

32. Tiampo, K.F., Bowman, D.D., Colella, H. and Rundle, J.B. (2008) The stress accumulation method and pattern informatics index: Complementary approaches to earthquake forecasting. Pure and Applied Geophysics, 165, 693-709. doi:10.1007/s00024-008-0329-5

33. Harmann, J. and Levy, K.J. (2006) The influence of seismotectonics on precursory changes in undergroundwater composition for the 1995 Kobe earthquake, Japan. Hydrogeology Journal, 4, 1307-1318. doi:10.1007/s10040-006-0030-7

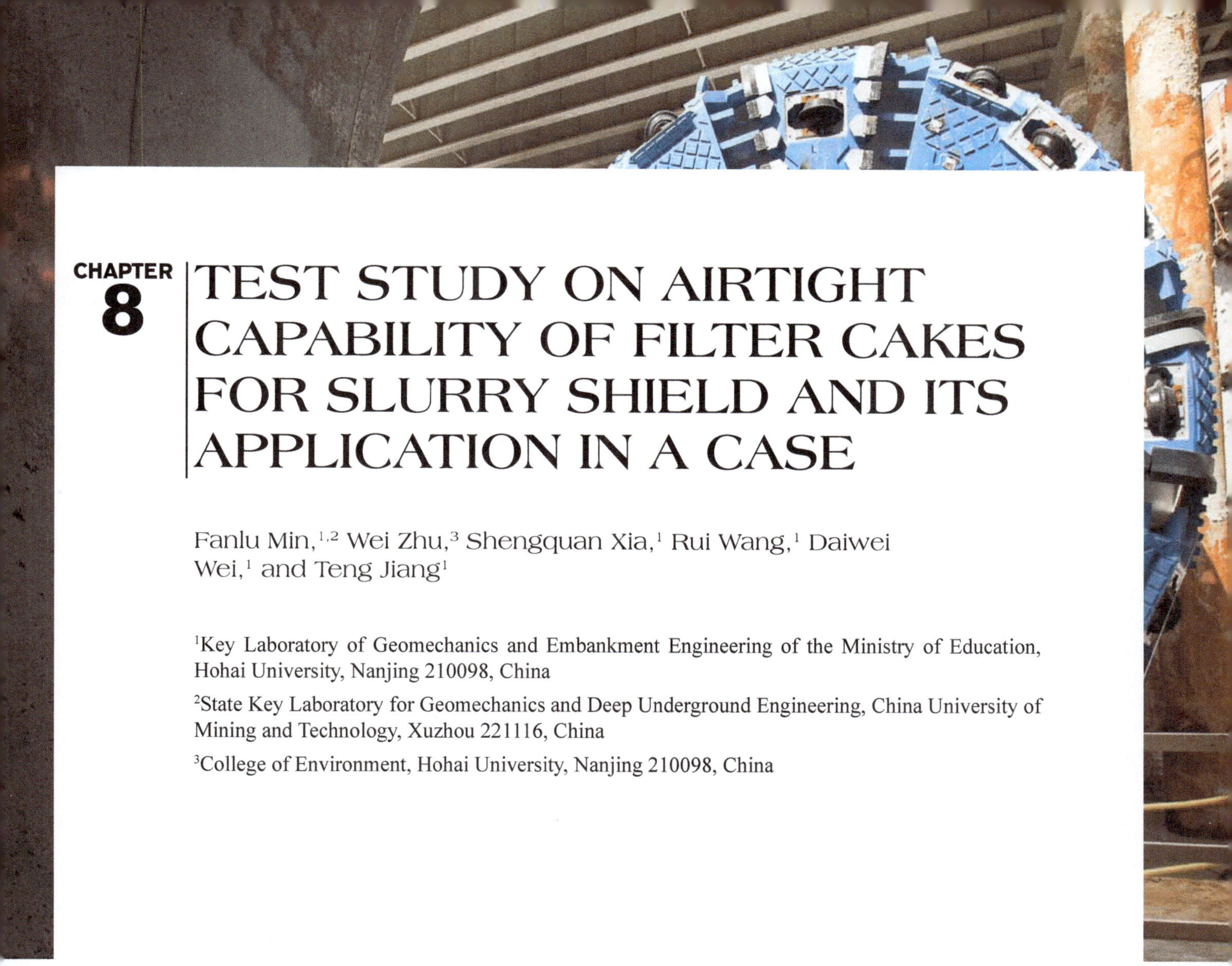

CHAPTER 8

TEST STUDY ON AIRTIGHT CAPABILITY OF FILTER CAKES FOR SLURRY SHIELD AND ITS APPLICATION IN A CASE

Fanlu Min,[1,2] Wei Zhu,[3] Shengquan Xia,[1] Rui Wang,[1] Daiwei Wei,[1] and Teng Jiang[1]

[1]Key Laboratory of Geomechanics and Embankment Engineering of the Ministry of Education, Hohai University, Nanjing 210098, China

[2]State Key Laboratory for Geomechanics and Deep Underground Engineering, China University of Mining and Technology, Xuzhou 221116, China

[3]College of Environment, Hohai University, Nanjing 210098, China

ABSTRACT

To learn the airproof capacity of filter cakes as opening chambers under air pressure, a series of tests were carried out. The variations of discharged water with air pressure and time were observed, and the relationship between airproof capacity of filter cakes and surrounding air pressure was analysed. The test results indicated that there were three stages as compressed air acting on filter cakes: completely not infiltration, a very small amount of infiltration, and penetration leakage. The certain air pressure between the first and second stages was called the airproofing value of filter cake. And a capillary bundle model was used to explain the mechanism of air tightness of filter cakes. In Nanjing Yangtze River Tunnel, a 5 cm thickness filter cake was formed in gravel sand, and its airproofing value was a little lower than 0.12 MPa. The air pressure used as opening chamber should be equal to the summation of water pressure in sand and airproofing value of filter cake. While the air pressure is larger than the summation, the filter cake would be gas permeable. The slurry formulation and airproofing value of filter cakes obtained in the tests were applied successfully in Nanjing Yangtze River Tunnel.

INTRODUCTION

Nowadays, shield tunneling method has been widely used in the construction of subway and river-crossing tunnels in China [1]. However, shield-tunneling machine malfunctions and highly worn cutting wheel or cutters were encountered frequently and nagged the construction enterprises seriously in the difficult geological conditions [2–4]. To restore the excavation, technicists have to open chambers to maintain equipment or exchange the cutters. Meanwhile maintaining the stability of tunnel face is the most vital problem as opening chambers in shield tunneling projects [5]. Generally, compressed air is used to support the tunnel face during opening chamber operations in some subjects. However, in some soils with high permeability, the tunnel face cannot be supported effectively without any accessory measures as the massive air leakage. This problem has attracted the attention of construction enterprises in the world and some accessory measures, such as filter cakes formed on the surface, have been

Citation: Fanlu Min, Wei Zhu, Shengquan Xia, Rui Wang, Daiwei Wei, and Teng Jiang, "Test Study on Airtight Capability of Filter Cakes for Slurry Shield and Its Application in a Case," Advances in Materials Science and Engineering, vol. 2014, Article ID 696801, 8 pages, 2014. doi:10.1155/2014/696801.

taken to reduce the permeability of soils as opening chamber under air pressure [2, 5].

In slurry-shield-driven project, the method of opening the working chamber under compressed air has been widely used, such as in Nanjing Yangtze River Tunnel [6–8]. A filter cake plays an important role on the working face stability in the normally tunneling [9, 10]. As opening an excavation chamber under compressed air, the air pressure is in direct action on the filter cake, and it is the key to the excavation face stability. However, few theoretical researches on the mechanism of air tightness of the filter cake are reported.

In shield tunnel construction, opening an excavation chamber under compressed air is an engineering challenge. Cases of excavation chamber opening have been reported in some countries, such as the Elbe 4 Tunnel in Germany, which was excavated in the slurry shield mode of the TBM composed of sand, marlite, and boulder in the chamber replacement center cutters [6]. The Red Line Metro in the St. Petersburg subway in Russia is under a pressure of 0.55 MPa, with a clay and fine sand formation opening an excavation chamber under compressed air [1]. The Holland Westerschelde Tunnel has a gas pressure of MPa in a depth of 60 m of a clay-layer open chamber [7, 8]. These aforementioned cases of opening an excavation chamber belong to engineering accident treatment. Thus, this study focused on describing the processes and points in chamber opening, such as symptoms of decompression sickness. However, theoretical and practical analyses on the mechanism of air tightness of filter cakes and an excavation face during chamber opening are limited.

A filter cake is a special porous medium whose airproof capability mainly varies with the movement of water and gas in a porous medium. Richards and Ogata advanced the concept of the intake value of a ceramic plate. The intake value is the air pressure that the saturated porous ceramic plate begins to drain; the air enters but not through this plate [11]. Fredlund and Rahardjo considered that air entry into unsaturated soils causes air pressure to break through the surface tension of the bound water on the surface of soil particles [12]. Ye et al. indicated that the threshold pressure of gas infiltrating in unsaturated soils causes the bound water to fill the pore channel of the clay [13].

This paper presents an experimental and a case study on the Nanjing Yangtze River Tunnel accessing the working chamber under compressed air to study the breathability of a filter cake under the air pressure. The concept of moderate air pressure and the filter cake air tightness mechanism were discussed.

EXPERIMENTAL METHODS AND MATERIALS

Experimental Apparatus

A self-developed filter cake air tightness device was used as the test apparatus in this paper. Figure 1 shows the schematic of the test apparatus. The apparatus mainly comprised a Perspex cylinder for pressure stabilization and filtering acquisition. The Perspex cylinder had an inner diameter of 8.4 cm and a height of 120 cm. It can withstand a maximum pressure of about 0.7 MPa. The control precision of the pressure stabilization system was 0.01 MPa. The accuracy of the filtering acquisition system was 0.1 g. Simulation of the shield working chamber-working face-stratum system was achieved by loading strata and slurry into the Perspex cylinder, whose end was sealed by a flange. An air compressor was used to provide the permeation pressure for slurry and air pressure for the air tightness of the filter cake. At the same time, the air pressure was controlled by the stabilization system. The pressure indicator monitored the pressure variation inside the apparatus. As placed at 1.2 m above the ground and filled with water in it, the pressure drainage apparatus provided a stable head as water or gas passed through the strata. And then the seepage discharge of water or gas was measured by electronic scales. The variation in seepage discharge with time was automatically recorded by the data acquisition apparatus.

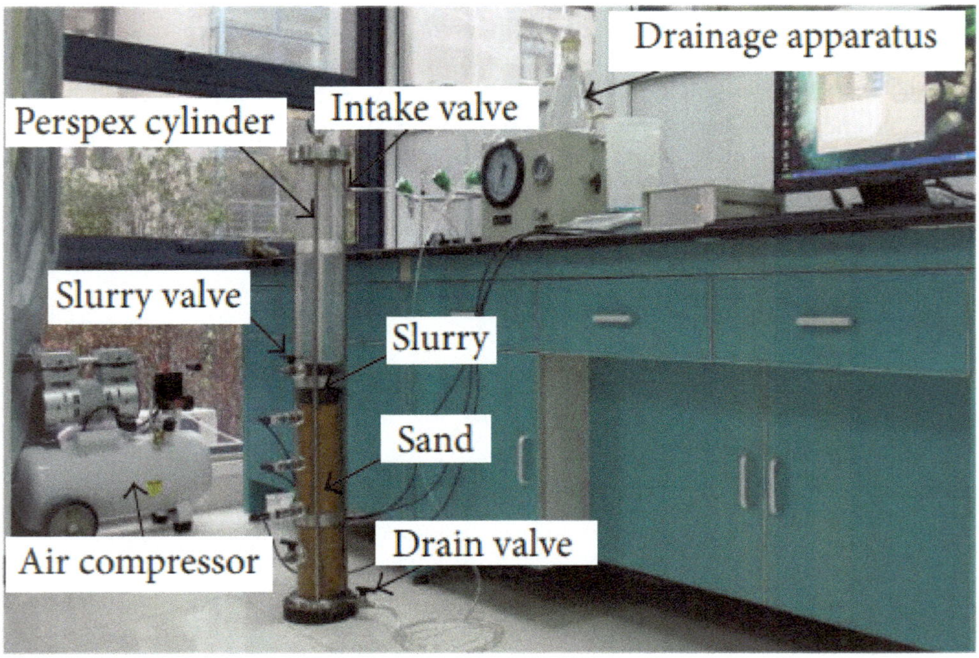

Figure 1: Filter cake air tightness test apparatus.

Experimental Materials

Slurry Materials

Two slurries were used in this experiment, namely, slurry 1 and slurry 2. Their properties are shown in Table 1. The major components of the slurry included bentonite, silt, and an appropriate amount of slurry additives. The bentonite was sodium bentonite from Nanjing Tangshan and the silt was obtained from Baima Lake in Jiangsu Province. The grain size distribution curve is shown in Figure 2.

Table 1: Properties of the tested slurries.

Identity of slurry	ρ (g/cm^3)	Funnel viscosity (s)	Bleeding rate (24 h) (%)
1	1.24	29	2
2	1.13	59	0

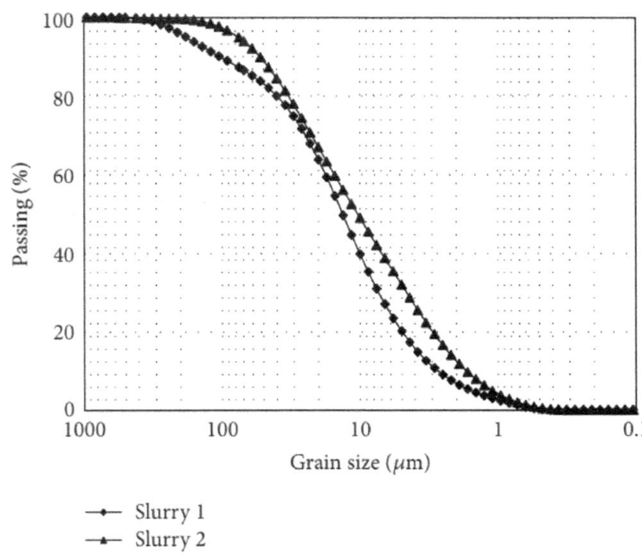

Figure 2: Grain size distribution curves for test slurries.

Stratum Materials

In this experiment, a uniform sand with particle sizes ranging from 0.25 mm to 0.5 mm was used as stratum materials. It was obtained by screening between the nearest two sizes of standard sieves, the 0.25 mm and 0.5 mm. The stratum was prepared by three times with a depth of 9 cm per layer, and the load height was 27 cm. The soil specimen was compacted to a dry density (ρ_d) of 1.5 g/cm^3 and the corresponding porosity (n) was 0.43. The hydraulic conductivity (k) of soils was measured by the constant head permeability tests, and its hydraulic conductivity was about 6.5 × 10^{-2}cm/s. To prevent sand draining, a 5 cm sand filter layer with particle sizes ranging from 2 mm to 5 mm and hydraulic conductivity of 4.2 cm/s was placed at the bottom.

Experimental Methods

Preparation of the Filter Cake Test. The filter layer was first loaded, hit, and compacted. The strata were then loaded, the layer was hit and compacted, and the surface was screed. Water was slowly flooded from the bottom to the top until the strata were completely saturated [14]. The whole saturation process required around 30 min. The slurry was slowly injected along the diversion rod into the strata and the experimental apparatus was sealed. The air compressor was turned on, and the air pressure of 0.1 MPa was acting on the slurry in the apparatus. The drain valve was then opened and the slurry permeated into the strata under the action of pressure. When the increment of filtering quantity was lower than 1 cm^3/min and stable, the filter cake was deemed to be formed. Finally, the intake and drain valves were closed, the slurry valve was opened, and the slurry was drained. The shape and thickness of the filter cake were then recorded.

Filter Cake Air Tightness Test. The slurry valve was closed, the intake valve was opened to exert air pressure to the filter cake, and then the drain valve was opened. The air pressure and variation in seepage discharge with time were then recorded. At the end of the experiment, the thickness of the filter cake was recorded. The thickness of every filter cake and infiltrated zone formed in the tests were measured three times by a ruler with a precision of 1 mm. The average value was recorded as the final thickness of filter cakes and infiltrated zone.

Two Sets of Experiment. (1) The filter cake formation test was performed four times using slurry 1 under 0.1 MPa pressure. Then, air pressure was exerted to each filter cake stepwise, with an initial pressure of 0.02 MPa and an increase of 0.02 MPa every 2 min until the variation in the air permeation quantity became very obvious. The seepage discharge of the filter cake under each stepwise increase in air pressure and the deformation of the filter cake were recorded. (2) The filter cake formation test was performed five times using slurry 2 under 0.1 MPa pressure. Different air pressures (0.1, 0.14, 0.15, 0.16, and 0.2 MPa) were then exerted to each filter cake continuously for 24 h. The variation in seepage discharge with time until air leakage appeared in the filter cake was recorded.

EXPERIMENTAL RESULTS

Form of the Filter Cakes

The forms of the filter cake of the two groups of nine experiments were recorded. The thickness of the filter cake and the infiltrated zone for each experiment are summarized in Table 2.

Table 2: Morphology of the filter cakes in the test.

Identity of filter cake	Filter cake thickness (mm)	Infiltrated zone thickness (mm)
1		
①	5.1	28.7
②	4.9	29.1
③	4.7	30.4
④	4.9	30.2
2		
①	5.8	28.1
②	5.9	26.7
③	6.1	25.9
④	6.3	26.4
⑤	6.1	27.3

The thickness of filter cake 1 and the infiltrated zone were approximately 4.9 and 29.6 mm, respectively. The thickness of the filter cake 2 and the infiltrated zone were approximately 6.0 and 26.9 mm, respectively. The thickness of the formed filter cake and infiltrated zone were essentially identical for the same slurry under the same conditions, which indicated that the experiment was of reusable operability and the experimental method was feasible.

Airproof Phenomena of the Filter Cake under Air Pressure

The variation in seepage flow of the filter cake with air pressure is shown in Figure 3. Below a specific pressure (≤0.12 MPa), the seepage flow was low and the cake was airtight. On the other hand, when the air pressure was higher than a specific pressure (≥0.14 MPa), the amount of discharged water became very large and the cake lost its air tightness. After the test, compressive deformation was generated.

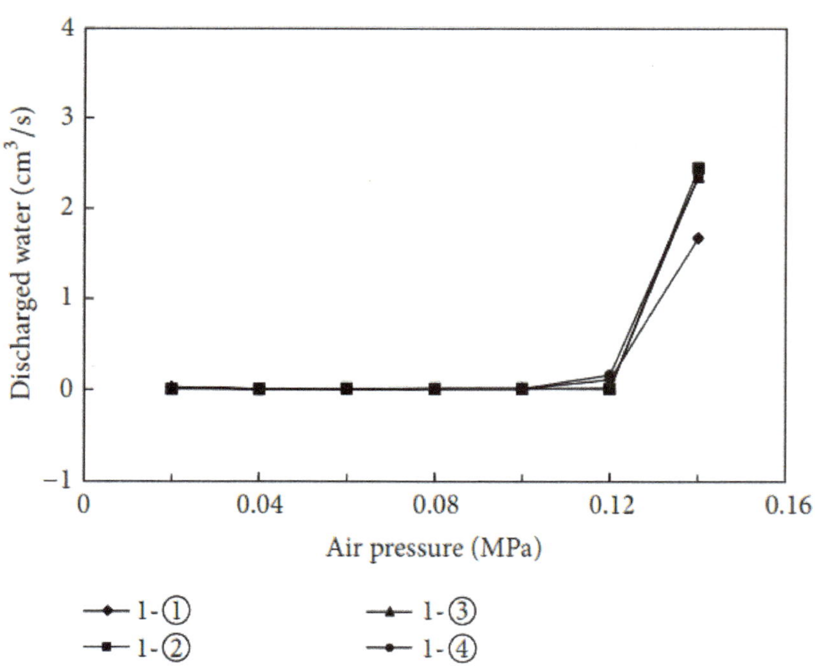

Figure 3: Relationship between discharged water and air pressure of filter cake 1.

Figure 4 shows the variation in the seepage flow of the filter cake with time. Under 0.1 and 0.14 Mpa air pressure, the seepage flow of the cake was very low (about 0.003 cm^3/s). After continuous observation for 24 h, the seepage flow decreased to 0 cm^3/s with time. During this period, the stratum was saturated and no air entered the cake. However, under 0.15, 0.16, and 0.20 MPa, the seepage flow of the filter cake obviously increased with time. A higher air pressure resulted in faster seepage flow.

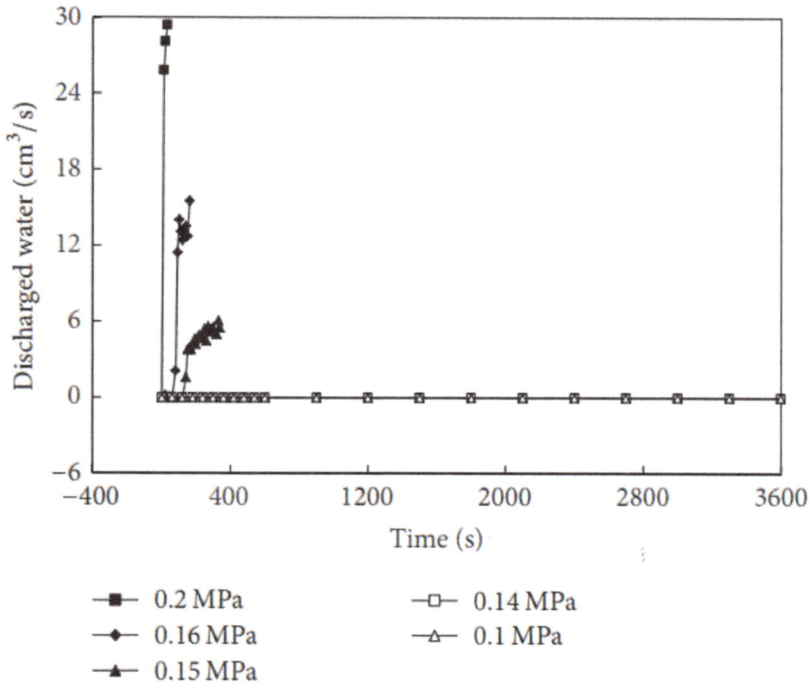

Figure 4: Relationship between discharged water and tested time of filter cake 2.

The physical processes of the filter cake under air pressure can be divided into four steps as shown in Figure 5. First, the air cannot enter the cake and the cake appeared to be undergoing a nondrainage process; the air pressure was low. Second, with increased air pressure, the amount of discharged water from the cake was low. Third, the air pressure increased until the air entered the cake, and the cake appeared to undergo a transient drainage process. Fourth, the air pressure increased to the airproofing pressure of the filter cake, the amount of discharged water immediately increased considerably, and the air entered the stratum through the filter cake. The essence of process is that cake became permeable from airproofing, and the required time depended on the magnitude of the air pressure.

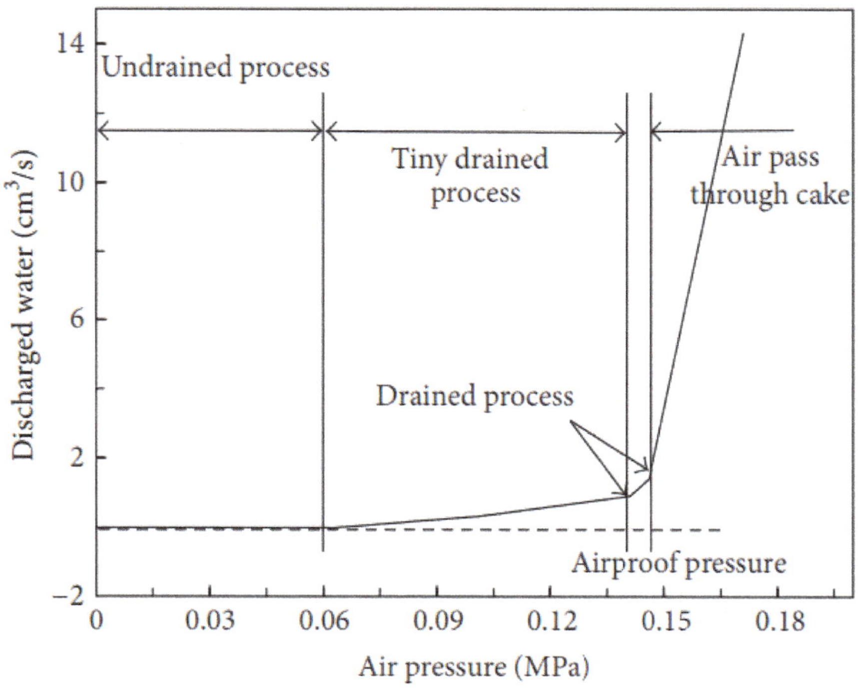

Figure 5: Filter cake permeability and drainage phenomena.

Airproofing Pressure of the Filter Cakes

Figure 3 shows that inflection points appeared in the curves. For filter cake 1, the inflection points appeared at 0.12 MPa air pressure for each test. Considering the air pressure at which the inflection point appeared as airproofing pressure, when the air pressure was below the airproofing pressure, the filter cake was airtight. On the other hand, when the air pressure was above the airproofing pressure, the filter cake was permeable. The test results indicated that the airproofing pressure also existed in filter cake 2. When the air pressure was below 0.14 MPa, the filter cake was airtight. The amount of discharged water was very low because of the compressive deformation of the cake under air pressure. By contrast, when the air pressure was above 0.15 MPa, the filter cake was permeable. The test results indicated that the airproofing pressure of filter cake 2 was between 0.14 and 0.15 MPa.

A CASE STUDY ON OPENING A SLURRY SHIELD EXCAVATION CHAMBER IN THE NANJING YANGTZE RIVER TUNNEL

General Conditions in the Nanjing Yangtze River Tunnel

The Nanjing Yangtze River Tunnel was excavated by the two-slurry shield mode of the TBMs. The outer diameter of the segments was 14.5 m [1]. The pressure cabin of the slurry shield mode of the TBM in the Nanjing Yangtze River Tunnel was divided into two chambers: excavation and air pressure. The top of the air pressure chamber had high air pressure, and the bottom was connected with the excavation chamber by suspensions. The two-chamber design was very flexible for inspection and repair works. Until August 2008, the TBM was excavated to 650–659 rings (K4 + 900 to K4 + 918), and the torque was rapidly increased to 15 MNm and sometimes to 20 MNm. The speed was lower than 5 mm/min [5]. Thus, the TBM had to be stopped to verify the explanation of the aforementioned exceptional events. The highest tide water level throughout the project from August 2008 to December 2008 was 8.37 m, and the water pressure at the centerline of the TBM was approximately 0.53 MPa.

Method of Opening the Chamber and Study on the Filter Cake

Selecting the Method of Opening the Chamber

The TBM was stopped 53.25 m under the water level of the Yangtze River (Figure 6). Thus, reinforcing the soils near the excavation face was not advisable, and opening the chamber under atmospheric pressure after soil improvement would not be conducted. The 3 m mode in the upside supported by compressed air and 11.96 m in the downside still filled with suspensions were selected. The cutting wheel would be circumrotated and the other cutters would be repaired.

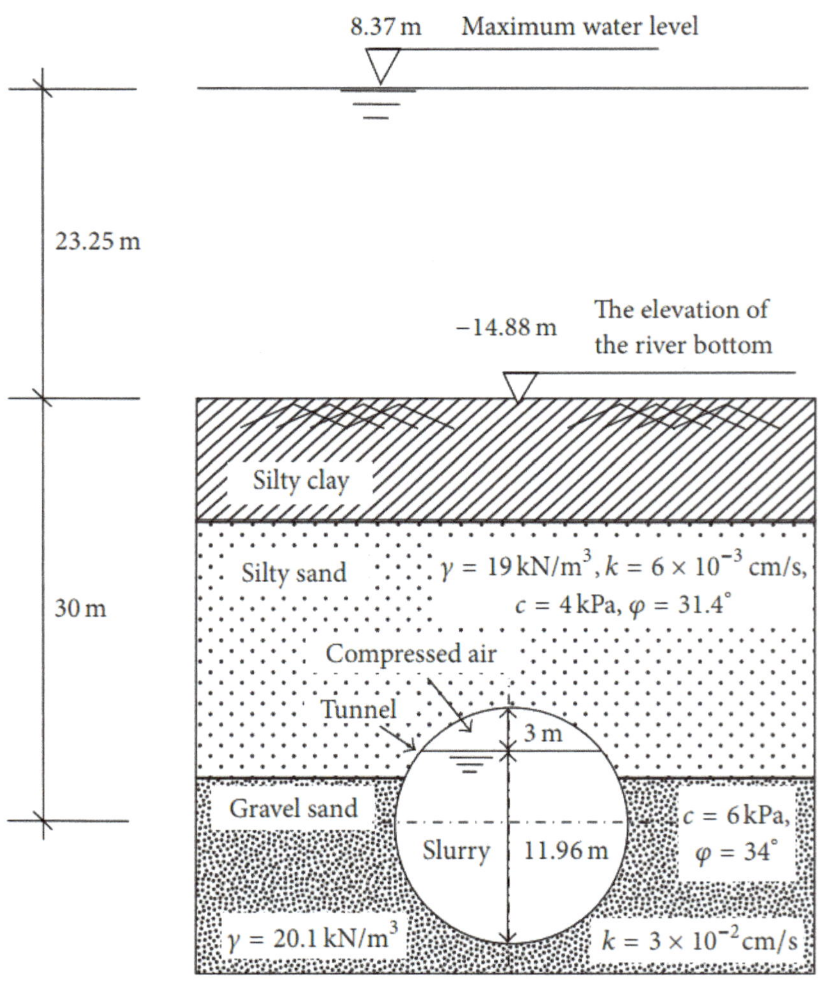

Figure 6: Sectional view of geological strata of the stop location of tunnel machine.

Filter Cake Formation Tests and Air Pressure Tests of the Filter Cake

Filter cake formation tests were conducted on the mixed strata under 0.3 MPa, and the slurry was the suspensions mixed with bentonite and natural clay ($\rho = 1.15$ g/cm^3, funnel viscosity = 25 s). A 5 mm thick filter cake was rapidly formed (Figure 7).

Figure 7: The 5 cm thickness filter cake formed in the slurry tests.

After filter cake formation on the surface, an air pressure test was conducted and a curve for the relationship between the air pressure and corresponding discharged water was obtained (Figure 8). Below a specific pressure, the water discharge amount was low. At an air pressure higher than the specific pressure, the discharged water immediately increased considerably. For a 5 mm thick filter cake formed in the test, no discharged water was measured because the air pressure was lower than 0.12 MPa. Thus, the compressed air cannot penetrate the filter cake. At 0.12 MPa pressure, the discharged water suddenly increased and some air penetrated the filter cake. Similar phenomena occurred when the pressure reached 0.13 and 0.14 MPa. Therefore, the critical pressure for this filter cake airproofing (the airproofing pressure) was 0.11 MPa. The threshold air pressure that can be adopted in the project was equal to the water pressure of the ground and the airproofing pressure of the filter cake formed in the test. As aforementioned, the threshold air pressure was 0.59 MPa (0.48 + 0.11 MPa). Aside from withstanding the pore water pressure, part of the air pressure (the airproofing pressure) can counteract the effective earth pressure in the tunnel face. The threshold air pressure was nearly equal to the total earth pressure at rest, that is, 0.59 MPa.

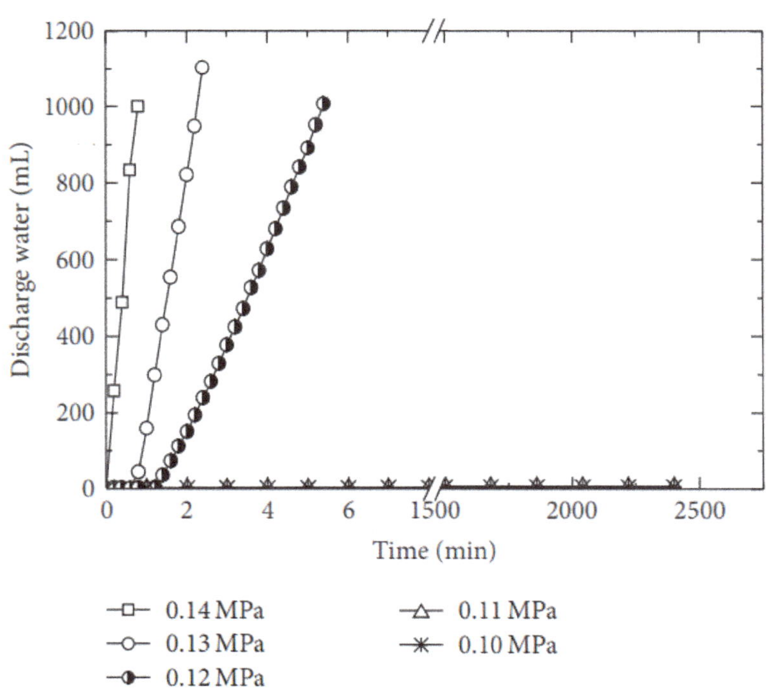

Figure 8: Results of the air pressure tests.

The Implementation of Opening Chamber under Air Pressure

Before opening the chamber, an air pressure test was performed in the site to verify the reliability of the filter cake formed on the surface. The level of the slurry in the working chamber decreased to 3 m after a dense filter cake was formed. The top of the chamber was filled with compressed air then. Considering the safety factor of the airproofing pressure of the filter cake, air pressure from 0.55 MPa to 0.57 MPa was set to support the tunnel face. This range of pressure was a little lower than the threshold airproofing pressure of 0.59 MPa but a little larger than the total active earth pressure of 0.547 MPa. The air pressure change in the excavation chamber was monitored to estimate the stability of the excavation surface. The monitoring data showed that the gas escape was rare and the pressure in the chamber was stable. The filter cake was sealed in this case. The technicians accessed the excavation chamber to repair the cutters, and they can work only from 2 h to 5 h in the chamber every time. After finishing the assignment, they exited the chamber and resumed the slurry level to the top. The same procedures were repeated in the next operation.

Figure 9 shows the images of filter cakes on the surface of the tunnel face. According to these photos and observations in the chamber, a dense and thick filter cake was formed on the tunnel face, with no collapse and air escape. The filter cake was also deemed stable.

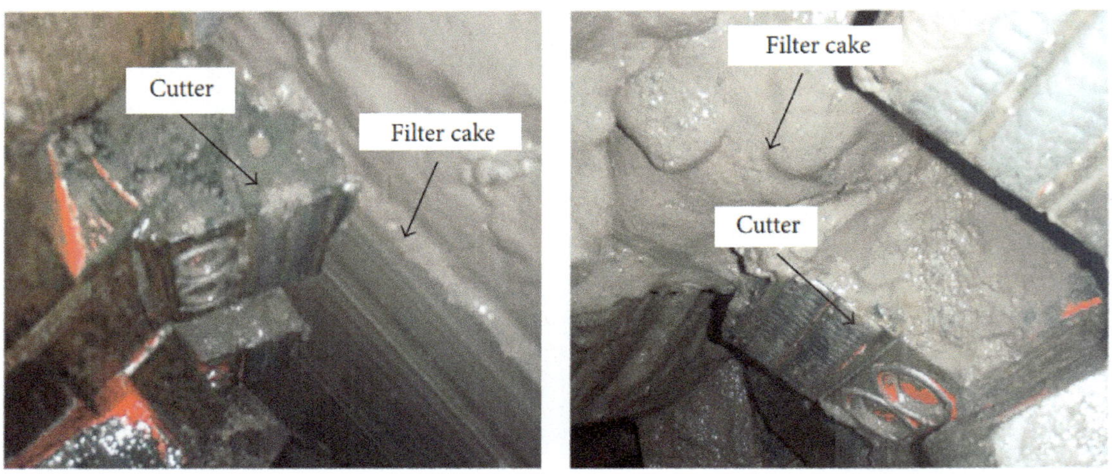

Figure 9: Photos of the filter cake screened in the excavation chamber.

DISCUSSION

Physical Meaning of the Airproofing Pressure of the Filter Cake

To study the nature of the airproofing pressure and pneumatically convert it to the effective stress in the mechanism, it should be considered from the surface tension of the water in the porous medium [15]. A capillary bundle model is very commonly used to study the movement of water in the porous medium [16, 17]. Pore connectivity in the media is observed as a capillary in the model [18]. The capillary phenomenon is formed by the joint action of the wettability of the capillary wall and the surface tension of water. The negative capillary pressure is borne by the meniscus and then passes through the surface tension to the wall [19].

When air pressure is applied to the saturated filter cake, a meniscus is formed to resist the pressure through the surface tension of the meniscus. The pressure is transmitted to the particles of the filter cake and then converted to the effective stress. Within a certain pressure range, the curvature of the meniscus increases with the pressure. When the curvature threshold is reached, the meniscus is pushed down by the pressure, which is the air pressure overcoming the surface tension of water in the pores of the filter cake. This pressure is called the airproofing pressure.

The airproofing pressure overcomes the surface tension of water in the pores of the filter cake. Assuming that the meniscus is a spherical surface, the stress analysis is shown in Figure 10. is pressure, is the coefficient of surface tension of water, and is the pore radius [14, 15].

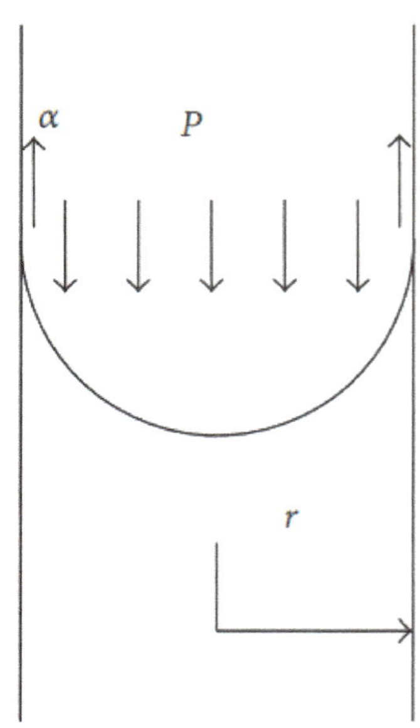

Figure 10: Stress analysis of the meniscus.

According to the balancing of forces in the vertical direction $P \cdot \pi r^2 = \alpha \cdot 2\pi r$. The pressure values $P = 2\alpha/r$ are then obtained. This equation shows that the airproofing pressure is related to the pore size; smaller pores result in higher airproofing pressure. This pressure magnitude is independent of the pressure time and the pressure exerted.

Figure 3 shows that when the pressure is less than the airproofing pressure, that is, the air pressure is equal to or less than 0.12 MPa, the air pressure is insufficient to overcome the surface tension of water in the pores of the airproofing pressure. The pressure is then passed through the interaction between the water and soil particles to the soil skeleton and becomes effective stress. The cake is compressed in the role of the effective stress. Figure 11compares the deformation and seepage discharge of filter cake when the airproofing pressure is reached, suggesting that the seepage discharge was mainly generated by the cake deformation. Figure 4 shows two curves of 0.1 and 0.14 MPa. The seepage discharge progressively decreases and finally approaches zero. The deformation of the filter cake also gradually becomes stable.

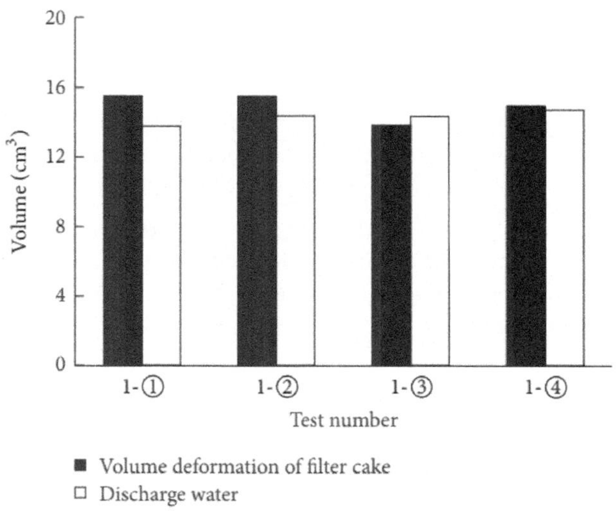

Figure 11: Comparison between the volume deformation and seepage discharge of filter cake.

When the pressure is higher than the airproofing pressure, such as the pressure reaches 0.14 MPa in Figure 3and 0.15 MPa in Figure 4, the air pressure overcomes the surface tension of the water in the pores of the filter cake, the gas penetrates the pores, seepage discharge significantly increases, and the filter cake is breathable. Figure 4 shows that, at different pressures (0.15, 0.16, and 0.2 MPa), the difference between the processes of air permeability of the cake indicates that a higher air pressure results in a shorter time within which the cake becomes breathable.

Airproofing Pressure of the Strata and Filter Cakes

The effects of the construction method under compressed air depend on the permeability of the strata; a lower stratum permeability results in better effects. When the permeability coefficient is higher than 10^{-2}cm/s, the construction method is difficult to use under compressed air. For high penetrative strata, a slurry is used to form a filter cake to decrease the permeability of the stratum and avoid the leakage of compressed air. Construction is then started under compressed air.

When forming a dense filter cake, moderate air pressure can counteract part of the effective earth pressure. Without a filter cake, compressed air can penetrate the pore and counteract the water pressure, but cannot counteract the effective pressure. The pore of the formed dense filter cake is so small that the air needs to overcome high resistance when passing through the channel. Before exceeding the resistance, the air cannot penetrate the pores but can only alter the curvature of the liquid in the channel and convert the air pressure to effective earth pressure. This mechanism explains why compressed air can counteract moderate effective pressure. In this case, the airproofing pressure of the filter cake is approximately 0.11 MPa. Considering a moderate safety factor, 0.1 MPa is used as the airproofing pressure, which is close to the effective earth pressure at rest (0.11 MPa). Therefore, filter cakes play an important role in balancing the earth pressure on the tunnel face. It also reduces air escape and enhances the stability of the tunnel face when opening the chamber.

Experience of Opening Chamber Successfully

In the present project, the air pressure less than the total earth pressure at rest and a little larger than the total active earth pressure was set to support the tunnel face when opening the chamber in the Nanjing Yangtze River Tunnel. In normal TBM excavation, the slurry pressure in excavation chamber was set 0.02 MPa larger than the total earth pressure at rest in general [18, 19]. However, the air pressure set in the present study was a little smaller than the slurry pressure. Lower cohesion of the sand could also maintain stability when supported by a pressure a little larger than the total active earth pressure because of the self-support stability of the soil caused by its own strength. Actually, the air pressure of 3 m on the top of the chamber established the slurry pressure of 12 m at the bottom. Increasing the air pressure has an important role on keeping the stability of the whole tunnel face. The slurry pressure was larger than the air pressure before it was replaced with compressed air. Whether the filter cake would fall off from the tunnel when supported by a little lower air pressure was a problem. However, the observations during physical access to the face showed that the filter cake had not fallen off and the face was stable.

CONCLUSIONS

(1) During chamber opening, a complete filter cake is formed after the infiltration of slurry, and a stage of airproof deformation of a filter cake occurs when compressed air acts on the surface of the filter cake. When the pressure of the compressed air is low, the air cannot pass through the filter cake and the amount of discharged water is low because of the deformation of the filter cake. When the pressure of compressed air is higher than a certain value, the air can pass through the filter cake and the discharged water rapidly increases. This pressure is defined as the airproofing pressure. This phenomenon is a draining process of pore water when its surface tension is overcome by air pressure from the exterior.

(2) The airproofing pressure of a filter cake has a certain magnitude that is independent of the pressure of compressed air and the acting time. Its magnitude only depends on the form of the filter cake. The airproofing pressure of the filter cake determines the magnitude of threshold pressure of the compressed air. If the difference between the compressed air pressure and water pressure on the working face is lower than the

airproofing pressure, the stability of the working face can be guaranteed by a construction method under compressed air. However, when the buried depth of the tunnel was too deep, the needed supporting pressure was very high, and sometimes the stability of the working face cannot be guaranteed only by the construction method under compressed air; other auxiliary measures should be used.

(3) Air pressure test results show that the airproofing pressure of the filter cake is 0.11 MPa. This result suggests that, below 0.11 MPa, air cannot penetrate the pores of the filter cake and the air pressure can withstand the effective pressure. For practical application, the air pressure must be lower than the airproofing pressure of the filter cake to support the tunnel face and maintain its stability.

Conflict of Interests

The authors declare that there is no conflict of interests regarding the publication of this paper.

ACKNOWLEDGMENTS

The authors would like to acknowledge the financial support of the National Natural Science Foundation of China (51408191), the National Program on Key Basic Research Project of China (2015CB057803/2012CB719804), the China Postdoctoral Science Foundation (2014M560388), the Project (SKLGDUEK1306) Supported by State Key Laboratory for Geomechanics and Deep Underground Engineering, China University of Mining & Technology, and the Natural Science Foundation of Jiangsu Province (BK2011025).

REFERENCES

1. F. L. Min, W. Zhu, X. R. Han, and X. C. Zhong, "The effect of clay content on filter-cake formation in highly permeable gravel," in Proceedings of the GeoShanghai International Conference—Geoenvironmental Engineering and Geotechnics: Progress in Modeling and Applications, pp. 210–215, June 2010.
2. L. Hou, Research on technology of replacing cutting cool with pressure of EPB shield in water-rich and pebbly sand stratum [M.S. thesis], Southwest Jiaotong University, Sichuan, China, 2006.
3. J. Zhao, Q. M. Gong, and Z. Eisensten, "Tunnelling through a frequently changing and mixed ground: a case history in Singapore," Tunnelling and Underground Space Technology, vol. 22, no. 4, pp. 388–400, 2007.
4. Á. Tóth, Q. Gong, and J. Zhao, "Case studies of TBM tunneling performance in rock-soil interface mixed ground," Tunnelling and Underground Space Technology, vol. 38, pp. 140–150, 2013.
5. W. Zhu, F. Min, S. Xia, and D. Wei, "A case study of tunnel face stability and opening the pressure cabin in nanjing yangtze river tunnel," in Proceedings of the 6th Japan-China Conference on Shield-Driven Tunneling, pp. 324–329, 2011.
6. C. Falk, "Pre-investigation of the subsoil developments in construction of the 4th Elbe Tunnel Tube,"Tunnelling and Underground Space Technology, vol. 13, no. 2, pp. 111–119, 1998.
7. J. Heijboer, J. van den Hoonaard, and W. van de Linde, The Westerschelde Tunnel: Approaching Limits, A. A. Balkema Publishers, Lisse, The Netherlands, 2004.
8. J. D. Nieuwenhuis and A. Verruijt, "Tunnelling problems in older sand formations," in Engineering Geology for Infrastructure Planning in Europe, vol. 104 of Lecture Notes in Earth Sciences, pp. 538–548, Springer, Berlin, Germany, 2004.
9. G. Anagnostou and K. Kovári, "The face stability of slurry-shield-driven tunnels," Tunnelling and Underground Space Technology incorporating Trenchless, vol. 9, no. 2, pp. 165–174, 1994.
10. S. H. Kim and F. Tonon, "Face stability and required support pressure for TBM driven tunnels with ideal face membrane—drained case," Tunnelling and Underground Space Technology, vol. 25, no. 5, pp. 526–542, 2010.
11. L. A. Richards and G. Ogata, "Materials for retainer plates and their use for retentivity measurements,"Soil Science Society of America Journal, vol. 20, no. 3, pp. 303–306, 1956.
12. D. G. Fredlund and H. Rahardjo, Unsaturated Soil Mechanics, China Building Industry Press, Beijing, China, 1997.
13. W. Ye, C. Wang, Q. Wang, and B. Chen, "Laboratory tests on the characteristics of air-permeation in unsaturated Shanghai soft soil," Journal of Engineering Geology, vol. 17, no. 2, pp. 244–248, 2009.
14. T. V. Duong, V. N. Trinh, Y. J. Cui, A. M. Tang, and N. Calon, "Development of a large-scale infiltration column for studying the hydraulic conductivity of unsaturated fouled ballast," Geotechnical Testing Journal, vol. 36, pp. 1–10, 2013.
15. R. H. Brooks and A. T. Corey, "Properties of porous media affecting fluid flow," Irrigation and Drainage, vol. 92, pp. 61–88, 1966.
16. A. Oukhlef, S. Champmartin, and A. Ambari, "Yield stress fluids method to determine the pore size distribution of a porous medium," Journal of Non-Newtonian Fluid Mechanics, vol. 204, pp. 87–93, 2014. ·
17. D. J. Fallow and D. E. Elrick, "Field measurement of air-entry and water-entry soil water pressure heads,"Soil Science Society of America Journal, vol. 60, no. 4, pp. 1036–1039, 1996.
18. S. Tian, G. Lei, S. He, and L. Yang, "Dynamic effect of capillary pressure in low permeability reservoirs,"Petroleum Exploration and Development, vol. 39, no. 3, pp. 405–411, 2012.
19. R. Rezaee, A. Saeedi, and B. Clennell, "Tight gas sands permeability estimation from mercury injection capillary pressure and nuclear magnetic resonance data," Journal of Petroleum Science and Engineering, vol. 88-89, pp. 92–99, 2012.

CHAPTER 9

INFLUENCE OF FRACTURE ROUGHNESS ON APERTURE FRACTURE SURFACE AND IN FLUID FLOW ON COARSE-GRAINED MARBLE, EXPERIMENTAL RESULTS

G. Dimadis, A. Dimadi, I. Bacasis

Department of Civil Engineering, Laboratory of Technical Geology, Aristotle University, Thessaloniki, Greece

ABSTRACT

Wetting fluid flow through rock discontinuities influence a great number of project among others: dam construction, underground projects, CO2 storage in underground schemes, geological disposal of radioactive wastes; Hydrocarbon storage caverns. Flow through fractures is considered to be laminar due to small aperture of the fracture walls and slow velocity. The fluid model called "Cubic law" describes the flow assuming parallel infinite plates. However, natural discontinuities on rock have roughness. In this experimental study an induced fracture on a sample of medium-grained marble was used, to determine the influence of roughness in water flow. This study is a preliminary part of research funding program for flow of CO2 through rocks (AUTH-GEOMechanics and Environment of CO2 geological Storage, Project No. 456,400).

Keywords

Cubic Law, Roughness, Hydraulic Aperture, Mechanical Aperture

INTRODUCTION

In relatively impermeable rocks, the fluid flow is essentially confined in joints and fractures of the rock mass. As a result the hydraulic conductivity of the rock mass is determined by the hydraulic conductivity of joints and fractures. The major factors govern the flow in fractured rocks are, a) fluid properties, b) void geometry and c) fluid pressure at the joint boundary (Olsson & Barton, 2001) (Kranzz, Frankel, Engelder, & Scholz, 1979). Void geometry is ruled by the variation of the aperture between the walls of the fracture and the contact areas (touching asperites) (Zimmerman & Bodvarsson, 1996) which induce the roughness of the fracture walls. In this study the parameter of void space is examined, by testing the effect of joint surface roughness on water flow rate through induced marble fracture.

The relationship between flow rate and hydraulic gradient ($\Delta h/\Delta l$) is given by Darcy's law or so called cubic law Equation (1) (Bear, 1979) (Witherspoon, Wang, Iwai, & Gale, 1980) even though introduced for planar parallel surfaces should not be

Citation: Dimadis, G. , Dimadi, A. and Bacasis, I. (2014) Influence of Fracture Roughness on Aperture Fracture Surface and in Fluid Flow on Coarse-Grained Marble, Experimental Results. Journal of Geoscience and Environment Protection, 2, 59-67. doi: 10.4236/gep.2014.25009.

generally used to estimate flow through rough surfaces (Oron & Berkowitz, 1998) (Park, Osada, Matsushita, Takahashi, & Ito, 2013; Zhao, Li, & Jiang, 2013; Zimmerman, Al-Yaarubi, Pain, & Grattoni, 2004; Brown, 1987). In conditions that fluid velocity is high and/or joint surface roughness is much, the Cubic Law cannot apply. Moreover it is crucial to calculate the appropriate (most accurate) fracture aperture, so as the predictions of cubic law to be accurate (Renshaw, 1995).

$$Q = \frac{-\rho g}{12\mu} b_h^3 w \frac{\Delta h}{\Delta l} \qquad (1)$$

where Q is the flow rate of water passing through two smooth parallel plates, Δh is the difference in hydraulic heads, w is the width of the flowing zone between the parallel plates in the direction normal to the flow vector, Δl is the length of sample, ρ is the fluid density, g is the acceleration due to gravity, μ is the kinematic viscosity of the fluid and b_h is the distance between the two plates (hydraulic aperture).

Roughness affection seems to be in length (Δl). Actually roughness increases the length that fluid has to cover from point A to point B. The longest route causes greater "friction effect" the solids would bring to the fluid, thus decreasing the fracture transmissivity (Bo Li, Jiang, Koyama, Jing, & Tanabashi, 2008). This becomes critical in small joint apertures, because it is translated in greater resistance to flow, and finally in lower conductivity (orders of magnitude) of rough joints (Oron & Berkowitz, 1998)

MATERIALS AND METHOD

Sample Preparation

A sample of marble, of medium-grained Interlocking crystals, with an induced fracture was used in this study. Total dimensions of the samples were: 186 × 45 × 105 mm (Figure 1). Artificial fracture was imposed by splitting (tension fracture) one sample in half using custom made metal splitting jaws (a kind of modified "Brazilian" loading method) in a UCS device. A thin scratch in the middle of all surfaces (parallel to the long axis) of the sample was created before splitting, so as to enhance the stress field while loading, as a result almost parallel surfaces of crack occurred.

We studied the influence of fracture roughness on water flow for different joint apertures. Fluid flow is assumed to occur only through the fracture, because the permeability of marble is considered negligible to the joint-flow magnitude. Eleven hydraulic gradients (Δh/Δl) were selected for each test Table 1. It must be mentioned that due to restrictions of the experimental setup in apertures of 750 μm and 1000 μm hydraulic gradients greater than 0.54 and 0.4 respectively, could not be achieved. Each test concerns the same induced mechanical aperture, six induced mechanical apertures were selected as seen in Table 2, these apertures could represent a natural fracture at stage of rest (0 - 500 μm) and stage after shear movement which causes dilation to the fracture surface (>1000 μm) (Yeo, de Freitas, & Zimmerman, 1998).

For each hydraulic gradient we conducted three flow measurements for 50 seconds (after the flow was stabilized) the average flow was calculated. Totally more than 150 measurements were conducted. To ensure that the induced mechanical apertures were as accurate as possible, each time we let the sample halves to rest on themselves (Em_1 = 0 μm) and then raised it on the desired aperture. The raise of the upper halve was manually controlled by four linear displacement indicators LDI (Wykeham Farrance, accuracy: ±2 μm) placed at each corner of the sample. For the same aperture to verify reproducibility in measurement at least 3 tests were performed.

Table 1. Relation of hydraulic gradient and sample length.

Relation between Hydraulic gradient and Δh/ΔL ratio											
Δh (mm)	2	7.5	15	25	35	50	75	100	125	150	190
Δh/ΔL[1]	0.01	0.04	0.08	0.13	0.19	0.27	0.4	0.54	0.67	0.81	1.02

Note: [1]In our study ΔL is steady (equal to sample length 189mm).

Table 2. Fracture mechanical apertures used in the study.

	Em_1	Em_2	Em_3	Em_4	Em_5	Em_6
Fracture mechanically induced aperture (μm)	0	200	250	500	750	1000

induced mechanical apertures were as accurate as possible, each time we let the sample halves to rest on themselves (Em_1 = 0 μm) and then raised it on the desired aperture. The raise of the upper halve was manually controlled by four linear displacement indicators LDI (Wykeham Farrance, accuracy: ±2 μm) placed at each corner of the sample. For the same aperture to verify reproducibility in measurement at least 3 tests were performed.

Experimental Set Up

Upper and lower surfaces of the samples were rigidly attached on acrylic plates with dimensions 10 × 65 × 260 mm, 4 thread rods at each corner were used for raising the upper half of the sample on z axis, while 4 Linear displacement indicators were used for exact measurement of level raising, a 5th indicator was used to check for linear movement along x axis. Two smaller acrylic plates with cavities in one side where used as entrance and exit chambers, which allowed water to enter/exit the fracture evenly, minimizing any radial movement. The hydraulic heads in entrance and exit where measured by two piezometers. Hydraulic gradient was applied each time manually using the piezometers. We employed constant head method by adding a water tank with a spillway to receive constant hydraulic head at specimen entrance, resulting in constant hydraulic gradient throughout each step.

At the exit the spillway of the water tank was used to lead the water coming out of the fracture in a container on a high precision laboratory scale (KERN N2400). Weighting the outcome water gave us the water flow of the crack by using the equivalence 1 ml = 1 g of water. Deaerated water was used throughout the experiment. Total view of the experimental apparatus in Figure 2. To avoid inserting anisotropy in flow as described in Li's study (B. Li, Wong, & Milnes, 2014) due to movement on x axis, 4 rigid metal plates were placed at both sides (y-z planes) of the sample, while raising it on desired mechanical aperture. We used strips of fiber-free elastomeric foam, as no-flow boundary.

Fracture Roughness Characterization

The sample with the rough faces fracture was scanned with a 3D scan in both faces and the exact micro-topography of the phases was electronically created (Figure 5), z axis measurement took place to calculate joint surface roughness. Characterization of joint surface roughness made by one empirical and one statistical method that are widely used in rock mechanics. JRC factor (joint roughness coefficient) introduced by Barton (Nick Barton, 1973) is a coefficient used to quantify rock joint roughness, it compares the joint surface duplicate taken with a Profile Gauge Tool for a length of 10 cm, to a table of 10 typical roughness profiles and JRC ranges from 0 (for totally planar faces) to 20 (for very rough surfaces) Figure 4. In our case compare of the joints profile gave as an approximation for JRC ≈ 17 (Figure 3 and Figure 4). The statistical method we used is the dimensionless parameter Z2 introduced by Myers (Myers, 1962). It is defined as:

$$Z_2 = \sqrt{\frac{1}{L}\int_0^L \left(\frac{dy}{dx}\right)^2 dx} \tag{2}$$

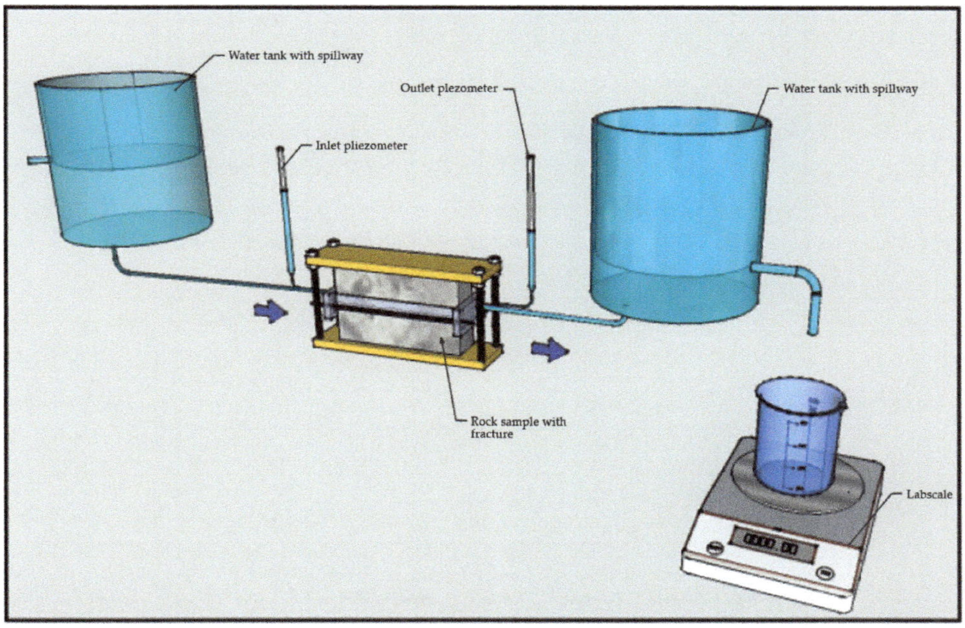

Figure 2. Experimental setup.

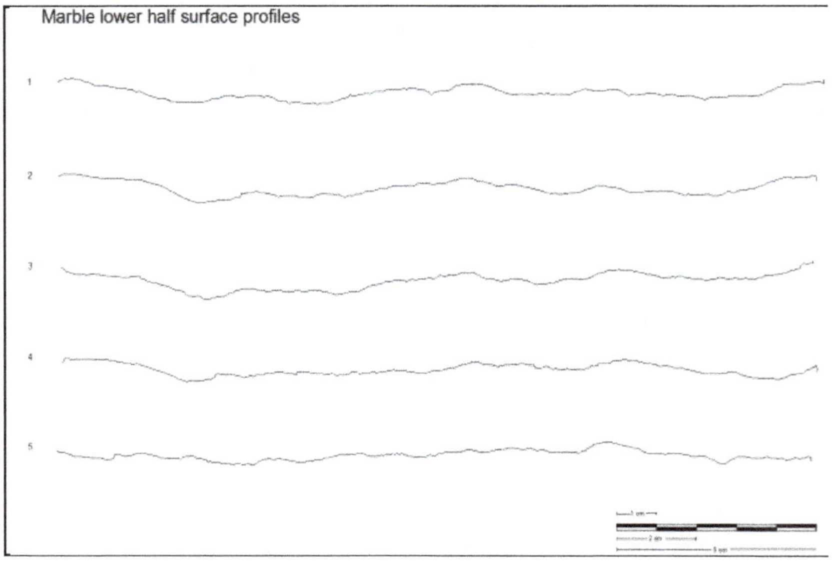

Figure 3. Profiles of lower half taken with profilometer.

where z(x) refers to profile of the fracture surface and L is the profile length. In practice it is calculated by the discretization of the surface profile:

$$Z_2 = \sqrt{\frac{1}{L}\sum\frac{(z_{i+1}-z_i)^2}{(x_{i+1}-x_i)^2}} \quad (3)$$

where x_i and z_i the coordinates of the surface profile and L the length of the profile (L = 189 mm).

For the rough fracture surface and 42 profiles along x axis Z2 factor was calculated and JRC coefficient was extracted using Tse's Equation (4) by (Tse & Cruden, 1979). The frequency of JRC results can be seen in Figure 6. Mean JRC was calculated as 16.38 very close to our approximation (≈17)

$$JRC = 32.2 + 32.47\log(Z_2) \quad (4)$$

RESULTS

The conducting aperture e of the fracture according to Barton (N. Barton, Bandis, & Bakhtar, 1985) is considered identical to parameter bh of cubic law Equation (1), so after measuring flow rates in our experiment, we back-calculated the conducting aperture that occurred in every flow.

Linear regression between hydraulic gradient and flow rate was observed so laminar flow exists (Figure 7).

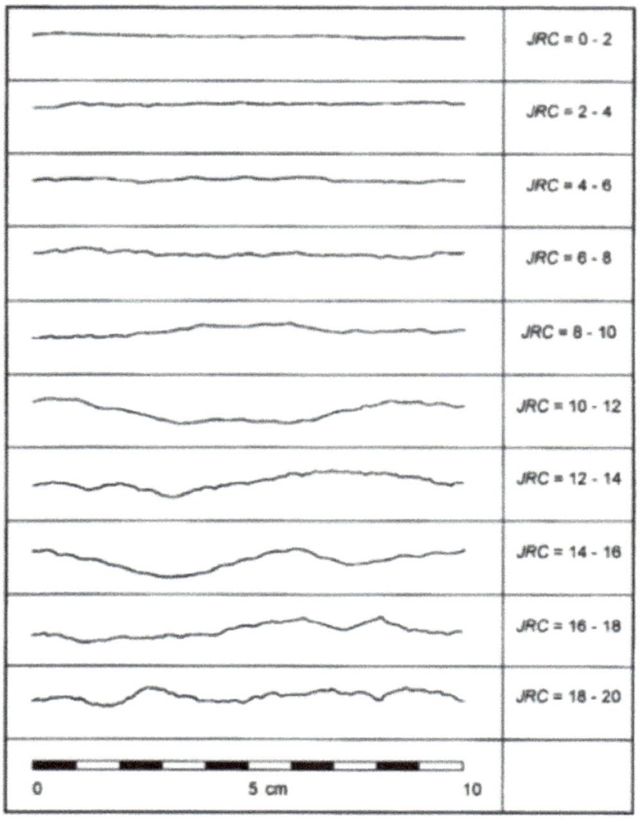

Figure 4. Typical profiles of fractures for JRC coefficient (N. Barton & Choubey, 1977).

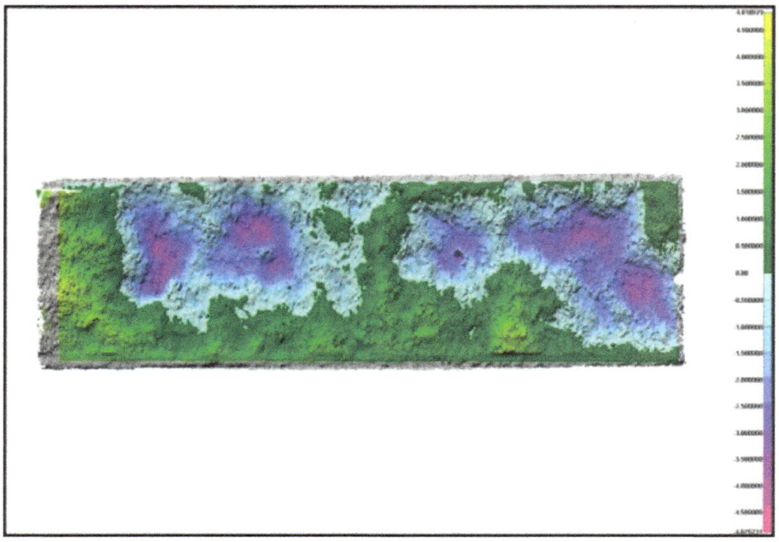

Figure 5. Topography of marble fracture (lower half).

We used the cubic law to back-calculate the hydraulic aperture e (Figure 8) for each induced mechanical aperture Emi 0 μm, 200 μm, 250 μm, 500 μm,750 μm and 1000 μm. The existence of flow for induced aperture Em1 = 0 μm (fracture closed under gravity only) and the back calculated hydraulic aperture of 122 μm (Figure 8(a)), is incompatible for closed fracture. This indicates that void spaces exist between the two walls of fracture due to roughness and water flows through them. Using the empirical Equation (5) of Bandis (Bandis, Makurat, & Vik, 1986) for JRC 16.3 and assuming conducting aperture e equal to back calculated hydraulic b_h (b_h = e = 122 μm), the real aperture E is approached as 363 μm (for fracture closed under self-weight).

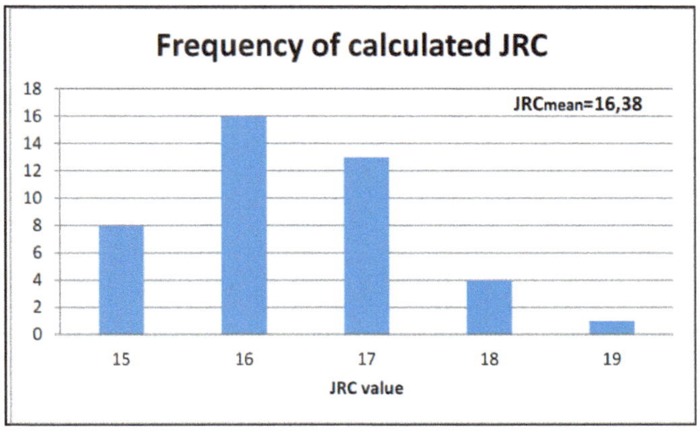

Figure 6. Calculated JRC coefficient for 42 profiles along x axis.

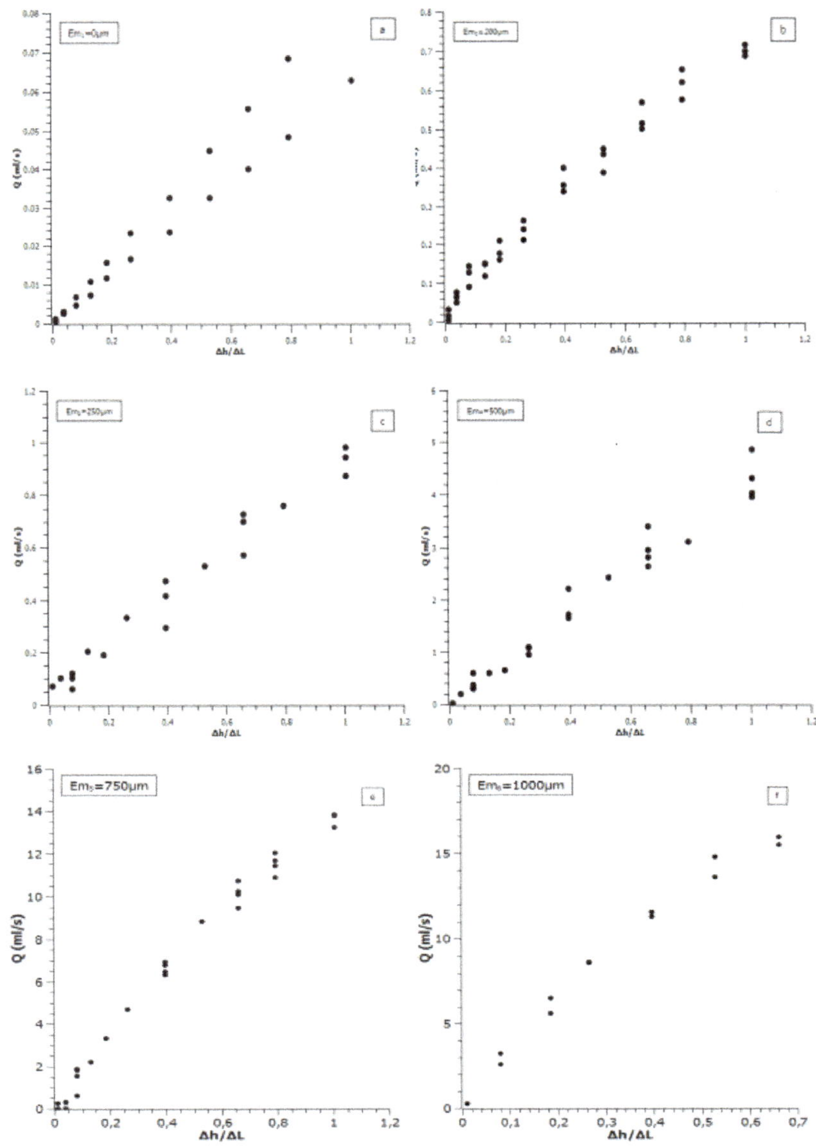

Figure 7. Measured flow rate "Q" for mechanically induced apertures: a) Em_1 = 0 μm; b) Em_2 = 200 μm; c) Em_3 = 250 μm; d) Em_4 = 500 μm; e) Em_5 = 750 μm; f) Em_6 = 1000 μm.

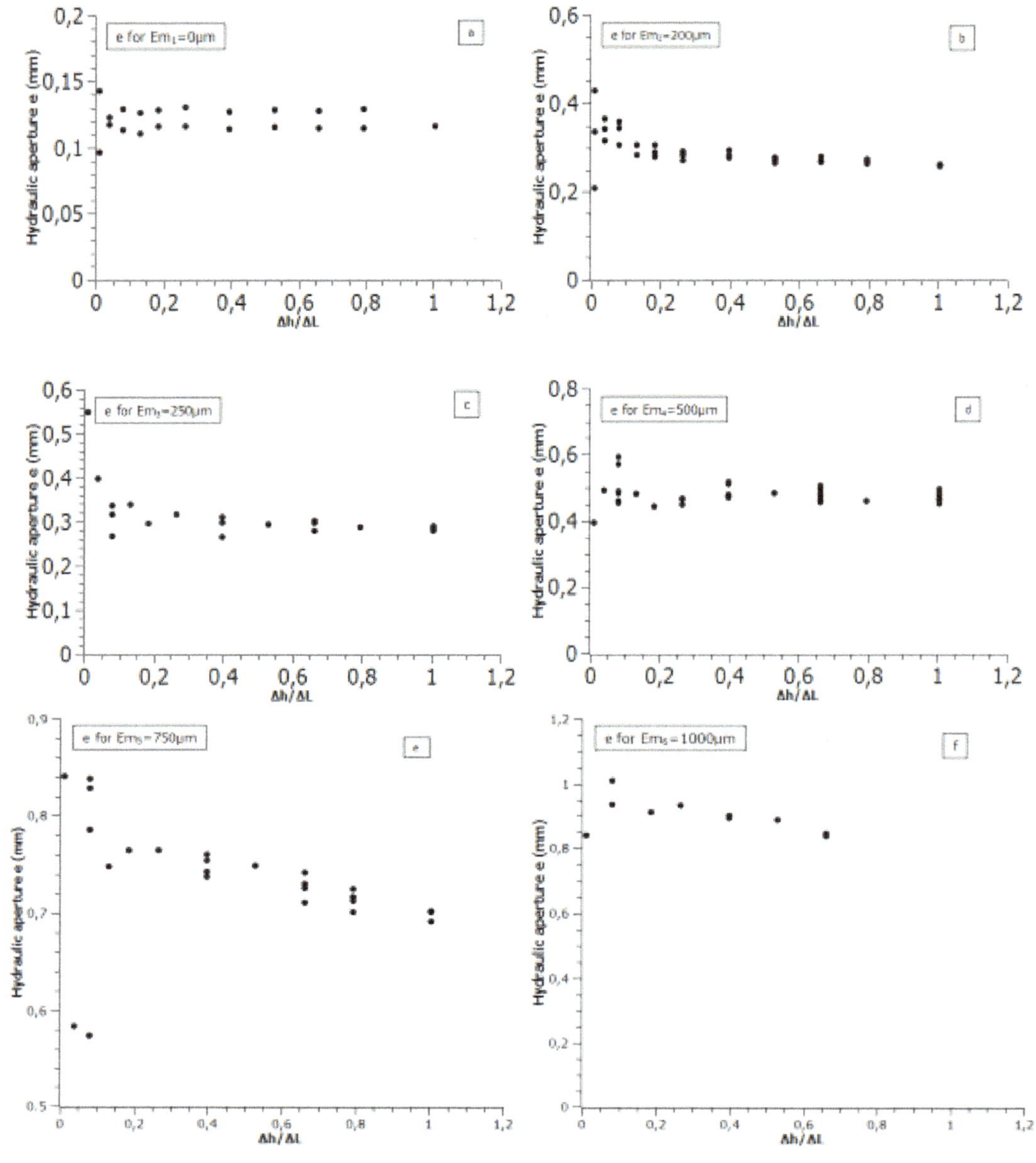

Figure 8. Back calculated hydraulic apertures for mechanically induced apertures: a) $Em_1 = 0$ μm; b) $Em_2 = 200$ μm; c) $Em_3 = 250$ μm; d) $Em_4 = 500$ μm; e) $Em_5 = 750$ μm; f) $Em_6 = 1000$ μm.

$$e = \frac{JRC^{2.5}}{\left(E/e\right)^2} (\mu m) \tag{5}$$

For every mechanically induced aperture the real aperture E was estimated (Table 3). The variation of induced mechanical aperture Em to real aperture E, is presented in Figure 9. Back calculated hydraulic apertures for mechanically induced apertures: a) $Em_1 = 0$ μm, b) $Em_2 = 200$ μm, c) $Em_3 = 250$ μm, d) $Em_4 = 500$ μm, e) $Em_5 = 750$ μm, f) $Em_6 = 1000$ μm.

CONCLUSIONS AND DISCUSSION

In this preliminary study we tested the influence of roughness and induced aperture on the flow regime by measuring flow rate in a fracture of marble. The roughness of the fracture walls was estimated by the index of JRC and factor Z2 which gave similar results (JRC ≈ 16.5). We observed that conducting aperture e (equal to hydraulic aperture b_h which was back calculated by cubic law) and E_m do not coincide, apart from induced aperture $Em_6 = 1000$ μm. This indicates that cubic law was not valid for apertures lower than 1000 μm.

During the experiment even for $Em_1 = 0$ μm flow existed. We calculated the real aperture E using Bandis's Equation (5) which gave E = 360 μm for $Em_1 = 0$ μm. Real aperture E became identical to induced aperture Emi only at 1000 μm. The great differences which were observed between real aperture E and induced aperture E_m indicate that the empirical Equation (5) can be improved. In our experiment values of conducting aperture, real aperture and induced aperture became identical only at 1000 μm.

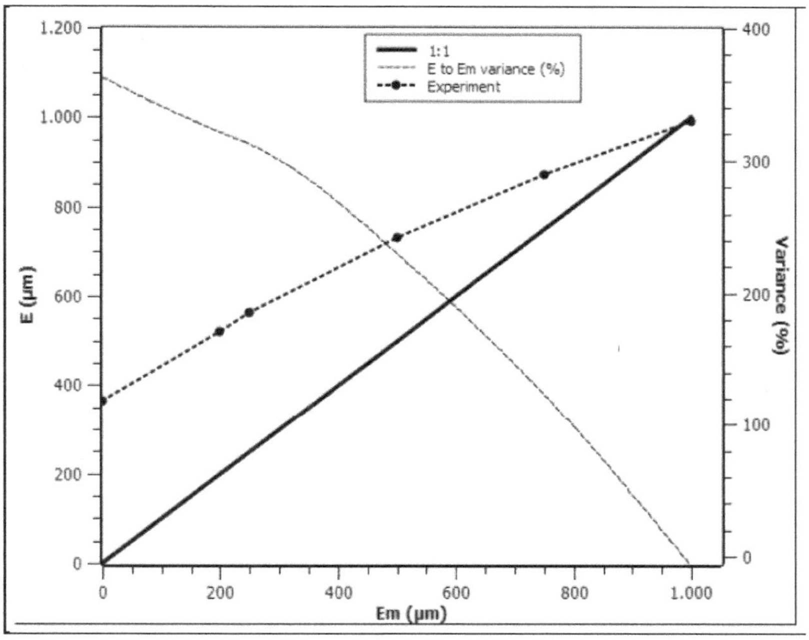

Figure 9. Real aperture E to Induced aperture Em variation.

Table 3. Estimated real apertures and induced apertures

Induced aperture Em_i	0	200	250	500	750	1000
Back calculated b_h (equal to e)	122	250	290	490	700	992
Real Aperture E	363	521	562	730	875	1000

Further tests will take place, in order to approach the relation between induced aperture Em and conducting aperture (e). Tests will be performed for lower apertures trying to minimize the gap between these two parameters. The value of real aperture for closed fracture (E_0) will be approached by putting the sample under normal stress, further flow tests will take place in these conditions to optimize Bandis' constitutive model (Equation (5)). The same procedure will be followed for samples with different fracture roughness (smooth till very rough). Later, flow experiments with acid solutions (simulating CO_2-brine conditions under pressure) will be performed, to examine the influence of acidity on fracture roughness and fracture transmissivity.

ACKNOWLEDGEMENTS

This research has been co-financed by the European Union (European Social Fund–ESF) and Greek national funds through the Operational Program "Education and Lifelong Learning" of the National Strategic Reference Framework (NSRF)-Research Funding Program: Thales. Investing in knowledge society through the European Social Fund.

REFERENCES

1. Bandis, S. C., Makurat, A., & Vik, G. (1986). Predicted and Measured Hydraulic Conductivity of Rock Joints. Publikas-jon-Norges Gotekniske Institutt, 164, 1-11.
2. Barton, N. (1973). Review of a New Shear-Strength Criterion for Rock Joints. Engineering Geology, 7, 287-332. http://dx.doi.org/10.1016/0013-7952(73)90013-6
3. Barton, N., Bandis, S., & Bakhtar, K. (1985). Strength, Deformation and Conductivity Coupling of Rock Joints. International Journal of Rock Mechanics and Mining Sciences & Geomechanics Abstracts, 22, 121-140. http://dx.doi.org/10.1016/0148-9062(85)93227-9
4. Barton, N., & Choubey, V. (1977). The Shear Strength of Rock Joints in Theory and Practice. Rock Mechanics Felsmechanik Mecanique Des Roches, 10, 1-54. http://dx.doi.org/10.1007/BF01261801
5. Bear, J. (1979). Hydraulics of Groundwater. MacGraw-Hill.
6. Brown, S. R. (1987). Fluid Flow through Rock Joints: The Effect of Surface Roughness. Journal of Geophysical Research, 92, 1337-1347. http://dx.doi.org/10.1029/JB092iB02p01337
7. Kranzz, R. L., Frankel, A. D., Engelder, T., & Scholz, C. H. (1979). The Permeability of Whole and Jointed Barre Granite. International Journal of Rock Mechanics and Mining Sciences & Geomechanics Abstracts, 16, 225-234. http://dx.doi.org/10.1016/0148-9062(79)91197-5
8. Li, B., Jiang, Y., Koyama, T., Jing, L., & Tanabashi, Y. (2008). Experimental Study of the Hydro-Mechanical Behavior of Rock Joints Using a Parallel-Plate Model Containing Contact Areas and Artificial Fractures. International Journal of Rock Mechanics and Mining Sciences, 45, 362-375. http://dx.doi.org/10.1016/j.ijrmms.2007.06.004
9. Li, B., Wong, R. C. K., & Milnes, T. (2014). Anisotropy in capillary invasion and fluid flow through induced

sandstone and shale fractures. International Journal of Rock Mechanics and Mining Sciences, 65, 129-140. http://dx.doi.org/10.1016/j.ijrmms.2013.10.004

10. Myers, N. O. (1962). Characterization of Surface Roughness. Wear, 5, 182-189. http://dx.doi.org/10.1016/0043-1648(62)90002-9

11. Olsson, R., & Barton, N. (2001). An Improved Model for Hydro-mechanical Coupling during Shearing of Rock Joints. International Journal of Rock Mechanics and Mining Sciences, 38, 317-329. http://dx.doi.org/10.1016/S1365-1609(00)00079-4

12. Oron, A. P., & Berkowitz, B. (1998). Flow in Rock Fractures: The Local Cubic Law Assumption Reexamined. Water Resources Research, 34, 2811-2825. http://dx.doi.org/10.1029/98WR02285

13. Park, H., Osada, M., Matsushita, T., Takahashi, M., & Ito, K. (2013). Devel-opment of Coupled Shear-Flow-Visualization Apparatus and Data Analysis. International Journal of Rock Mechanics and Mining Sciences, 63, 72-81. http://dx.doi.org/10.1016/j.ijrmms.2013.06.003

14. Renshaw, C. E. (1995). On the Relationship between Mechanical and Hydraulic Apertures in Rough-Walled Fractures. Journal of Geophysical Research, 100, 629-636. http://dx.doi.org/10.1029/95JB02159

15. Tse, R., & Cruden, D. M. (1979). Estimating Joint Roughness Coefficients. In-ternational Journal of Rock Mechanics and Mining Sciences & Geomechanics Abstracts, 16, 303-307. http://dx.doi.org/10.1016/0148-9062(79)90241-9

16. Witherspoon, P. A., Wang, J. S. Y., Iwai, K., & Gale, J. E. (1980). Validity of Cubic Law for Fluid Flow in a Deformable Rock Fracture. Water Resources Research, 16, 1016-1024. http://dx.doi.org/10.1029/WR016i006p01016

17. Yeo, I. W., de Freitas, M. H., & Zimmerman, R. W. (1998). Effect of Shear Displacement on the Aperture and Permeability of a Rock Fracture. International Journal of Rock Mechanics and Mining Sciences, 35, 1051-1070. http://dx.doi.org/10.1016/S0148-9062(98)00165-X

18. Zhao, Z., Li, B., & Jiang, Y. (2013). Effects of Fracture Surface Roughness on Macroscopic Fluid Flow and Solute Transport in Fracture Networks. Rock Mechanics and Rock Engineering, 47, 2279-2286 http://dx.doi.org/10.1007/s00603-013-0497-1

19. Zimmerman, R. W., Al-Yaarubi, A., Pain, C. C., & Grattoni, C. A. (2004). Non-Linear Regimes of Fluid Flow in Rock Fractures. International Journal of Rock Mechanics and Mining Sciences, 41, 384. http://dx.doi.org/10.1016/j.ijrmms.2003.12.045

20. Zimmerman, R. W., & Bodvarsson, G. (1996). Hydraulic Conductivity of Rock Fractures. Transport in Porous Media, 23, 1-30. http://dx.doi.org/10.1007/BF00145263

CHAPTER 10

AN ENERGY-BASED SAFETY EVALUATION INDEX OF BLAST VIBRATION

Mingsheng Zhao,[1] Dong Huang,[2] Maosen Cao,[3] En-an Chi,[4] Jun Liu,[5] and Qiang Kang[6]

[1]Guizhou Xinlian Blasting Engineering Group Co. Ltd., 78 Xinhua Road, Fuzhong Building 17, Guiyang, Guizhou 550002, China

[2]Key Laboratory of Mountain Hazards and Surface Process and Institute of Mountain and Environment, Chinese Academy of Science and Ministry of Water Conservancy, Chengdu, Sichuang 610041, China

[3]College of Mechanics and Materials, Hohai University, 1 Xikang Road, Nanjing, Jiangsu 210098, China

[4]Guizhou Xinlian Blasting Engineering Group Co. Ltd. and Mining College, Guizhou University, 78 Xinhua Road, Fuzhong Building 17, Guiyang, Guizhou 550002, China

[5]Key Laboratory of Ministry of Education for Geomechanics and Embankment Engineering and Institute of Engineering Safety and Disaster Prevention, Hohai University, 1 Xikang Road, Nanjing, Jiangsu 210098, China

[6]Institute of Geotechnical Engineering, Hohai University, 1 Xikang Road, Nanjing, Jiangsu 210098, China

ABSTRACT

The combined peak particle velocity (PPV) and frequency safety criterion for blast vibration is widely used in blasting engineering. However, some field investigations are inconsistent with this criterion. On the basis of field investigations, it is found that there are two failure modes of structures subjected to blasting seismic waves, that is, first-excursion failure and cumulative plastic damage failure. Moreover, the nature of structural responses under blast vibrations is a process of energy input, transformation, and dissipation. Therefore, an energy-based dual safety standard is proposed in this work to more comprehensively explain all failure modes of structures under blast vibrations. To this end, structures are simplified into elastic-plastic single degree of freedom (SDOF) systems with bilinear restoring force models, and energy responses of SDOF systems are then determined using the Newmark-β method. From the energy responses, the maximum instantaneous input energy and hysteretic energy are selected as the basis of the dual safety criterion, because they can reflect first-excursion failure and cumulative plastic damage failure, respectively. Finally, field investigations in a blasting site in Zunyi, Guizhou province, China, are used to prove that compared to the PPV-frequency criterion the proposed energy-based dual safety criterion is more capable of assessing the damage potential of blast vibrations.

INTRODUCTION

Bench blasting with medium to deep blast holes is used increasingly in China in the current mining industry, infrastructure construction, and buildings demolition in cities. Blasting expedites the progress of construction and improves productivity. However, adverse effects of blasting (such as ground vibration, flying rocks, noise, etc.) are inevitable and should be minimized to a permissible level. Among all the adverse effects, ground vibration is a prime concern, attracting considerable research effort.

Based on extensive field investigations, early studies of structural responses and damage from blast-induced ground vibration focused on establishing empirical allowable ground vibration levels in terms of either peak particle velocities (PPV) [1–4] or PPV together with the principal ground vibration frequency (the so-called PPV-frequency criterion) to limit structural damage [5, 6]. As these two criteria can be

Citation: Mingsheng Zhao, Dong Huang, Maosen Cao, En-an Chi, Jun Liu, and Qiang Kang, "An Energy-Based Safety Evaluation Index of Blast Vibration," Shock and Vibration, vol. 2015, Article ID 698193, 9 pages, 2015. doi:10.1155/2015/698193.

easily implemented, they have been adopted in the national standards of many countries [7]. For example, the PPV-frequency criterion is recommended by the latest Chinese standard (GB6722-2003) [8]. Therefore, in research into blast-induced ground vibration, efforts are still exerted to precisely predict PPV under different circumstances [9–11] and to propose a more accurate PPV-frequency criterion for specific categories of buildings [7]. In this criterion, however, the shapes of wave forms, the duration of dynamic loading, and the dynamic responses of structures are not taken into account. Thus the PPV-frequency criterion still has limitations in controlling permissible ground vibrations. Because energy is capable of representing the resultant action of the amplitude, frequency, and duration of blast-induced ground vibrations, energy-based failure criteria have been proposed to assess the damage to buildings subjected to blast vibrations. For example, equivalent energy proposed by Wenhua et al. [12], response energy by Guosheng et al. [13], equivalent peak energy by Li et al. [14], maximum instantaneous input energy by Zhang et al. [15], and total input energy by Wang et al. [16] have been, respectively, suggested as failure criteria for buildings. These works more comprehensively reflect the failure process of buildings subjected to blast vibrations. It should be borne in mind that the nature of structural responses under blast vibrations is a process of energy input, transformation, and dissipation. The structural responses depend not only on input blasting seismic waves, but also on the parameters of structures, such as natural period, damping ratio, yield strength factor, and stiffness reduction factor [17, 18]. Therefore, the accuracy of failure criteria that are simple additive weighted energy based on either different frequency bands or different intrinsic mode function (IMF) components of blasting seismic waves is still questioned.

From the viewpoint of energy analysis in elastoplastic structures, energy is continuously transferred from seismic waves to structures in natural earthquakes. A proportion of energy is stored as recoverable kinetic energy and elastic strain energy, and the rest is dissipated by damping and hysteretic energy of plastic deformation. As the damping ratios of structures vary over a small range, hysteretic energy is regarded as the most important energy index of the cumulative plastic damage of structures [19, 20]. By analogy, hysteretic energy can also be used to indicate the cumulative damage of structures subjected to blast vibrations. Research into speed pulse earthquakes [21, 22] has shown that the maximum instantaneous input energy (MIIE) transferred into structures occurs at the position of speed pulses and is very likely to cause the maximum displacement and the first-excursion failure of structures. Researchers [23–25] have demonstrated that MIIE is closely related to the maximum displacement of a structure, and thus MIIE can indicate the first-excursion failure of structures under earthquake.

There are two primary types of structural failure under blast vibrations, namely, first-excursion failure and cumulative plastic failure. These two failures result from both the characteristics of blast vibrations and the dynamic responses of structures [15, 16]. MIIE fails to show the cumulative effect of the duration of ground vibrations on the nonlinear plastic deformations of structures [25], and hysteretic energy scarcely reflects first-excursion failure of structures caused pulse-like ground motions [26, 27]. Therefore, a dual failure (or safety) criterion is often used in seismic design in earthquake engineering [28, 29]. Investigations of blast vibrations have shown that the failure mechanisms of structures subjected to blast vibrations and natural earthquakes are similar [14, 15, 30]. In this paper, therefore, a dual safety criterion is proposed to more comprehensively assess the damage potential of structures under blast vibrations. To this end, (1) buildings affected by blast vibrations are simplified into elastoplastic single degree of freedom (SDOF) systems with bilinear hysteretic restoring forces, and the corresponding structural energy responses are analyzed using the Newmark-β method; (2) MIIE and hysteretic energy are proposed as a dual safety criterion for structures, as they reflect first-excursion failure and cumulative plastic failure, respectively.

HYSTERETIC ENERGY OF STRUCTURES

Concept of Hysteretic Energy

In a SDOF system, the relative energy equation of a unit mass structure at time t is

$$\int_0^t m\ddot{U}(t)\dot{U}(t)\,dt + \int_0^t c\left(\dot{U}(t)\right)^2 dt + \int_0^t F(U(t))\dot{U}(t)\,dt = \int_0^t -m\ddot{U}_g(t)\dot{U}(t)\,dt, \tag{1a}$$

where $\ddot{U}(t), \dot{U}(t)$ and $U(t)$ are the relative acceleration, velocity, and displacement between the structure and the ground, respectively; FU(t) is the restoring force; c is the damping coefficient; m is the mass of the block in SDOF; $\ddot{U}_g(t)$ is the ground acceleration induced by blast vibration. Assuming a bilinear hysteretic restoring force model and a dividing mass m at both sides of (1a), (1a) becomes

$$\int_0^t \ddot{U}(t)\dot{U}(t)\,dt + \int_0^t 2\xi\omega\left(\dot{U}(t)\right)^2 dt + \int_0^t p\omega^2\dot{U}(t)\,dt = \int_0^t -\ddot{U}_g(t)\dot{U}(t)\,dt, \tag{1b}$$

where $\xi = c/(2m\omega)$ is the structure damping ratio, typically assumed as a constant; $\omega = \omega_0\sqrt{1-\xi^2}$ is the natural frequency of the damped structure and $\omega_0 = \sqrt{k/m}$ is the natural frequency without damping; $\omega \approx \omega_0$ when $\xi \ll 1$; p is the stiffness ratio in the bilinear hysteretic restoring force model, p = 1 for the elastic stage and $p = k_i/k$ for the plastic stage, where k is the initial elastic stiffness and k_i is the stiffness after yielding.

Equation (1b) is alternatively expressed as

$$E_K + E_D + E_H + E_E = E_I, \tag{2}$$

where $E_K = \int_0^t \ddot{U}(t)\dot{U}(t)dt$ is the kinetic energy of the structure; $E_D = \int_0^t 2\xi\omega(\dot{U}(t))^2$ is the damping energy dissipation of the structure; $E_H + E_E = \int_0^t p\omega^2\dot{U}(t)dt$ is the deformation energy, that is, the sum of hysteretic energy E_H and elastic strain energy E_E, and E_H=0 for an elastic system; $E_I = \int_0^t -\ddot{U}_g(t)\dot{U}(t)dt$ is the total input energy of the system. Research [20] has shown that the proportion of E_E in the deformation energy of an elastic-plastic structure is so small that can be ignored, and thus E_H can represent the deformation energy.

Restoring Force Model of Structures

Restoring force is structural resistance against deformation. The curve of the relation between the restoring force and the nonlinear deformation of a structure under cyclic loading is called the characteristic curve of restoring force. For convenience, many researchers have simplified and abstracted the characteristic curves of restoring force and proposed practical restoring force models such as bilinear, trilinear, and semidegenerate models. Researchers have found that restoring force models have little effect [31] or negligible effect [32] on hysteretic energy. Therefore, the bilinear restoring force model, which is the simplest and the most widely used one in elastoplastic dynamic analysis, is selected in this work to study the hysteretic energy responses of structures under blast vibrations, as shown in Figure 1.

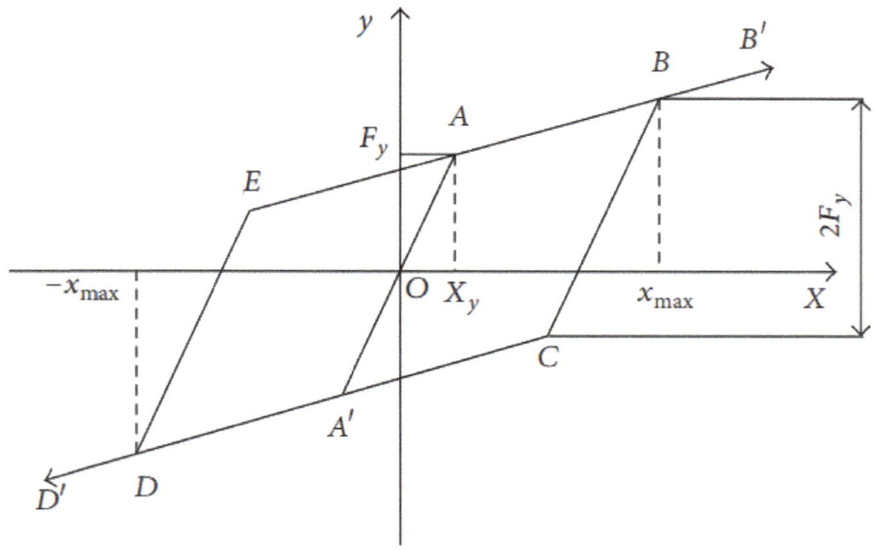

Figure 1: Bilinear restoring force model.

Figure 1 shows the bilinear restoring force model, which has symmetric envelopes, loading, unloading, and reloading rules in positive and negative branches. Note that the bilinear model follows the kinematic hardening rule. According to the deformation history x, the model is classified into five states: (1) In the initial elastic state (i.e., along the line AOA') when $|x| \le x_y$, the stiffness is $k=k_1$. (2) In the positive plastic state (i.e., line AB) when $x>x_y$, the stiffness decreases, such that $k=k_2 = \alpha k_1$, where α is the stiffness reduction factor after yielding. (3) In the reversed elastic path BC when $x \le x_{max}$ and $x \ge (x_{max} - 2x_y)$, x_{max} is determined at $x=0$. In this state, the stiffness is $k=k_1$. When $x>x_{max}$, the structure reaches the positive plastic state ABB' with the stiffness of $k=k_2 = \alpha k_1$. (4) In the negative plastic state (i.e., path CD) when $x < (x_{max} - 2x_y)$, the stiffness is $k=k_2 = \alpha k_1$. (5) In the positive elastic path DE when $-x_{max} \le x \le (-x_{max} + 2x_y)$, the corresponding stiffness is $k=k_1$. When $x < -x_{max}$, the structure reaches the negative plastic state CDD' with the stiffness of $k=k_2 = \alpha k_1$.

Calculation of Hysteretic Energy

In research into obtaining the energy responses of elastoplastic structures, it has been suggested that when the parameter β=1/4, the Newmark-β method is unconditionally stable in solving the incremental equations of motion (EOM) with positive and negative stiffness [20, 26–28, 32]. Therefore, this numerical method is employed, and the incremental quantities of EOM are given by

$$\Delta U = \left(-\Delta \ddot{U}_g + \frac{1}{\beta \Delta t}\dot{U}_n + \frac{1}{2\beta}\ddot{U}_n + 4\xi\omega_0\left(\frac{1}{2\beta}\dot{U}_n + \left(\frac{1}{4\beta} - 1\right)\ddot{U}_n \Delta t\right)\right) \cdot \left(p\omega_0^{\,2} + \xi\omega_0\frac{1}{\beta\Delta t} + \frac{1}{\beta \Delta t^2}\right)^{-1},$$

$$\Delta \dot{U} = \frac{1}{2\beta\Delta t}\Delta U - \frac{1}{2\beta}\dot{U}_n - \left(\frac{1}{4\beta} - 1\right)\ddot{U}_n \Delta t,$$

$$\Delta \ddot{U} = \frac{1}{\beta \Delta t^2}\Delta U - \frac{1}{\beta\Delta t}\dot{U}_n - \frac{1}{2\beta}\ddot{U}_n, \qquad (3)$$

Where $\ddot{U}_n$, $\dot{U}_n$, and U_n are the acceleration, velocity, and displacement at the previous time step, respectively.

Following the above algorithm, the measured data of an open-pit mine is selected as an example to demonstrate the energy responses of a structure subjected to blast vibrations. Figure 2 shows the measured blast vibration velocity signal with a sampling frequency of 2000 Hz. The affected structure is assumed to be a unit mass elastoplastic SDOF system with a damping ratio of $\xi = 0.05$ and a natural period of T=0.1 s. The yield strength coefficient of the bilinear restoring force model is 0.3 (equal to the ratio of x_y/x_{max}), and the stiffness reduction factor after yielding (i.e., α) is 0.02. Thereafter, in accordance with (1b), the incremental quantities in (3) are integrated step by step to obtain the total input energy (E_I), the kinetic energy (E_k), the deformation energy (E_E+E_H), and the damping energy dissipation (E_D) of the SDOF system. The time histories of these energies are shown in Figure 3. Further, the time histories of the percentages of E_k, E_E+E_H and E_D to the total input energy E_I are shown in Figure 4.

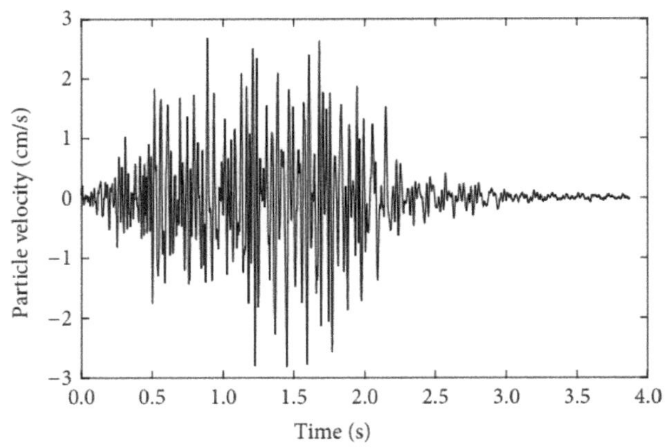

Figure 2: A velocity waveform measured in an open pit induced by a blast.

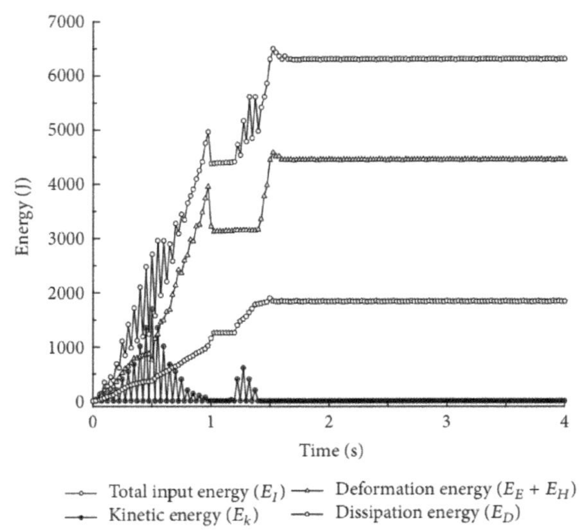

Figure 3: Time histories of energy responses in elastoplastic structure.

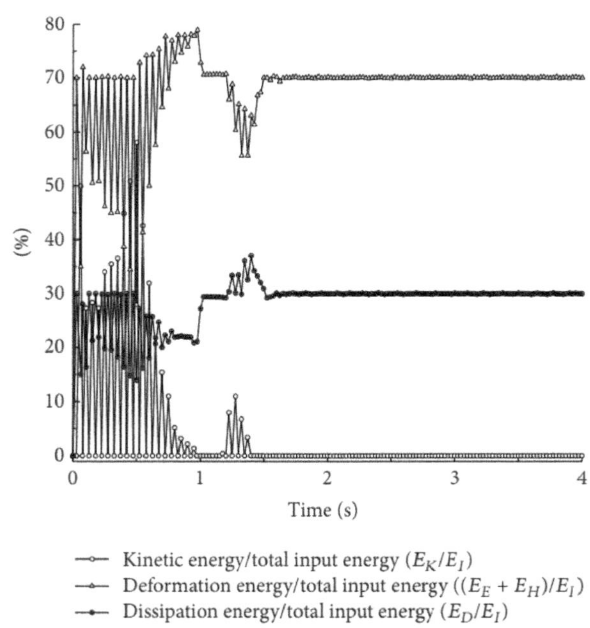

Figure 4: Time histories of percentages of structural energies to total input energy.

From Figures 3 and 4, the following can be found: (1) Responses of input energy, damping energy, and hysteretic energy of a structure are gradual process of accumulation. These three energies all increase with time and attain their maximum values at the end of the blast vibration. (2) The energy of an elastoplastic SDOF structure under blast vibrations consists of input energy, kinetic energy, deformation energy, and damping energy. With an increase in vibration duration, kinetic energy and elastic strain energy of deformation energy undergo mutual transformation and eventually tend to 0. At the end of the blast vibration, all the input energy is dissipated by hysteretic energy of plastic deformations and damping energy. (3) With an increase of the duration of blast vibration, the percentages of hysteretic energy and damping energy to total input energy tend to be stable, at around 70% and 30%, respectively, as shown in Figure 4. As hysteretic energy dominates the energy dissipation in the hardening stage of structural nonlinear responses, it can be used for assessing the cumulative plastic damage of structures under blasting vibration.

MAXIMUM INSTANTANEOUS INPUT ENERGY OF STRUCTURES

Definition of Instantaneous Input Energy

The input energy between two adjacent zero-velocity points in a structural energy response is defined as instantaneous input energy, ΔE. Due to zero-velocity between the two adjacent points, the incremental kinetic energy is null, and thus (1b) can be written as

$$\int_t^{t+\Delta t} 2\xi\omega\left(\dot{U}(t)\right)^2 dt + \int_t^{t+\Delta t} p\omega^2\dot{U}(t)\,dt = \int_t^{t+\Delta t} -\ddot{U}_g(t)\dot{U}(t)\,dt, \tag{4}$$

where time interval Δt is the time needed in a half cycle of structural vibration and Δt is not a constant value, because of the nonstationary random characteristics of blasting seismic waves. The energy equation can be expressed by (5) according to (2); that is,

$$\Delta E_D + \left(\Delta E_H + \Delta E_E\right) = \Delta E_I. \tag{5}$$

Good correlation has been found between MIIE $\Delta E_{I\,max}$ and the maximum displacement of a structure over a short period of vibration [32–34]. Because the dominant frequencies of blasting seismic waves are higher than those of natural earthquakes, structural responses to blasting seismic waves occur mainly in over short period range [35, 36]. Accordingly, the MIIE of elastoplastic structural responses obtained by using bilinear restoring force models in this paper can accurately indicate the first-excursion failure of a structure subjected to blasting seismic waves.

Calculation of Instantaneous Input Energy

According to the definition of instantaneous input energy (IIE), points with zero-velocity are selected from the time history of velocity as shown in Figure 2. The differences of input energy E_I in the time intervals between each two adjacent selected points are calculated to establish the time history of IIE, as shown in Figure 5. The maximum value in Figure 5 is the MIIE.

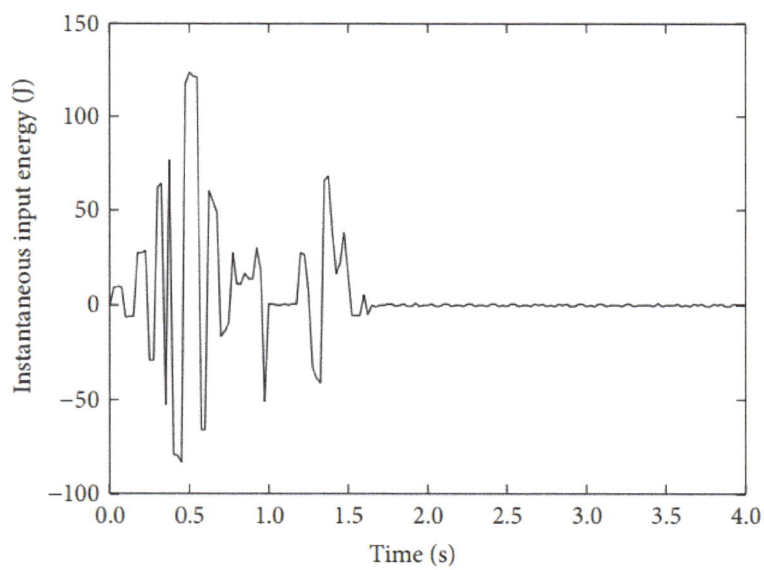

Figure 5: Time history of instantaneous input energy.

Figure 5 shows that the MIIE affects structures for a very short time, less than 1.5 s. However, the dissipation of input energy due to damping is a process of accumulation (from 0 s to 4 s), as shown in Figure 3. Therefore, if the IIE is too large, a structure may fail in a short time due to large deformations, prior to the excitation of damping. Therefore, the first failure of structures caused by MIIE under blasting seismic waves plays an important role in situations of a small distance from an explosive charge center. According to the definition and the calculation of MIIE, MIIE is mainly affected by PPV and the dominant frequencies of blasting seismic waves as well as by the natural periods of structures. Nevertheless, the duration of blasting seismic waves has no effect on MIIE. Consequently, MIIE cannot reflect the cumulative effect of blasting seismic waves on structure failure.

Following the above analysis, MIIE can reflect the first-excursion failure of structures under blasting seismic waves, and hysteretic energy is an important index of the cumulative plastic damage in structures. Therefore, MIIE and hysteretic energy are proposed as a dual safety criterion for structures subjected to blast vibrations.

DYNAMIC PARAMETERS OF STRUCTURES

The primary structural parameters that affect MIIE and hysteretic energy are the natural periods and damping ratios of structures. Damping ratios are usually assumed to be constant in structural dynamic analysis, ranging within 0.02~0.05 for typical building structures. A damping ratio of 0.05 has been used in analyzing structural responses under blasting seismic waves [13, 15, 16, 35], so the damping ratio in this paper is also assumed to be 0.05 in assessing the damage potential caused by blast vibrations. There are two methods to determine the natural periods of structures, namely, theoretical calculation and empirical formula, and the latter is preferred in engineering applications [13]. Most buildings analyzed in this paper are brick-concrete structures. Based on the derivation of uniform cantilever beams and the measurement of dynamic characteristics of multistory brick-concrete structures [13], the natural period of a brick-concrete structure is closely related to its height. In the general field, the natural period is estimated as

$$T_0 = 0.0168\,(H + 1.2)\,, \tag{6}$$

where H is the height of a structure.

APPLICATION OF THE ENERGY-BASED SAFETY EVALUATION METHOD

Experimental Details

The proposed energy-based dual safety criterion was used in a series of bench blasts. The bench excavation elevation was 8 m. Drilling equipment consisted of 351 down-the-hole drills and hydraulic drilling machines. There were three hole diameters, 90 mm, 115 mm, and 140 mm. Quincuncial hole arrangement and plastic Nonel tubes were used. Ten-period detonators were used with millisecond delays in each hole, three-period detonators were used with a millisecond delays every 2-3 holes, and five-period detonators were used in millisecond delays between rows. Two types of explosives were utilized, including the mixed loading ANFO and the mixed loading emulsion explosive. The total charge amount each time was not greater than 12,000 kg, and the maximum charge amount per delay was about 160 kg.

The buildings in the surroundings of the blasting area are shown in Figure 6. It can be seen that residential buildings are extensively distributed around the blasting area. These buildings are typically two-story brick-concrete structures built at the beginning of this century. The floors in those buildings are constructed of precast panels that are basically not connected by ring beams. Therefore, the global seismic performance of these buildings is poor. Some residential buildings around the blasting area were not covered by land requisition schemes and hence were still in use during the blasting construction period. As a result, it was necessary to assess and control the effects of blast vibrations on these residential buildings.

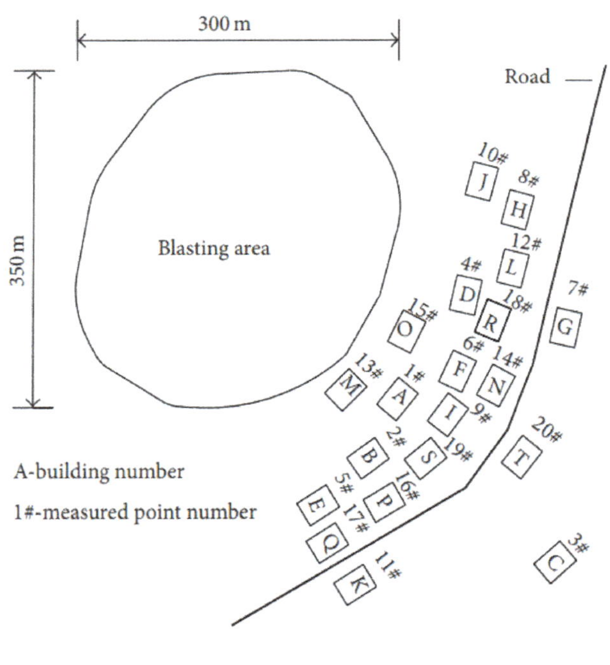

Figure 6: Buildings surrounding blasting area.

During blasting, a great many investigations were carried out to assess the possible damage to the residential buildings caused by blast vibration. To prove that the proposed dual safety criterion based on MIIE and hysteretic energy was superior to the PPV-frequency criterion, the measured data at the foundations of 20 residential buildings were selected for demonstration. The basic parameters and the conditions prior to blasting of these 20 residential buildings are given in Table 1. The main parameters extracted from the measured results are listed in Table 2. These results are discussed by the aforementioned two criteria later.

Table 1: The main parameters of the surrounding buildings of the blasting area.

Building number	Types of buildings	Building height (m)	Status before blast
A	Brick-concrete	7.3	Intact
B	Brick-concrete	7.5	Intact
C	Brick-concrete	7.1	Intact
D	Brick-concrete	7.3	Intact
E	Brick-concrete	7.0	Intact
F	Brick-concrete	7.1	Intact
G	Brick-concrete	7.4	Intact
H	Brick-concrete	7.8	Intact
I	Brick-concrete	7.6	Intact
J	Brick-concrete	7.6	Intact
K	Brick-concrete	12.5	Intact
L	Brick-concrete	7.2	Intact
M	Brick-concrete	7.3	Intact
N	Brick-concrete	7.5	Intact
O	Brick-concrete	7.5	Intact
P	Brick-concrete	7.4	Intact
Q	Brick-concrete	7.7	Tiny crack on wall
R	Brick-concrete	11.4	Intact
S	Brick-concrete	7.2	Tiny crack on wall
T	Brick-concrete	7.0	Intact

Table 2: The results measured at the buildings.

Measured point number	Maximum charge per delay (kg)	Distance from explosion source (m)	PPV (cm/s)	Dominant frequency (Hz)	Duration (s)
1#	120	55.12	3.035	29.608	1.6
2#	85	83.44	1.693	11.576	0.7
3#	140	273.87	0.702	24.691	1.2
4#	160	64.32	2.839	27.454	2.5
5#	110	88.53	1.427	21.277	2.2
6#	60	91.65	1.510	21.677	2.5
7#	80	152.39	0.827	9.412	1.0
8#	60	98.11	1.164	13.519	2.5
9#	110	103.25	1.137	31.250	1.6
10#	140	96.38	2.731	9.519	0.7
11#	120	162.84	0.608	8.333	2.1
12#	110	103.96	1.126	17.478	2.2
13#	60	10.16	5.916	48.868	0.5
14#	132	124.48	1.619	14.286	1.0
15#	144	26.12	4.801	41.544	0.6
16#	148	120.49	1.423	9.957	4.7
17#	120	113.11	0.905	10.659	4.8
18#	115	101.26	1.221	14.804	3.8
19#	135	113.45	1.092	9.241	3.4
20#	144	166.49	0.874	12.868	4.3

Analysis Based on the PPV-Frequency Safety Standard

The latest national safety regulations for blasting (GB6722-2003) were issued in 2003 [8]. The regulations adopted the PPV-frequency safety standard (as shown in Table 3) to assess the damage potential for buildings subjected to blast vibrations.

Table 3: The safe level standard of blast vibrations in China (GB6722-2003).

Types of structures	Allowable PPV (cm/s)		
	<10 Hz	10–50 Hz	50–100 Hz
Soil cave houses, adobe houses, and rubble stone structures	0.5–1.0	0.7–1.2	1.1–1.5
Common brick and large nonseismic designed masonry buildings	2.0–2.5	2.3–2.8	2.7–3.0
Reinforced concrete framed structures	3.0–4.0	3.5–4.5	4.2–5.0
Common historical buildings and interests	0.1–0.3	0.2–0.4	0.3–0.5
Hydraulic tunnels		7–15	
Traffic tunnels		10–20	
Mine roadways		15–30	

The dominant frequencies of blasting seismic waves were in the range 10–50 Hz at the measuring points 1#, 4#, 13#, and 15# (see Table 2), where all the tested buildings tested were brick-concrete structures. According to Table 3, the maximum allowable PPV was 2.8 cm/s. All the measured velocities at these four points have exceeded 2.8 cm/s (see Table 2). After blasting, no damage was found in residential buildings A and D corresponding to measuring points 1# and 4#, respectively. Buildings M and O at measuring points 13# and 15# were nearer to the blasting area and had already been abandoned prior to blasting. After blasting, it was found that in building M that a large area of floated coat had chipped off, "X" shaped cracks had formed on the windowsills and doorways, and many cracks occurred on the walls and the floors. In building O after blasting, it was observed that the floated coat had chipped off, cracks appeared on the walls, and a horizontal crack occurred at a staircase. The dominant frequency of the blast seismic wave at measuring point 10# was 9.519 Hz, and the maximum allowable PPV was 2.5 cm/s according to Table 3. However, the measured PPV was 2.7 cm/s, and no damage was found in the corresponding residential building J at measuring point 10#.

All the PPVs of the other measuring points were lower than the corresponding maximum allowable PPVs. Nevertheless, in building P at measuring point 16#, the floating coat had chipped off and cracks appeared on the windowsills; in building Q at point 17#, original tiny cracks had broadened; in building S at point 19#, the floating coat had chipped off, original tiny cracks had widened, and new "X" shape cracks had developed in the doorways.

In summary, the measured PPVs at 15 buildings out of 20 were lower than the maximum allowable value, and no damage was found in 12 of those buildings. Although the measured PPVs in the other 5 buildings exceeded the allowable value, only 2 of them were damaged. The findings from 14 of the buildings can be explained by the safety level standard as shown in Table 3, indicating that the safety level standard based on extensive investigations of blasting vibrations could provide valuable suggestions. However, the findings from the other 6 buildings were beyond the current national safety level standard. These differences are ascribed to the facts that (1) the current standard only considers the PPV and the dominant frequency of a blasting seismic wave, without accounting for the duration of the blasting seismic wave and the associated structural responses and (2) the current standard permits reduction of the allowable PPVs for old and originally damaged buildings but has no clear quantitative specifications for the reduction.

Analysis Using the Energy-Based Safety Evaluation Method

For comparison with the analysis based on the PPV-frequency safety standard, the measured data from 20 buildings was also analyzed using the proposed dual safety criterion based on MIIE and hysteretic energy. Because all the residential buildings were brick-concrete structures, their natural vibration periods were obtained by (6). In addition, it was assumed that, for all 20 buildings, the damping ratios were 0.05, the yield strength coefficient of the bilinear restoring force model was 0.3, and the stiffness reduction factor after yielding was 0.02. Following the procedures illustrated in Sections 2 and 3, the MIIE and the hysteretic energy of the 20 structures under blasting seismic waves were calculated and are shown in Table 4.

Table 4: Energy responses of the residential buildings.

Measured points number	Building number	Natural vibration period (s)	Maximum instantaneous input energy (J)	Hysteretic energy (J)
1#	A	0.1428	123.12	2745.44
2#	B	0.1462	45.98	854.35
3#	C	0.1394	17.33	356.32
4#	D	0.1428	131.44	2678.21
5#	E	0.1378	33.65	677.93
6#	F	0.1394	34.13	768.86
7#	G	0.1445	20.66	459.19
8#	H	0.1512	31.28	679.37
9#	I	0.1478	35.76	712.21
10#	J	0.1478	119.76	2338.91
11#	K	0.2302	35.67	697.44
12#	L	0.1411	40.13	754.35
13#	M	0.1428	220.25	2987.19
14#	N	0.1462	39.77	732.87
15#	O	0.1462	215.11	2853.33
16#	P	0.1445	155.54	3022.22
17#	Q	0.1495	113.17	2930.09
18#	R	0.2117	145.76	2871.15
19#	S	0.1411	103.22	2853.87
20#	T	0.1378	143.23	2973.12

MIIE is mainly related to the PPV and dominant frequency of a blasting seismic wave and the natural periods of vibrated structures but is unable to reflect the adverse effect of the duration of blasting vibration. In contrast, hysteretic energy can account for the effects of the vibration duration. For example, Table 4 shows that the MIIEs at measuring points 13# and 15# are 220.25 J and 215.11 J, respectively, much higher than those at the other points. However, the hysteretic energies at points 13# and 15# are 2987.19 J and 2853.33 J, respectively, smaller than that at point 16# (3022.22 J). The maximum MIIEs at measuring points 16#, 17#, and 19# are 155.54 J, 113.17 J, and 103.22 J, respectively, and the corresponding hysteretic energies are 3022.22 J, 2930.09 J, and 2853.87 J, respectively. Except for the above-mentioned 5 points, the MIIEs and the hysteretic energies at all the other points are below 200 J and 3000 J, respectively. Moreover, no damage was found in the buildings at these 15 points. Prior to blasting, some tiny cracks had already occurred on the walls of residential buildings Q and S at measuring points 17# and 19#, respectively, and all the other 18 buildings were intact, as shown in Table 1. It is concluded, therefore, that (1) MIIE beyond 200 J caused the first-excursion failures of the intact residential buildings with brick-concrete structures, such as buildings M and O at measuring points 13# and 15#, respectively; hysteretic energy beyond 3000 J resulted in cumulative plastic damage, such as in building P at point 16#; (2) the threshold values of the MIIE and the hysteretic energy for the residential buildings with original tiny cracks were, respectively, lower than those of intact residential buildings. In this test, the hysteretic energies at points 17# and 19# were closer to the threshold value of the cumulative plastic failure of intact buildings. Therefore, the damage found in these two buildings can be tentatively regarded as cumulative plastic failure.

In summary, the dual safety criterion based on MIIE and hysteretic energy takes into account the combined effects of blast vibration characteristics and associated structural responses. Therefore, this criterion can explain phenomena that are inconsistent with the current PPV-frequency safe level standard and is more comprehensive than the current standard.

CONCLUSIONS

Because the PPV-frequency safety level standard does not take into consideration the shape and the duration of blasting seismic waves as well as the corresponding dynamic responses of structures, the standard is limited in its ability to assess the potential for damage to structures subjected to blast-induced ground vibrations. To address these shortcomings, in this paper the following steps took place. (1) Structures are simplified into elastoplastic SDOF systems with bilinear restoring force models. The energy responses of SDOF systems subjected to blast vibrations are then determined using the Newmark-β method. Those energy responses more comprehensively show the combined effects of blasting seismic waves and associated structural responses on the damage to structures. (2) A dual safety criterion based on MIIE and hysteretic energy is proposed to explain the damage to structures under blast vibrations. Note that the damage incurred by some structures is inconsistent with the current PPV-frequency standard but can be understood well using the proposed criterion.

MIIE affects structures for a very short time, during which damping is not excited. Therefore, MIIE causes first-excursion structural failure. MIIE is mainly affected by PPV and the dominant frequencies of blasting seismic waves as well as the natural period of structures but cannot reflect the cumulative effect of blasting seismic waves on structural failure. In contrast, hysteretic energy is an important index for showing cumulative plastic damage in structures. Typically, structures subjected to blasting seismic waves with long durations are more likely to fail through cumulative plastic damage. The case study presented in this paper shows that the proposed dual safety criterion based on MIIE and hysteretic energy can more comprehensively and reasonably explain findings from field investigations.

Conflict of Interests

The authors declare that there is no conflict of interests regarding the publication of this paper.

ACKNOWLEDGMENTS

This work was supported by the grants from the Guizhou Science and Technology Plan Project (no. SY2010365), Key Industrial Research Project Supported by Science Technology Department of Guizhou Province (no. (2013) 3013), Guizhou Province Outstanding Young Scientific Talents Training Program Funded Projects (no. (2013) 30) and the National Natural Science Foundation of China (nos. 41301009 and 51174076), the International Cooperation Program of the Ministry of Science and Technology of China (Grant no. 2013DFA21720), and the National Science & Technology Support Program during the Twelfth Five-Year Plan Period (Grant no. 2011BAK12B01). These grants are gratefully acknowledged.

REFERENCES

1. U. Langerfors, H. Westerberg, and B. Kihlstrom, "Ground vibration in blasting," Water Power, vol. 10, no. 9, pp. 335–424, 1958.
2. A. T. Edwards and T. D. Northwood, "Experimental studies of the effects of blasting on structures," The Engineer, vol. 210, pp. 538–546, 1960.
3. W. I. Duvall and D. E. Fogleson, "Review of criteria for estimating damage to residences from blasting vibrations," Report of Investigation BM-RI-5968, Bureau of Mines, College Park, Md, USA, 1962.
4. T. D. Northwood, R. Crawford, and A. T. Edwards, "Blast vibrations and building damage," The Engineer, vol. 215, pp. 973–978, 1963.
5. C. H. Dowding, Blast Vibration Monitoring and Control, Prentice-Hall, Englewood Cliffs, NJ, USA, 1985.
6. C. H. Dowding, Construction Vibrations, Prentice-Hall, Upper Saddle River, NJ, USA, 1996.
7. P. K. Singh and M. P. Roy, "Damage to surface structures due to blast vibration," International Journal of Rock Mechanics and Mining Sciences, vol. 47, no. 6, pp. 949–961, 2010.
8. GB6722-2003, "China Safety regulations for blasting," 2003.
9. M. Khandelwal and T. N. Singh, "Evaluation of blast-induced ground vibration predictors," Soil Dynamics and Earthquake Engineering, vol. 27, no. 2, pp. 116–125, 2007.
10. H. Ak and A. Konuk, "The effect of discontinuity frequency on ground vibrations produced from bench blasting: a case study," Soil Dynamics and Earthquake Engineering, vol. 28, no. 9, pp. 686–694, 2008.
11. R. Nateghi, "Evaluation of blast induced ground vibration for minimizing negative effects on surrounding structures," Soil Dynamics and Earthquake Engineering, vol. 43, pp. 133–138, 2012.
12. H. Wenhua, X. Quanjun, and S. Wei, "Application of wavelet transform in evaluation of blasting vibration damage," Engineering Blasting, vol. 7, no. 1, pp. 24–28, 2001.
13. Z. Guosheng, F. Yingguang, and X. Guoyuan, "Safety criterion of blasting vibration for structure based on wavelet packet analysis," Chinese Journal of Geotechnical Engineering, vol. 7, no. 1, pp. 24–28, 2001.
14. H.-T. Li, X.-G. Yang, W.-B. Lu, D.-Q. Shu, and J.-W. Zhou, "Safety assessment for structures under blasting vibration based on equivalent peak energy," Chinese Journal of Geotechnical Engineering, vol. 33, no. 5, pp. 821–825, 2011.
15. Y. Zhang, B. Li, Y. Wang, and X. Li, "Model and calculation of momentary input energy of blast-induced vibration based on Hilbert spectrum," Mining Research and Development, vol. 29, no. 4, pp. 88–90, 2009.
16. Z. Wang, X. Liang, Y. Chen, F. Li, and G. Liu, "Study of safety evaluation method of blasting vibration based on input energy," Chinese Journal of Rock Mechanics and Engineering, vol. 29, no. 12, pp. 2492–2499, 2010.
17. B. Liu, M. Xiao, and M. Lai, "Distribution of earthquake total input energy of structures," Journal of Chongqing Jian Zhu University, vol. 18, no. 2, pp. 100–109, 1996.
18. A. Erberik and H. Sucuoğlu, "Seismic energy dissipation in deteriorating systems through low-cycle fatigue," Earthquake Engineering and Structural Dynamics, vol. 33, no. 1, pp. 49–67, 2004.
19. G. Manfredi, "Evaluation of seismic energy demand," Earthquake Engineering & Structural Dynamics, vol. 30, no. 4, pp. 485–499, 2001.
20. C.-H. Zhai, L.-L. Xie, and Z.-F. Wu, "Influence analysis on hysteretic-energy based on 1999 CM-Chi, Taiwan Earthquake," Journal of Harbin Institute of Technology, vol. 38, no. 1, pp. 59–62, 2006.
21. H. Norio, T. Tomaya, and I. Norio, "Damaging properties of ground motions and response behavior of structures based on momentary energy response," in Proceedings of the 12th World Conference on Earthquake Engineering (WCEE '00), Auckland, New Zealand, 2000.
22. Z. Qiong, W. Qiang, and C. Shaolin, "Characteristics of instantaneous input energy of ground motion containing velocity pulses," Industrial Building, vol. 39, supplement 1, pp. 186–190, 2009.
23. R. Hu and Y. Wang, "A study on relation of instantaneous energy of earthquake strong motion and maximum displacement response of structures," Journal of Building Structures, vol. 21, no. 1, pp. 71–76, 2000.
24. N. Makris and C. Black, "Dimensional analysis of inelastic structures subjected to near-fault ground motions," Tech. Rep. 03-05, Earthquake Engineering Research Center, University of California, Berkeley, Calif, USA, 2003.
25. K. Chen, Z.-F. Liu, and P.-S. Shen, "Study on the duration spectra of structural momentary input energy response," Engineering Mechanics, vol. 28, no. 1, pp. 19–25, 2011.
26. X. Ming-Kui, B. Shao-Liang, and L. Ai-Ming, "Maximum displacement estimation of seismic structures based on hysteresis energy," Journal of Chongqing University, vol. 26, no. 3, pp. 133–137, 2003.
27. R. Hu and Y. Wang, "Evaluation of long-period structural maximum displacement response on momentary input

energy," Journal of HUST: Urban Science Edition, vol. 23, no. 4, pp. 100–104, 2006.

28. Y.-J. Park and A. H.-S. Ang, "Mechanistic seismic damage model for reinforced concrete," Journal of Structural Engineering (ASCE), vol. 111, no. 4, pp. 722–739, 1985.
29. Y. Wang, "A review of seismic response spectra, time history analysis and energy method," Journal of Building Structures, vol. 21, no. 1, pp. 21–27, 2000.
30. Z. Duo and Y. Shaoguo, "Difference and connection of earthquake vibration and blasting vibration influence on building," Open Mining Technology, vol. 16, no. 3, pp. 68–69, 2010.
31. M. Sheng and Q. Luo, "Comparison of hysteretic energy spectra in Northridge and ChiChi earthquake,"Journal of Tongji University (Natural Science), vol. 36, no. 10, pp. 1314–1319, 2008.
32. Q. Liu, R. Zhou, and Y. Liu, "Research on relation between instantaneous energy spectrum of earthquake ground motion and displacement response of structures," Journal of Earthquake Engineering and Engineering Vibration, vol. 29, no. 5, pp. 46–51, 2009.
33. W. Changfeng and Z. Dongsheng, "Research on instantaneous energy of bilinear system," Journal of Lanzhou Railway University (Natural Sciences), vol. 20, no. 4, pp. 54–59, 2001.
34. P. Fajfar, T. Vidic, and M. Fischinger, "On energy demand and supply in SDOF systems," in Nonlinear Seismic Analysis and Design of Reinforced Concrete Buildings, P. Fajfar and H. Krawinkler, Eds., Elsevier, 1992.
35. S.-H. Chen, H.-X. Wei, A.-K. Zhang, and X.-H. Mao, "Computational model and safety criterion of blasting vibration effect based on wavelet packet techniques," Explosion and Shock Waves, vol. 30, no. 4, pp. 377–382, 2010.
36. T.-H. Ling and X.-B. Li, "Influence of maximum decking charge on intensity of blasting vibration,"Journal of Central South University of Technology (English Edition), vol. 13, no. 3, pp. 286–289, 2006.

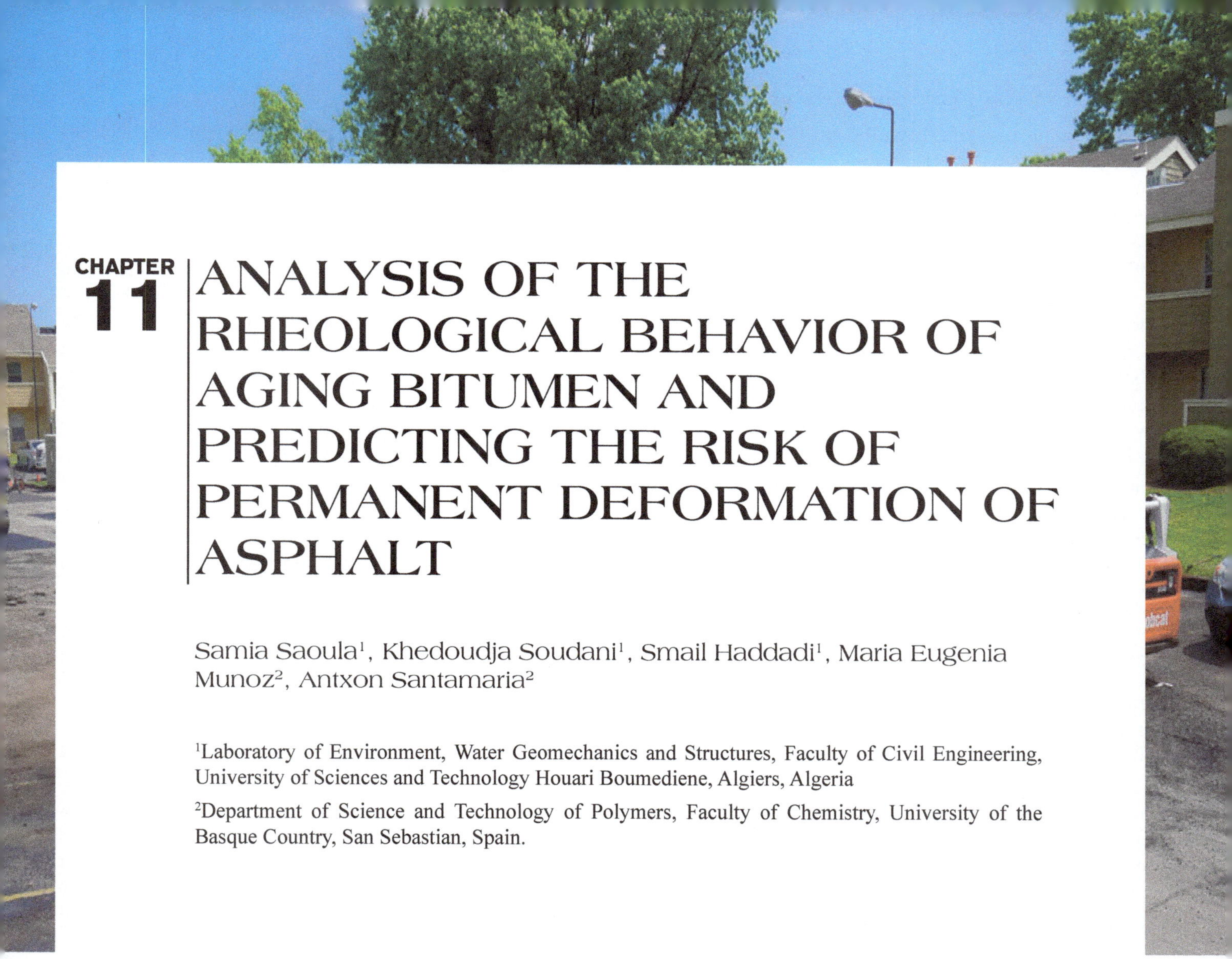

CHAPTER 11

ANALYSIS OF THE RHEOLOGICAL BEHAVIOR OF AGING BITUMEN AND PREDICTING THE RISK OF PERMANENT DEFORMATION OF ASPHALT

Samia Saoula[1], Khedoudja Soudani[1], Smail Haddadi[1], Maria Eugenia Munoz[2], Antxon Santamaria[2]

[1]Laboratory of Environment, Water Geomechanics and Structures, Faculty of Civil Engineering, University of Sciences and Technology Houari Boumediene, Algiers, Algeria

[2]Department of Science and Technology of Polymers, Faculty of Chemistry, University of the Basque Country, San Sebastian, Spain.

ABSTRACT

The aging of bitumen has a significant impact on the mechanical behavior of asphalt. This aging is carried out at the mixing operation with the aggregates generally at 163°C in the case of pure mixtures or at temperatures higher than in the case of the modification by polymers or industrial waste. This paper presents experimental results of a study of the rheological behavior of a class of bitumen 40/50 to the artificial aging. Three aging temperatures were selected: 163°C, 173°C and 183°C. From obtained results, a study of the risk of deformation of asphalt is performed, based on the results of correlation between the behavior of the coated and his binder deducted SHRP specifications and technical advice French. These results show that the stiffness of the aged binders increases with aging temperature. Moreover, there is no risk of rutting and fatigue cracking for thermal coated projected.

INTRODUCTION

Bituminous binders are widely used for road paving applications [1]. The ageing of the bitumen during storage, mixing, transport and laying on the road, as well as in service life, are the most important problems presented by the use of bitumen in pavements [2].

Indeed, Bitumen ageing is a very complex process resulting in hardening of bitumens and embrittlement, both in application and in service, which contributes greatly to the deterioration of paving applications [3].

Important ageing related modes of failure are traffic and thermally induced cracking, and ravelling and shorten the lifespan of pavement [4,5].

In bitumen aging, two types of mechanisms are involved. The main aging mechanism is an irreversible one, characterised by chemical changes of the binder, which in turn has an impact on the rheological properties [6]. The second mechanism is a reversible process called physical hardening [7,8]. Physical hardening may be attributed to molecular structuring [5].

Several methods have been proposed to replicate the effect of aging and, therefore, to foresee bitumen behavior during application and service life. To simulate the age hardening occurring during plant mixing and laydown the most utilised

Citation: S. Saoula, K. Soudani, S. Haddadi, M. Munoz and A. Santamaria, "Analysis of the Rheological Behavior of Aging Bitumen and Predicting the Risk of Permanent Deformation of Asphalt," Materials Sciences and Applications, Vol. 4 No. 5, 2013, pp. 312-318. doi: 10.4236/msa.2013.45040.

test is Rolling Thin Film Oven Test (RTFOT, ASTM D-2872). The age hardening is evaluated by observing how viscosity, penetration and softening point change with aging [3]. Aging, which occurs in asphalt roadway pavements during construction and service life, affect the rheological properties of bitumen [5,8].

Rheology involves the study and evaluation of the time—temperature dependent response of materials that are stressed or subjected to an applied force. Rheological properties of bitumen consist of age hardening, temperature susceptibility, shear susceptibility, stiffness, penetration, ductility, and viscosity [9].

Over the past few years, rheology has been extensively used in order to classify and evaluate bituminous binders according to their performance properties. This has led to a better knowledge of bitumen behaviour that occurs when subjected to different thermal and mechanical conditions, as seen during road construction and service in the field [3].

One of the key material properties of bitumen is the complex modulus. This property is related to major distress modes such as traffic induced permanent deformation and rutting, as well as fatigue and low temperature cracking. The complex modulus is dependent upon temperature and loading frequency [10].

This paper presents experimental results of a study of the rheological behavior of a class of bitumen 40/50 to the artificial aging. Three aging temperatures were selected: 163°C, 173°C and 183°C.

In this work, test aging was evaluated by means of complex modulus (G^*) and phase angle (δ), determined by dynamic rheological analysis.

From the results obtained a study of the risk of deformation of asphalt has been achieved, based on the results of correlation between the behavior of the coated and his binder deducted SHRP specifications and technical advice French.

MATERIALS

The bitumen of class of 40/50 (B) used in this study is commercialized by NAFTAL. The main physical properties of the bituminous material are the following: penetration (at 25°C 1/10 mm) = 27, softening point temperature = 54.9°C and penetration index (PI) = −1.33.

Ring and ball temperature is considered as a value attached to the permanent deformation behavior (resistance to rutting) at high operating temperatures [11]. The binder B has an average value of the ring and ball temperature. Knowing that in Algeria the temperature of 50°C is easily reached in the roadway this will increase certainly the risks of rutting in hot time [12], the risk of rutting in hot weather is not ruled out. The penetration index (PI) shows that the bitumen (B) has a high susceptibility to temperature change.

EXPERIMENTAL METHODS

Penetration test according to ASTM D-5 was used to determine the consistency of a bitumen binder according to the following procedure [13]: bitumen is placed in a standard cup at 25°C. A needle is then allowed to fall into the bitumen under a 100 g load for 5 s. The length the needle penetrates into the bitumen during this time is expressed in units of 0.1 mm. Soft bitumens exhibit high penetrations and harder bitumens low penetration values.

The test of softening point temperature was performed according to ASTM D36. The heated bitumen sample is placed in each of two brass rings. The rings are then placed in a bath of demineralised water/glycerol and a ball bearing is placed on each. The water/glycerol is heated at a rate of 5°C/min until the binder deforms due to its decreasing viscosity with increasing temperature. The softening point temperature was taken as the temperature at which the sample became soft enough to allow the ball, enveloped in the sample material, to fall a distance of 25.4 mm. This was recorded as the ring and ball softening point temperature.

The susceptibility of bitumen temperature, that is to say the measurement of the change in consistency depending on the temperature, is characterized by a number called Penetration Index noted PI [14,15].

Pfeiffer and Van Doormaal in 1936 [16] have defined the thermal susceptibility of bitumens based on penetration index and stated that this index can take values ranging from −3 for asphalt very sensitive to +7 for bitumens with low thermal sensitivity.

In addition, the temperature susceptibility of the bitumens has been calculated in terms of penetration index (PI) using the results obtained from penetration and softening point tests. Temperature susceptibility is defined as the change in the consistency parameter as a function of temperature. A classical approach related to PI calculation has been given in the Shell Bitumen Handbook as shown with the following equation [17]:

$$PI = \frac{1952 - 500 \times \log(pen_{25}) - 20 \times SP}{50 \times \log(pen_{25}) - SP - 120} \quad (1)$$

where, pen_{25} is the penetration at 25°C and SP is the softening point temperature of bitumens. Nowadays, bitumen of straight run have negative PI generally between −1 and −2, semi-blown bitumens thermally less susceptible have PI almost zero or slightly positive (from −0.2 to +0.6). Industrial bitumen sealing have positive PI (>+1) [15]. For asphalt "sol-like", penetrability indices were negative by cons for asphalt "gel-like" they are positive [16].

According to ASTM D2872-04, simulation of aging of bitumen was done by Rolling Thin-Film Oven Test (RTFOT). Accordingly, a moving film of bitumen was heated in an oven for 85 min at 163°C.

The aging of bitumen was determined from changes in physical and rheological properties of bitumen, as measured before and after the oven treatment.

The most commonly used method of fundamental rheological testing of bitumen was by means of dynamic mechanical analysis (DMA) using oscillatory-type test. It generally conducted within the region of linear viscoelastic response.

These oscillatory tests were undertaken using controlledstress rheometer (Haake RS1, Germany). It applied oscillating shear stresses and strains to samples of bitumen sandwiched between parallel plates at different loading frequencies and temperatures. A temperature sweep was applied over the range 40°C - 100°C at a fixed frequency of 10 rad/s and variable strain. The strain was chosen to be as small as possible to ensure measurement in the linear region, but large enough to allow sufficient torque readings. A frequency sweep was applied over the range 0.01 - 10 rad/s at variable temperatures (40°C to 80°C).

The binders rheological properties thus determined are appreciated by operating different curves plotted (isotherms, isochronous, Black and master).

RESULTS AND DISCUSSIONS

Aging Binders

The simulation of the aging RTFOT shows an acceptable aging, according to the European specifications EN 12591 (**Table 1**).

Aged binders do not pose a risk of thermal fatigue cracking as their softening point does not belong to the interval (64 to 68) where the risk of cracking are not negligible [18].

Table 1. Binder characteristics aging.

	B	B_163°C	B_173°C	B_183°C
Penetration (0.1 mm)	27	24.8	19.8	18.3
Softening point (°C)	54.9	59.7	59.9	60.1
PI	−1.33	−0.55	−1.00	−0.05
Δ Softening point (°C)	-	4.8	5	5.2

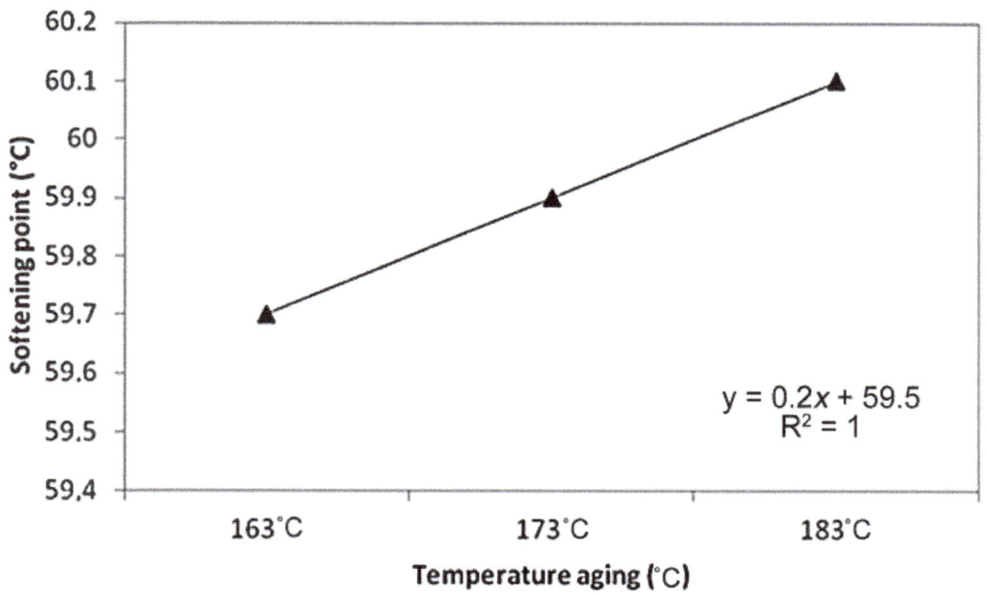

Figure 1. Softening point according to the aging temperature.

Another indicator can be used to estimate the risk of thermal fatigue cracking; it is the evolution of the softening point during testing RTFOT. This value is limited to 8°C (EN 12591) for class 35/50.

Increasing the ring and ball softening binder studied does not exceed 8°C showing that there is no risk of thermal fatigue cracking.

The value of temperature ball and rings obtained on the asphalt before RTFOT may fear, risk of permanent deformations if the evolution of the binder in the coating was low. But its strong growth during RTFOT suggests that these risks are limited, at least if the formula of the mix is correct. This fact is verified by the modulus measured by rheological tests.

The evolution of the softening point according to the temperature of aging is in a proportional manner (**Figure 1**). Indeed, the softening point increases of 0.2°C for a 10°C change in temperature aging.

The evolution of the thermal susceptibility PI (**Table 1**) after RTFOT not improved thermal sensitivity (negative values).

Rheological Behavior

1) Isotherms The isotherms of the complex modulus are plotted on **Figure 2**.

The complex modulus increases with frequency. For high temperature slope of the curve of the module is marked. In addition, the complex modulus isotherms show a small overlap area for high temperatures. This means that dispersion temperature and time are not equivalent in this region.

2) Isochrones Isochrones of the complex modulus and the phase angle at 10 rd/s, before and after RTFOT, are shown in **Figure 3**. The aging of the binder resulted a marked increase in stiffness binders with high operating temperatures thus reducing the thermal sensitivity. In addition, the decrease in phase angle increases the elastic contribution to the module [19]. The main ageing mechanism of bitumen is the loss of volatiles and oxidation, which leads to having higher viscosity and a bitumen that is stiffer than fresh bitumen [20].

3) Black curves

Figure 4 presents black curves of the binders before and after RTFOT. The analysis of these results shows that in the low frequency domain and high temperatures, the complex modulus G* decreases strongly, while the phase angle approaches 90° reflecting completely fluid behavior.

Higher values of phase angle indicate a tendency towards more viscous behaviour, whilst lower values indicate more elastic response [21].

The uniqueness of the curve Black binders is relatively good. Curves are not too "waves" due to poor implementation time/temperature. Discontinuities in the curve Black, often referred to as "waves" [19,22] reflect structural changes, based on changes in temperature, due to:

- Either the merger of certain chemical species such as certain paraffin;
- Or the modification of some physical links between asphaltene agglomerates.

Another phenomenon may be responsible for the formation of these waves [23] in the case of very paraffinic binders, relaxation due to dissolution/crystallization of paraffins is superimposed on the glass transition associated with the maltenes.

Activation of two different phenomena explains the dispersion increasingly important experimental points on the curve Black when the content of crystallizable fractions (paraffins) increases bitumen. This phenomenon frequently observed in the aged binders in the coating or in situ occurs only rarely on new asphalt as is. Aging results in an increase in the structure reducing thermal and kinetic sensitivities. It may be that these binders are microscopically heterogeneous and contain asphaltenes highly interactive teaming agglomerates more or less stable with temperature and perhaps even the frequency.

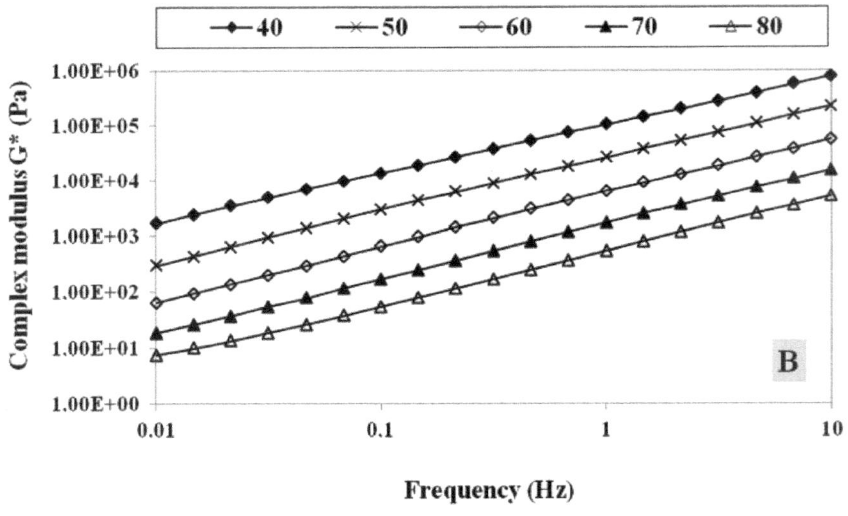

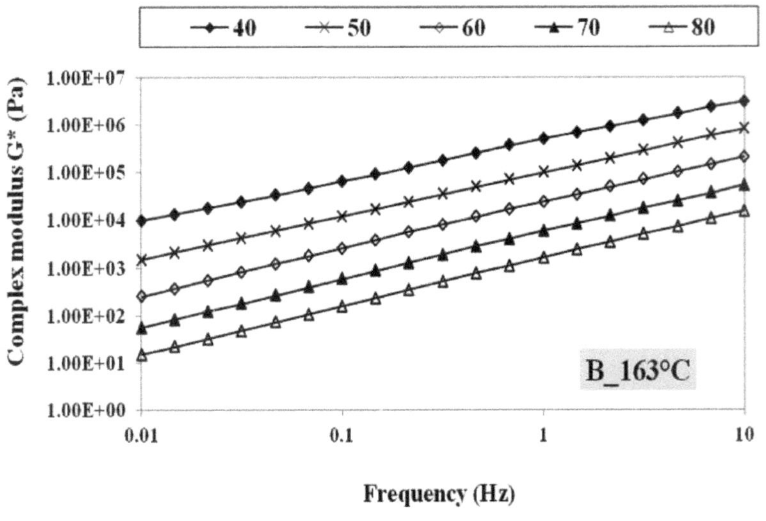

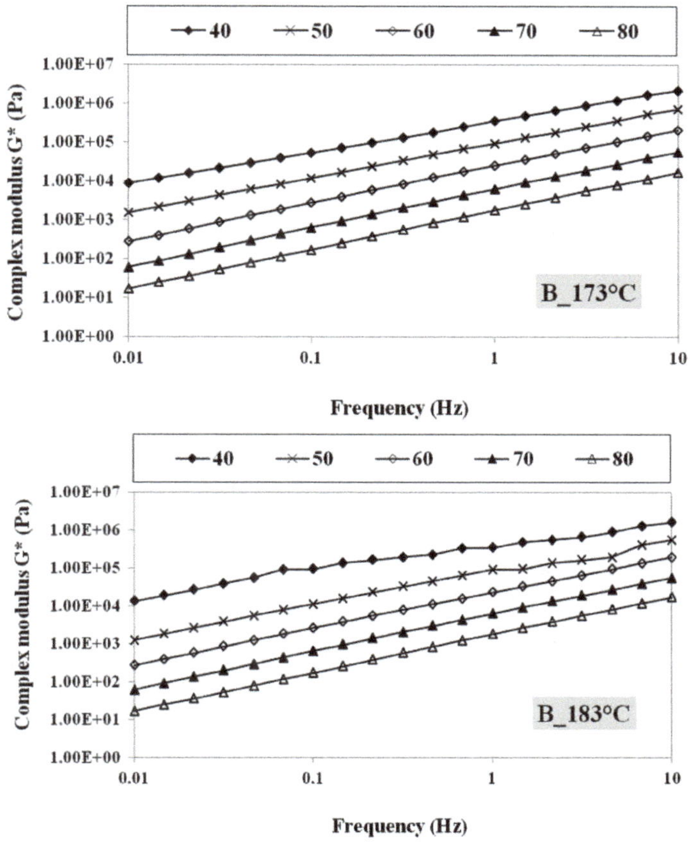

Figure 2. Isotherms of the complex modulus.

4) Master curves Master curves of complex modulus (G^*) can be traced to a reference temperature, by translating the different isotherms according to the frequency axis [24]. This property is called "time-temperature superposition principle (TTSP)" [25] as the translation procedure isotherms give a single master curve and continue only for the complex modulus. **Figure 5** shows the master curve plotted for the bitumen at a temperature of 25°C.

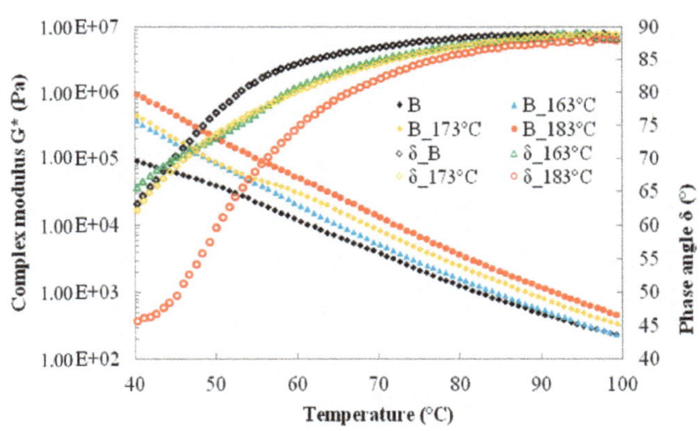

Figure 3. Isochrones of the complex modulus and phase angle.

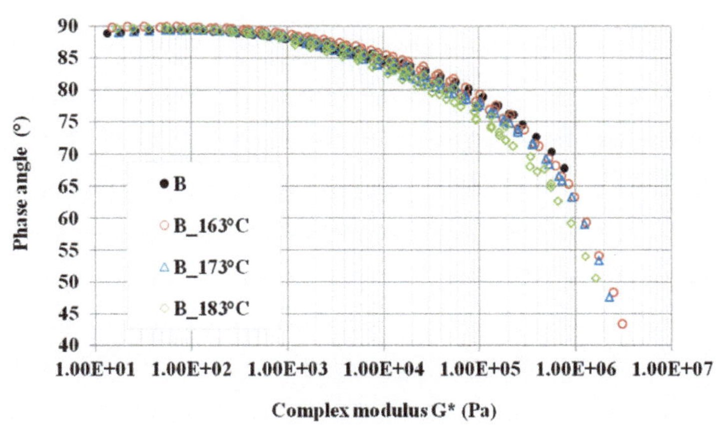

Figure 4. Black curves.

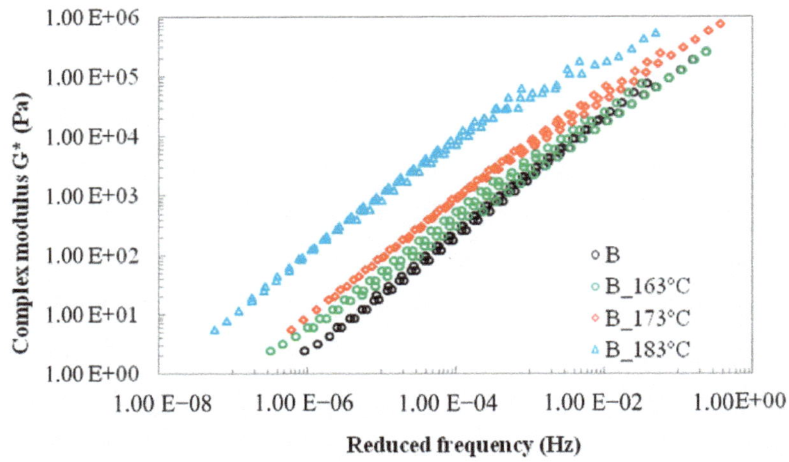

Figure 5. Master curve at 25°C.

An examination of the complex modulus master cu-rves obtained before and after RTFOT for different aging temperatures, show that the value of each isothermal shifts to values higher modules, especially for the aged bitumen to 183°C. This increase in modulus results in a cure.

Note that the curves tend to join from a value of 10^{-1}. Indeed, these isotherms also become less steep, for low tem-peratures indicating that the kinetic and thermal sensitivities decrease [15].**Table 2** gives the values of the constants C1 and C2 of Law WLF.

Study of the Risk of Deformation of the Asphalt through the Performance of Bitumen Rheologi-cal

The rheological characteristics measured on binders can be used to predict [19], given the Algerian climate, the type strain following:

- Permanent deformation (rutting);
- Thermal fatigue cracking;

In fact, according to research arising from the program SHRP (Strategic Highway Research Program) [26] the binder is responsible for 40% of rutting phenomenon, at 60% of the fatigue phenomenon.

Risk of rutting To reduce the risk of rutting, the French technical advice [19], suggest the value of the complex modulus G^* at 7.8 Hz measured at a temperature T1 as an indicator [27-29] (in the case of this study T1 = 64.6°C). It must be greater than the critical value of 1.10^4 Pa.

This condition is checked for the binder studied (Table 3). The ratio $G^*/\sin \delta$ is taken to SHRP specifications [19,30] it is measured at 60°C and 1.6 Hz. It presents the impact of the module with the phase shift value. SHRP specifications set indirectly and by means of the temperature, a minimum value is 1 KPa report before RTFOT and a value of 2.2 KPa after RTFOT. The ratios calculated for the four binders are higher than these limits.

Glita & Conan (1996) [31] proposed the ratio $G^*/\sin \varphi$ measured at 5 Hz and 60°C as a characteristic of bitumen (test conditions similar to that of the rutting test on asphalt), the greater its value, better resistance to creep.

Comparing the ratio given for each asphalt most aged binders are more rigid.

The risk of rutting evaluated by three criteria, although different, give the same trend: the binders studied do not present a risk of rutting.

The higher Temperature aging indices (TAI) was proposed [32] for indicate the effect of RTFO aging in rheological properties with respect to un-aged condition at high temperatures. The higher TAI values indicate a higher aging degree induced to bitumen. The indices use the measured rutting parameters at high in-service temperatures as follow:

$$\text{High TAI} = \frac{(G^*/\sin\delta)_{RTFOT}}{(G^*/\sin\delta)} \tag{2}$$

The index values show that when the temperature of aging increases, the rutting resistance is better.

Risk of thermal fatigue cracking To assess the risk of thermal fatigue cracking, French technical advice proposes as a requirement the value of the phase angle measured at 40°C and 7.8 Hz before RTFOT which must be greater than 45° [33]. This criterion (**Table 4**) is verified for the binder studied.

CONCLUSIONS

The aging of the binder to the different temperatures studied is acceptable.

Thermal fatigue cracking was estimated from two physical criteria and a criterion based on rheology. Each criterion shows that the base binder is no risk of thermal fatigue cracking. However, it should always be considered a good formulation to avoid joining the zone of un-certainty and move forward to cracking.

Table 2. Values of the coefficients C1 and C2 of Law WLF.

Law WLF coefficients	B	B_163°C	B_173°C	B_183°C
C1	9.01	13.53	15.55	10.11
C2	68.30	110.96	148.39	51.21

Table 3. Valuation to risk of rutting.

Criterion of rutting	B	163°C	173°C	183°C
G^*(7.8 Hz, 64.6°C) [Pa]	4.9×10^4	15.8×10^4	16.7×10^4	17×10^4
$G^*/\sin\delta$ (1.6 Hz, 60°C) [Pa]	1.3×10^4	2.18×10^4	3.2×10^4	5.4×10^4
$G^*/\sin\delta$ (5 Hz, 60°C) [Pa]	2.8×10^4	11.5×10^4	12.2×10^4	12.8×10^4
$(G^*/\sin\delta)_{RTFOT}/(G^*/\sin\delta)$	-	1.67	2.46	4.15

Table 4. Valuation to thermal fatigue cracking.

Criterion of thermal fatigue cracking	B	163°C	173°C	183°C
δ (7.8 Hz, 40°C) [°]	68.1	55.05	53.43	49.15

The rheological analysis showed a marked increase in the stiffness of binders to high operating temperatures thus reducing thermal sensitivity. The binders studied do not present a risk of rutting. In so far as these risks may not be favoured by a high traffic load or an inadequate formulation of the mix.

The good behaviour at high temperatures should be accompanied by an adequate response of the binder at low temperatures, at which cracking effects can happen.

Decreasing the elasticity of the binder at low temperatures facilitates to avoid cracking, since the absorbed energy is more easily dissipated. In summary, a low elasticity is suitable at low temperatures to facilitate dissipation, but a high elasticity is convenient at high temperatures to avoid viscous flow of the bitumen binder.

This study showed any interest in the choice of mixing temperature of bitumens and modified bitumen in the asphalt. Indeed, the rigidity of the binder, so the asphalt, depends on this temperature. However, as is often the case in road, everything is a compromise, and especiallythe definition of basic requirements. To asphalt formulators and pavement designers choose the most appropriate temperature.

REFERENCES

1. D. Whiteoak, "Shell Bitumen Handbook," Shell Bitumen UK, Riversdell Hause, 1990.
2. M. Zargar, et al., "Investigation of the Possibility of Using Waste Cooking Oil as a Rejuvenating Agent for Aged Bitumen," Journal of Hazardous Materials, Vol. 233- 234, 2012, pp. 254-258. doi:10.1016/j.jhazmat.2012.06.021
3. D. Mastrofini and M. Scarsella, "The Application of Rheology to the Evaluation of Bitumen Ageing," Fuel, Vol. 79, No. 9, 2000, pp. 1005-1015. doi:10.1016/S0016-2361(99)00244-6
4. I. Gawel and K. Baqinska, "Effect of Chemical Nature on the Susceptibility of Asphalt to Aging," Petroleum Science and Technology, Vol. 22, No. 9, 2004, pp. 1261-1271.
5. X. H. Lu and U. Isacsson, "Effect of Aging on Bitumen Chemistry and Rheology," Construction and Building Materials, Vol. 16, No. 1, 2002, pp. 15-22. doi:10.1016/S0950-0618(01)00033-2
6. A. B. Brown, et al., "Steric Hardening of Asphalts," Proceedings of the Association of Asphalt Paving Technologists, Vol. 26, No. 26, 1957, pp. 486-494.
7. H. U. Bahia and D. A. Anderson, "Glass Transition Behaviour and Physical Hardening of Asphalt Binders," Journal of the Association of Asphalt Paving Technologists, Vol. 62, 1993, pp. 93-129.
8. S. Wu, et al., "Rheological Properties for Aged Bitumen Containing Ultraviolate Light Resistant Materials," Construction and Building Materials, Vol. 33, 2012, pp. 133- 138.doi:10.1016/j.conbuildmat.2012.01.019
9. S. H. Firoozifar, et al., "The Effect of Asphaltene on Thermal Properties of Bitumen Chemical," Engineering Research and Design, Vol. 89, No. 10, 2011, pp. 2044-2048.doi:10.1016/j.cherd.2011.01.025
10. M. Yilmaz, et al., "Investigation of Complex Modulus of Base and EVA Modified Bitumen with Adaptive-Network-Based Fuzzy Inference," System Expert

Systems with Applications, Vol. 38, No. 1, 2011, pp. 969-974. doi:10.1016/j.eswa.2010.07.088

11. B. J. Dongmo-Engeland, "Characterization of Rutting Deformation of Bituminous Pavement," Ph.D. Thesis, National Institute of Applied Sciences of Lyon, Lyon, 2005.
12. S. Saoula, et al., "The Study of the Improvement of Mechanical Performance of Asphalt Modified by the Optimization of Mixing Time of EVA Bitumen," AIP Conference Proceedings, Vol. 1400, 2011, pp. 262-267. doi:10.1063/1.3663125
13. N. Dehouche, et al., "Influence of Thermo-Oxidative Aging on Chemical Composition and Physical Properties of Polymer Modified Bitumens," Construction and Building Materials, Vol. 26, No. 1, 2012, pp. 350-356. doi:10.1016/j.conbuildmat.2011.06.033
14. M. Harlin, "Bitumen: Properties, Characteristics, Specifications and Testing Laboratory," National School of Bridges and Roads, Paris, 1992.
15. H. Di Benedetto and J. F. Corte, "Bituminous Road Materials 1: Description and Properties of Components," Lavoisier, 2004.
16. O. Solomatnikova, "Rheological Behavior and Cohesive and Adhesive Properties of Bituminous Binders," Memory of Master of Applied Science, University of Laval, Quebec, 1998.
17. A. Topal, "Evaluation of the Properties and Microstructure of Plastomeric Polymer Modified Bitumens," Fuel Processing Technology, Vol. 91, No. 1, 2010, pp. 45-51. doi:10.1016/j.fuproc.2009.08.007
18. S. Gazeau, et al., "Application of Differential Scanning Calorimetry (DSC) to the Characterization of Ethylene Copolymers Modified Bitumen," Proceedings of the 1st Eurasphalt and Eurobitume Congress, Strasbourg, 7-10 May 1996.
19. G. Ramond, et al., "Caractéristiques de Bitumes Utilisés en Algérie," Bulletin de Liaison Ponts et Chaussées, Vol. 225, 2000, pp. 3-11.
20. G. Holleran, et al., "Rejuvenation Treatments for Aged Pavements," Transport Research Board, 2006.
21. I. Widyatmoko and R. Elliott, "Characteristics of Elastomeric and Plastomeric Binders in Contact with Natural Asphalts," Construction and Building Materials, Vol. 22, No. 3, 2008, pp. 239-249. doi:10.1016/j.conbuildmat.2005.12.025
22. G. Ramond and M. Epinat, "Contribution à l'Élaboration d'un Mode Opératoire de Mesure du Module Complexe," Bitume Actualités, No. 101, 2000, pp. 37-42.
23. F. Olard, "Thermomechanical Behavior of Bituminous Mix Asphalt with Low Temperatures, Relationship between Binder Properties and Asphalt," Ph.D. Thesis, INSA de Lyon, France, 2003. [Paper reference:1]
24. F. Olard, et al., "Rhéologie des Bitumes: Prédiction des Résultats des Tests de Fluage BBR à Partir des Résultats de Module Complexe," Bulletin des Laboratoires des Ponts et Chausses, No. 252-253, 2004, pp. 3-15.
25. G. D. Airey, "Rheological Evaluation of Ethylene Vinyl Acetate Polymer Modified Bitumens," Construction and Building Materials, Vol. 16, No. 8, 2002, pp. 473-487. doi:10.1016/S0950-0618(02)00103-4
26. J. J. Lesage, et al., "Vers une Approche Rhéologique des Liants Bitumineux et Prédictive des Performances Mécaniques des Enrobés," Analusis, Vol. 23, 1995, pp. 492- 496.
27. Y. Marciano, et al., "Les Spécifications Superpave: une Contribution Shell au Décryptage. Carte de France des PG (Performance Grade)," Revue Générale des Routes et Aerodromes, No. 748, 1997, pp. 55-59.
28. C. Such and G. Ramond, "Les Spécifications SHRP sur les Bitumes et la Température de Ramollissement Bille et Anneau," Bulletin de Liaison Ponts et Chausses, No. 200, 1995, pp. 3-12.
29. C. Such and G. Ramond, "Les Spécifications SHRP une Tentative de Décryptage," Revue Générale des Routes et Aérodromes, No. 730, 1995, pp. 57-59.
30. M. Barral, et al., "Novel Bituminous Mastics for Pavements with Improved Fire Performance," Construction and Building Materials, Vol. 30, 2012, pp. 650-656.doi:10.1016/j.conbuildmat.2011.12.055
31. S. Glita and J. Conan, "Rheological Analysis of 7 Hard 10/20 Bitumens" Proceedings of the 1st Eurasphalt and Eurobitume Congress, Strasbourg, 7-10 May 1996.
32. S. Dessouky, et al., "Effect of Pre-Heating Duration and Temperature Conditioning on the Rheological Properties of Bitumen," Construction and Building Materials, Vol. 25, No. 6, 2011, pp. 2785-2792. doi:10.1016/j.conbuildmat.2010.12.058
33. G. Ramond and M. Pastor, "Rappel des Notions de Base et des Principes d'Exploitation du Module Complexe," Report No. 135 D96, Central Laboratory of Roads and Bridges, France, 1996.

CHAPTER 12

NEW APPROACH OF GEOMECHANICAL PROPERTIES BY SCALE EFFECT AND FRACTAL ANALYSIS IN THE KEDOUGOU-KENIEBA INLIER (SENEGAL-WEST AFRICA)

Déthié Sarr[1], Meissa Fall[1], Papa Malick Ngom[2], Mamadou Gueye[3]

[1]Laboratoire de Mécanique et Modélisation, UFR Sciences de l' Ingénieur, Université de Thiès, Thiès, Sénégal

[2]Département de Géologie, FST-UCAD, Université Cheikh Anta Diop de Dakar, Dakar, Sénégal

[3]Institut des Science de la Terre, FST-UCAD, Université Cheikh Anta Diop de Dakar, Dakar, Sénégal

ABSTRACT

This paper shows geometric aspects of rock masses in the Senegalese side of Kédougou-Kéniéba inlier. The studies are done in one part on sedimentary rocks (represented by sandstones) with stratification and on the other part in igneous rocks (represented by basalts). Geometric studies are the fractal configuration and the scale effect. Scale effect is studied by comparing the results of mechanical tests done in the laboratory and on outcrops. For all samples, laboratory parameters are higher than values of field. In this paper, scale effect is characterized by the decrease of mechanical parameters from laboratory samples to field. The scale coefficient is defined and characterizes the relation between mechanical properties. More importantly, the scale effect is more significant. This scale effect depends largely on the rock anisotropy. The anisotropy depends on the fracturation and the filling. The scale effect takes into account the fracturation, the filling minerals and their diversity which define the complexity of the rock. The anisotropy is complex; the impact of scale effect traduced by an increase of scale coefficient is the highest. Configuration of discontinuities also defines fractal geometry. This fractal aspect appears on the asperities, the spacing and the apertures of discontinuities. The fractal dimension is different from a parameter to another. All these two parameters estimate the stage of fracturation of the rock in a geological system and depend one on another if they are studied in the same element.

INTRODUCTION

The West African Craton has been the object of many geological studies which concluded substantial knowledge of the Precambrian in Senegal. This research is mainly focused on the domain of the basic principle of geology and mines.

Citation: D. Sarr, M. Fall, P. Malick Ngom and M. Gueye, "New Approach of Geomechanical Properties by Scale Effect and Fractal Analysis in the Kedougou-Kenieba Inlier (Senegal-West Africa)," Geomaterials, Vol. 3 No. 4, 2013, pp. 145-155. doi: 10.4236/gm.2013.34019.

In this paper, we are interested in the geometrical aspects of rocks. Geometrical aspects are represented by fractal geometry and scale effect, which permit to analyze the appearance in term of microscopic and macroscopic dimension.

Fracture systems in the south eastern Senegal are discontinuous structures or discrete breaks within a rock mass that are developed by response to stress. Fractures include faults, joints, and filled structures such as veins. These figures appear in the area on a wide range of scales from microns to several hundreds of meters. In recent years, it has been suggested that various technics of fractals may usefully be applied to the study of spatial distribution of fractures, because of similarities between small and large-scale fracture arrays, which have been found over many orders of magnitude [1]. This work deals with the consequence of the structuration of the rock mass in their geometry. The fractal geometry and scale effect allow estimating the occurrence of the fracturation and other parameters to the studies of discontinuities. For these estimations, we introduce a more complex aspect using the filling materials and define the "scale coefficient".

GEOLOGICAL FRAMEWORK

Three main directions of discontinuities characterize the Proterozoic rock masses. In the Birimian domain, the discontinuities show a very large motion due to the complexity of the geological history. The structures correspond mainly to the boundary of the rock formations. In these domains, discontinuities show some shear that are dextral NE-SW or senestral NW-SE [2]. The fractures are sometimes conjugated. In the Kédougou-Kéniéba Inlier (KKI), [3-5] bring out three main Eburnean tectonic phases: 1) A D_1 phase, fairly high crustal level tectonic activity with a tendency toward thrusting; 2) A D_2 phase related to the major senestral shear zones and the emplacement of granitic intrusions [6-7]. This second phase is responsible of large sub meridian senestral shear corridors, where the most important are: The Main Transcurent Shear Zone (MTZ), separating Mako and Dialé Supergroups [3-8] and the Senegal-Mali fault (SMF), along the western Falémé bank [9] (**Figure 1**); 3) A D_3 Eburnean tectonic phase has been also characterized by [5] in the Birimian formations of the KédougouKéniéba Inlier (KKI).

Some authors show that the Birimian of Senegal show relative amount of faults D_2 and D_1. D_2 is identified in all the Birimian (Mako super-group with volcanic rocks and Dialé-Daléma supergroup with sedimentary ones associate to volcanic ones) and D_1 is only presents in the DialeDalema [4,10-12]. While these deformations are considered to be represented in the scale of the West African domain, D_1 is absent in some sector. In the new denomination by [2], D_1 and D_2 are respectively called Dt_1 and Dt_2. They belong to a continuum transcurent tectonic system Dt. So, this transcurent tectonic occurs in two stapes (Dt_1 and Dt_2). This consideration is different of the first one, which defended that D_1 is tangential and D_2 is transcurent [4]. The hypothesis of a continuum transcurent tectonic defined by [2] is also supported by the fold produced when schistosity occurred and which would be folded again. The two stages of folds have the same axial surface oriented (N (20 to 40) - (10 to 30) NW).

Studies done in the sector of Mako on the latitude of Mako [12] show that this domain is crossed by more important framework of discontinuities as those explained by [2-4]. This framework of discontinuities is caused by the described phenomena, the cracks due to the crystallization of the magma and at least the repercussion of the Hercynian orogenesis. Tectonic fractures were mapped and measured in the field area of Mako and Ségou. The sedimentary rocks show less anisotropy represented essentially by cracks and rarely by faults. They correspond to sandstones mainly associate to pelites rocks [13].

In the Ségou-Madina kouta basin, the domain of Segou is characterized by fracturations, which appear like cracks. These cracks show three directions representing three families of discontinuities NE, ENE and ESE associated to erratic joints.

FRACTAL THEORY

The fractal term is used to describe a natural geometry where studied objects are recurrent, fragmentary and elements are seen in all scale [14,15]. Mathematicians define fractals using three equations [16]:

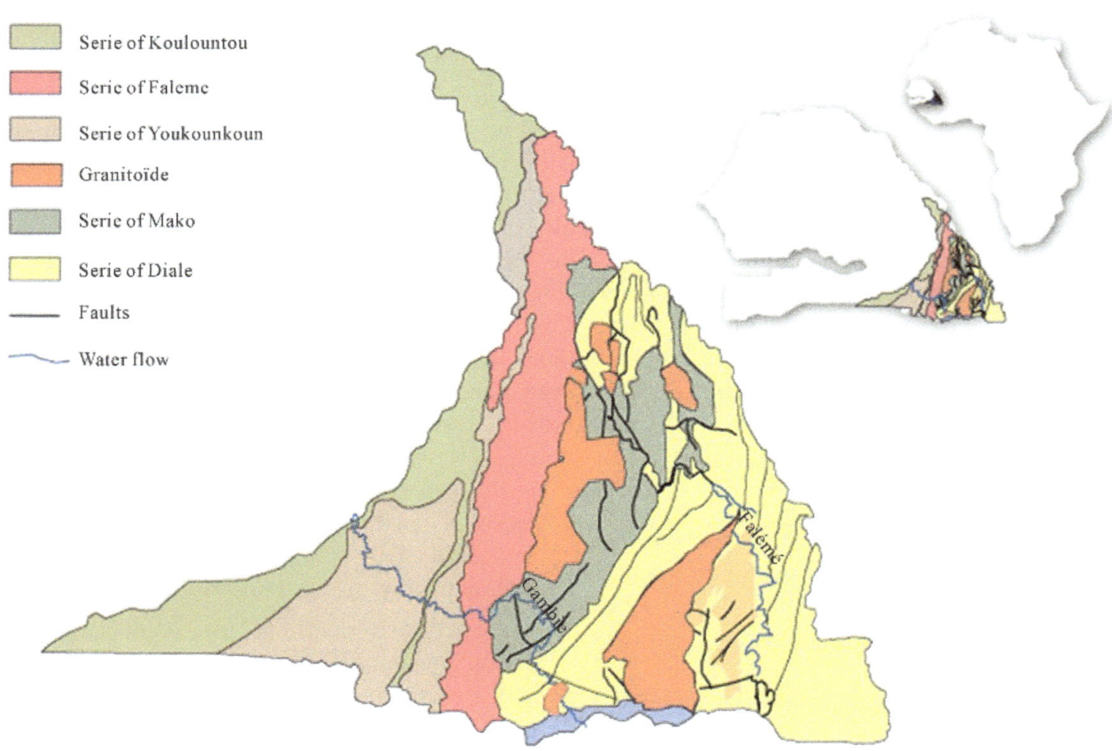

Figure 1. Structural map of the Kédougou-Kéniéba inlier [11 modified by 2 modified].

$$N(a) \approx a^{d_f} \tag{1}$$

N is the number of portions resulting to iteration, a the size of a portion and d the fractal dimension.

$$M(r) \approx r^{d_f} \tag{2}$$

M (r) is the weight of a portion and r is its linear extension.

$$\rho(r) \approx \frac{M(r)}{r^{d_0}} = r^{d_f - d_0} \tag{3}$$

ρ(r) is the density.

As Euclidian geometry, fractal geometry is also characterized by a dimension called fractal dimension d_f. The fractal dimension is fractionary and in difference of Euclidian dimension it is no characteristic of an object. Its value varies between 1 and 3 ($1 < d_f < 3$) and different to 2. It is a dimension deduced of the number of portions N(s) after iteration and the length (s) of one portion using the equation

$$d_f = \frac{\log N(s)}{\log s} \tag{4}$$

In Earth Sciences, the fractal analysis can be used to quantify many elements of discontinuities. It's the case of spacing, aperture, asperities and length of discontinuities or any other parameter, which allows describing the geometry of discontinuities [17,18]. For easier use, the fractal geometry is described by using process of iteration as the Sierpinsky, the flocon of Koch or the Cantor dust. It is very difficult to determine the fractal dimension by using the definition of fractals. So, in many case, it is done by graphical representation. This graphical resolution consists of plotting, the function log N(s) versus log(s). The obtained curve is linear and it slope corresponds to the fractal dimension. In Earth Sciences and fracture characterizations, the box-counting method is the most used method describing fractal geometry. These geological elements are non-euclidian shapes. The boxcounting method consists of dividing the defined domain in box and analyzing the box cross by discontinuities [18]. Fractal analysis is critical issue due to the complexity of it use. Multifractal theory is developed to take account this complexity which characterizes multiple elements.

THEORY OF SCALE EFFECT OF THE ROCKS

Scale effect is studied by some authors in the literatures [19-22]. Some of them subdivided the scale effect into two components [19] represented by the shape scale which take account the variation of the ratio D/L and the size scale in which this ratio is constant and the size of the specimen growth. For that, values obtained in tests show that shape scale have effect in results of studies but no effect for size effect. About [20], the scale effect is more remarkable for brittle rocks than ductile rocks. The scale effect define by [20] corresponds of shape scale define by [19]. Samples with artificial fractures were also tested to characterize the scale effect on the specimen's behavior [21]. These authors generated artificial discontinuities on specimens of rocks. Two types of discontinuities are represented. Ones are no embedded and others are embedded. They show that a large framework of fracture on specimen can caused a scale effect while it is considered to be homogeneous. The scale effect is also a characteristic of the in situ rock masses. It influence scale effect in both size and time [23]. For in situ behavior, these authors show that joint which are present in rocks causes scale effect. These joints are characterized by their frequencies; connectivities and spacing caused by rock burst in deep. Another parameter which can be characterized by scale effect in geology and rock engineering is the permeability [24]. The scale effect depends largely on the heterogeneity and the connectivity of discontinuities. Sometimes, discontinuities are no connected. From these authors, there are no correlated parameters characterizing the conductivity of discontinuities on variable directions for the same on the field. So the scale effect is largely discussed in the literature and some functions are put on to model this parameter. So, from laboratory test to field parameters, Hoek-Brown criterion and Zhang and Einstein function can be used. The first permits to define the uniaxial compressive strength and the second the Young moduli. The Zhang and Einstein [25] is defined as:

$$E_m = E_i\left(10^{0.0186RQD-1.91}\right) \tag{5}$$

Scale effect appears also in some results, which characterize the roughness of joint. That's why from laboratory parameters to field ones, it is necessary to adapt values. This is valid also for Joint Compressive Strength (JCS). These last elements are defined by taking about scale effect by:

$$JRC_n = JRC_0\left[\frac{L_n}{L_0}\right]^{-0.02JRC_0} \tag{6}$$

The consequence of scale effect is also noted on the result of uniaxial compressive strength. For this, uniaxial compressive strength of a sample is given by taking about the diameter of the study sample and the typical diameter of 50 mm by the function

$$\sigma_{cd} = \sigma_{c50}\left(\frac{50}{d}\right)^{0.18} \tag{7}$$

In all cases, scale effect is traduced by variations of mechanical parameters when the dimensions of the samples vary. In this paper, it is observed when pass from values of laboratory test to values of field. It depends on the discontinuities of the rocks, type of filling materials. When pass from intact rock to rock with multiple fractures, some of which are filled, these values of mechanical parameters fall down.

STRUCTURAL ANALYSIS OF DISCONTINUITIES ON STUDIED ROCKS

The structural studies carried on the domain of Mako show a very important number of discontinuity families. In this domain, the igneous outcrop rocks (basalt) show also a very important number of fractures, which are represented in the stereogram. This disposition reflects the presence of folds (Figures 2-4). At the west side of Sékoto, the poles of fractures are organized along little circles. So, in this area, the folds are conic (**Figure 2**). These conic folds show some axes oriented north south.

In the south-west of Bafoundou, the poles of fractures have an organization along a great circle. This kind of organization is characteristic of cylindrical folds (**Figure 3**). The pole of the great circle meaning cylindrical fold is located in the east side mining that the axe of these folds is east.

This projection (**Figure 4**) shows a complex appearance. There are poles disposed along a great circle characterizing cylindrical fold and others along little circle traducing conic folds. Axes of cylindrical folds are oriented south-east while ones of conic folds are oriented north-south.

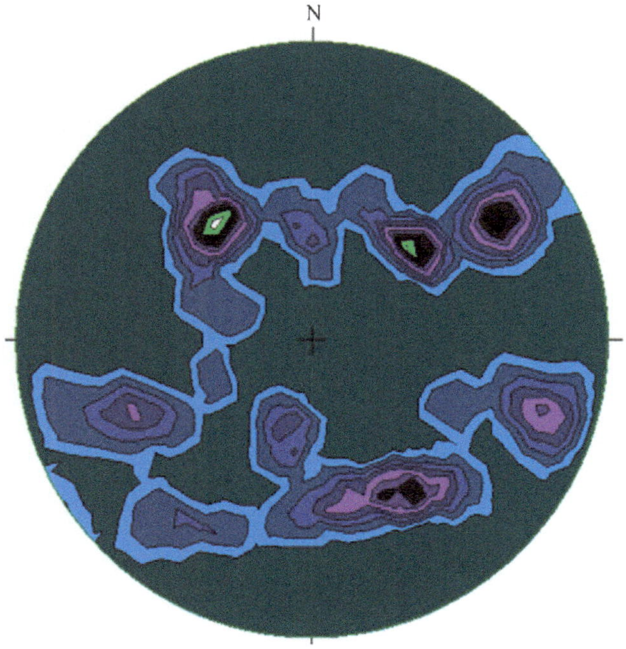

Figure 2. Contoured poles of fractures in the basalts (lower hemisphere equal angle projections): conic folds in the Mako area.

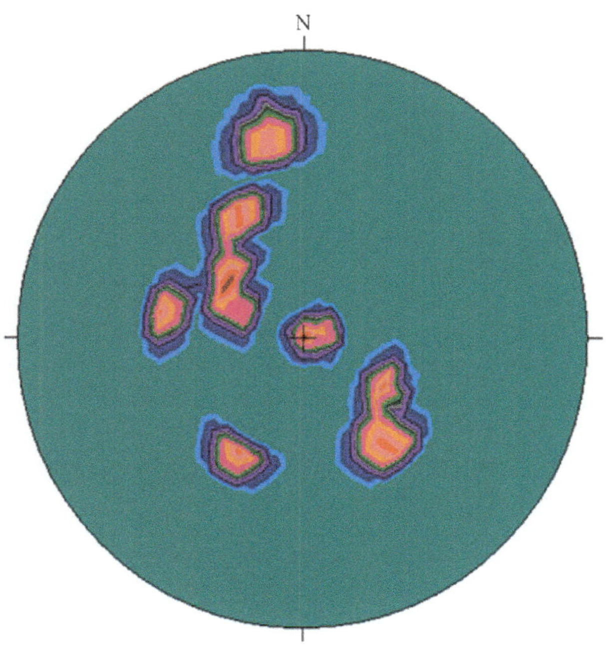

Figure 3. Contoured poles of fractures in the basalts (lower hemisphere equal angle projections): cylindrical folds southwest of Bafoundou.

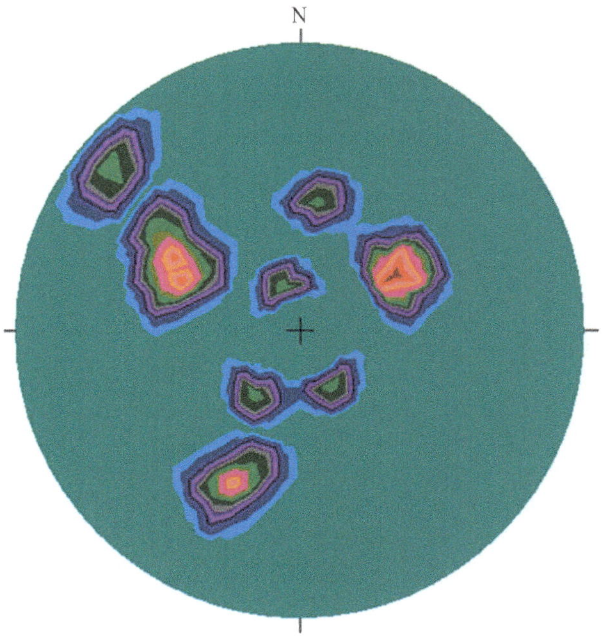

Figure 4. Contoured poles of fractures in the basalts (lower hemisphere equal angle projections): superposed cylindrical and conic fold.

Associated to a dome-basin system, they traduce a few different generations of folds that are or not same nature (Figures 2-4). This is due to the complex history of the Mako area in one part traduced by the three eburnean tectonic phases and in another part by the repercussion of Panafrican and Hercynian orogenesis.

Oldest sedimentary rocks of Senegal constitute the domain of Southern east Senegal in the Segou-Madina kouta basin. Those rocks show a simple fracturation represented by cracks and rarely faults which are often parallel to cracks (**Figure 5**).

On the stereogram, the fracturation of Segou doesn't show a particular disposition. They are grouped into three families (NE, ENE and ESE) while slight variation of fracturation can be seen. These families of discontinuities are associated with erratic joints accidentally present on the field (**Figure 5**).

GEOMETRIC AND SPATIAL STUDIES

Geometric parameters that caused fracturation and their iteration in the rock are fractal geometry and scale effect. Fractal geometry and scale effect depend on each other because both of them depend on stage and scale of the studies.

Scale Effect

Studies are conducted from laboratory to field. From one to another scale, rocks show important number of joint. This is a consequence of the scale effect. The increase of the number of discontinuities is due to the fact that in rocks microdiscontinuities are associated to microdiscontinuities.

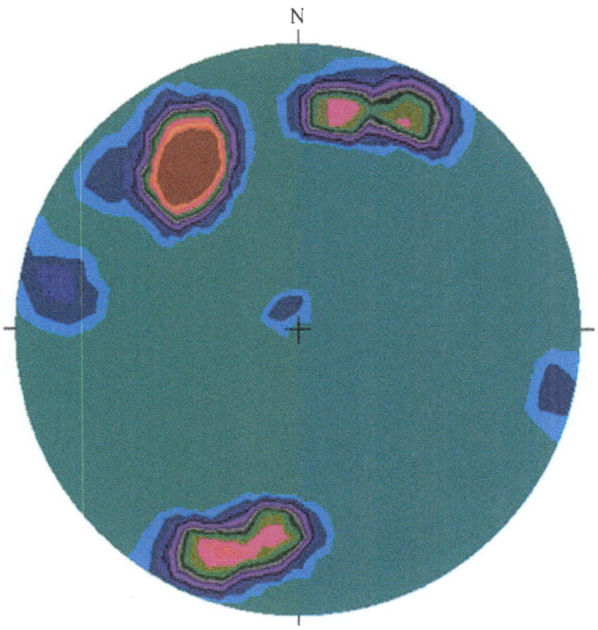

Figure 5. Contoured poles of fractures in the sandstone of Ségou area lower hemisphere equal angle projections.

When the sizes of the samples increase, the number of macrodiscontinuities present on the rock increases. The increasing of the number of discontinuities in the specimens affects the mechanical parameters. So, the smallest volume with all directions of discontinuities is called EVR (Elementary Representative Volume). If we have a volume of material equivalent or greater than this volume, mechanical parameters depending on the anisotropy don't change. They become stable around one value. This phenomenon appears very well in results of studies of all rocks, sedimentary or igneous. It can be remarked by lowest values of uniaxial compressive strength and Young moduli for rock masses than for laboratory samples. But the scale effect is more sensible for basalts than for sandstones (Figures 6 to 8). Structural studies at Mako (domain of basaltic rocks) and Segou (domain of sandstones rocks) show that the network of discontinuities is most complex for basalts. In fact, those sedimentary rocks of Segou are not the area of intensive tectonic history. Discontinuities are represented essentially by cracks. But basalts are affected by D_1 and D_2 phases of deformation and the Hercynian orogenesis impact. A large number of directions of discontinuities are present. Also, some of these discontinuities are filled with crystallization. Filling minerals are very different.

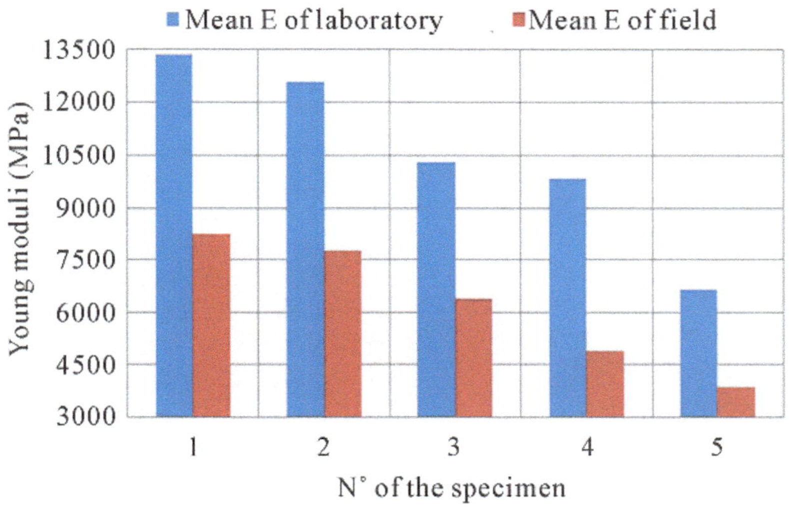

Figure 6. Graphical representation of Young moduli for different kind of basalts in laboratory samples and field ones (1: healthy basalts, 2: basalts with unfilled fractures, 3: basalts with tension cracks filled with quartz, 4: basalts with multiple fractures, 5: basalts with fractures filled with calcite, chlorite, epidote).

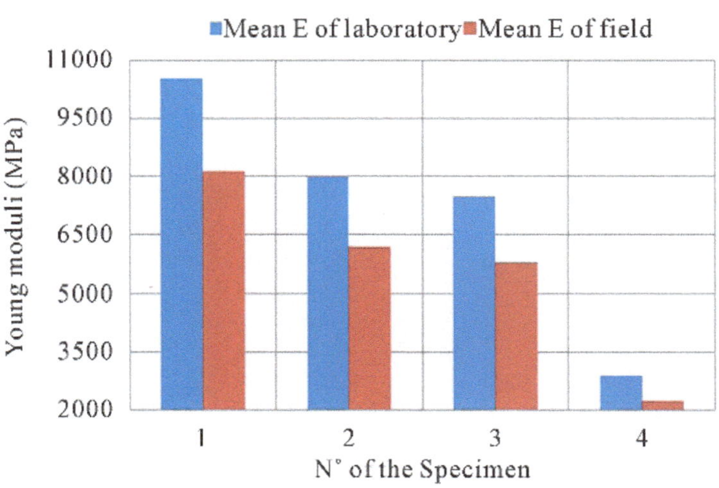

Figure 7. Graphical representation of Young moduli for different kinds of specimens of sandstones (1: white sandstones; 2: red sandstones; 3: purple sandstones; 4: sandstones with intercalation of pelites).

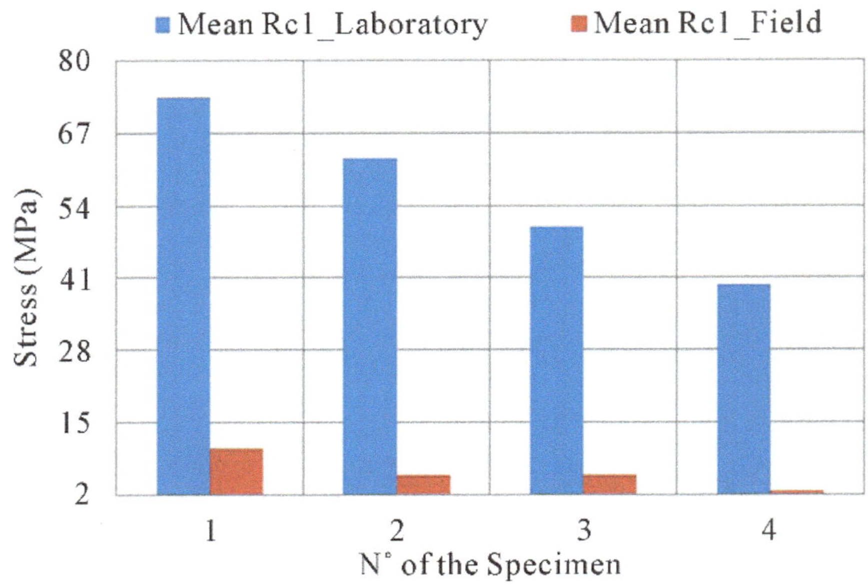

Figure 8. Graphical representation of uni-axial compression stress for different kinds of specimens of sandstones (1: white sandstones; 2: red sandstones; 3: purple sandstones; 4: sandstones with intercalation of pelites).

Some, which are hard, and the others are soft. These facts show that scale effect depends on the fracturation of the rock and the filling minerals of discontinuities.

For basalts (**Figure 6**), scale effect is very characteristic by the number of family of discontinuities and also the number of filling minerals and their natures. In fact, with these elements, Young moduli of the basalts in laboratory and in the outcrops vary remarkably. This variation is caused by the variability of the product of filling which can be very complex. So, it is very difficult to have a complete element of volume, which can represent the ERV. So, this kind of materials is characterized by a very heterogeneous nature and also, the heterogeneity can vary exceptionally from a locality to another.

In these studies, we introduce a parameter called scale coefficient (ech). This coefficient is obtained by the function:

$$ech = \frac{E_{lab} - E_{mass}}{E_{lab}} \times 100 \quad (8)$$

E_{lab} is the Young modulus for laboratory sample; E_{mass} is Young modulus of rock mass.

Basalts show large interval of variation of the scale coefficient (between 38% and 50.4%). that is caused by the complexity of the structuration of the samples (**Table 1**).

Healthy basalts, unfilled fractures basalts and basalts with tension cracks filled with quartz show same values of scale coefficients (38% to 38.3%). That is due to the fact that these cases correspond to a low complexity of structuration.

For basalts with fractures filled with calcite, chlorite and epidote, the scale coefficient increase (42.4%) because of an increase of the level of complexity. The complexity is a consequence of the duality of the material of filling. While calcite is hard, epidote and chlorite are soft.

For basalts with multiple fractures, increasing of the scale coefficient (ech) is most important (50.4%). This is due to a very important level of complexity. The material shows many directions of fractures and some of them are filled with soft materials, others by hard materials and the last ones are unfilled.

Therefore, the dimension of the ERV increases with an increasing of the scale coefficient, which denote of the increasing of the level of complexity of the rock.

Sandstones show also characteristics of scale effect. From laboratory to field values, the Young moduli decrease (**Figure 7**). This evolution is quite slight because of a low to medium level of complexity. In a structural part of view, the sandstones are crossed by cracks and there are no filling materials which can increase the complexity of the rock mass. Specimens present relatively the same kind of behaviors.

If we consider the scale coefficients (ech) for sandstones, their values are low and vary around 22% for all samples (**Table 2**). This is due to the fact that the number of discontinuities taking part to the behavior of this rock mass is limited to three and there are no filling materials which reinforce the complexity of the sandstones. While variation is exceptionally low, they are in disfavor of the sandstones with intercalation of pelites.

The scale effect affects also the uniaxial compressive strength (**Figure 8**). For sandstones, the action of scale effect is most remarkable on uniaxial compressive strength than on Young moduli. Another coefficient that's method of calculation is similar to the ech and called compression scale coefficient ech' can be defined using the uniaxial compressive strength for sandstones. This variable is calculated using the following function which whom R_{lab} is the uniaxial compressive strength (UCS) of laboratory; R_{field} is the UCS of the outcrop.

$$ech' = \frac{R_{lab} - R_{field}}{R_{lab}} \times 100 \quad (9)$$

For sandstones, this parameter takes high values more than 85%. This coefficient increases remarkably when the rock is most heterogeneous and takes values higher than 93% (**Table 3**).

Fractal Geometry

Fractal geometry defines a geometric pattern which is repeated at ever-smaller scales to produce irregular shapes and surfaces which can't be represented by classical geometry. Configurations of objects are small size, recurrent and can be identified in large range of scale. In Geosciences, this conception is used to quantify a large number of phenomena like spacing of discontinuities, escarpment of asperities and aperture of discontinuities. Due to the tectonic history of Mako, some of filled frac-tures and microfractures are folded like the outcrop system. These folds appear on preparation as undulations. For outcrops, folds can't be identified by simple observations. Determination of orientation of fractures permits to characterize these folds by stereographic representation (**Figure** 9(a)). In bloc scale, folds can be easily identified (**Figure** 9(b)) and it is also valid for preparations (**Figure** 9(c)).

Table 1. Evaluation of Scale coefficient (ech) for basalts.

Parameters	Healthy basalts	Basalts with unfilled fracture	Basalts with tension cracks filled with quartz	Basalts with multiple fractures	Basalts with fractures filled with calcite, chlorite, epidote
N° of Specimen	1	2	3	4	5
E (MPa)	13.351	12.566	10.300	9833	6647
E_m (MPa)	8236	7752	6387	4878	3826
ech (%)	38.3	38.3	38.0	50.4	42.4

Table 2. Evaluation of Scale coefficients (ech) for sandstones.

Parameters	White sandstones	Red sandstones	Purple sandstones	Sandstones with intercalation of pelites
N° of Specimen	1	2	3	4
E (MPa)	10.540	7999	7500	2897
E_m (MPa)	8140	6178	5793	2237
ech (%)	22.76	22.76	22.76	22.78

Table 3. Evaluation of compression scale coefficient (ech') for sandstones.

Parameters	White sandstones	Red sandstones	Purple sandstones	Sandstones with intercalation of pelites
N° of Specimen	1	2	3	4
R_{c1} (MPa)	73.31	62.37	50.25	39.83
R_{c2} (MPa)	10.36	5.62	5.5	2.52
ech' (%)	85.87	90.99	89.05	93.67

This fractal aspect appears also in the values of Joint Roughness Coefficients (JRC). For results of these studies, JRC values of unfilled microscopic and macroscopic fractures are highest than filled ones. For geological systems, more fractures are former, the more they are filled. Based on this fact, undulated discontinuities (**Figure 9**) are former. They are folded by posterior orogenesis system, which caused most deformation of the geological materials. Latest events are characterized by fractures, which shear the filled discontinuities. These shears are dextral or senestral like the field ones.

Microdiscontinuities show also same arrangements than macroscopic ones. According to the studied event and the number of iterations, the fractal dimension change. This characteristic is studied in this paragraph. The filled fractures show also conjugate aspects on outcrops and in laboratory scale with same angles and sinuosities (**Figure 10**).

The Birimian of Kédougou-Kenieba (example of Mako supergroup) inlier shows some shear fractures or faults. In the scale of the field, shears are in many cases represented by stumbles separated by ravines. For microscopic scale, these senestral shears are identified by studing fractures, which crossed another, filled one and which caused displacements (**Figure 11**). The displacements correspond to the one determined in the field in locality where sample have been taken.

Figure 12 shows filled fractures in the field scale (Figure 12(a)) and on preparation scale (**Figure 12**(b)). These fractures are filled with chlorite, epidote and calcite. Due to the same nature of filling products and the form of structures at the field and on the preparation, we deduce that they are contemporaries. Spacing of fractures measured for these discontinuities are 5.4 centimeters for rock masses and 175 micrometers for preparations. Before iteration, we consider an element with one discontinuity and one spacing in rock mass scale. After iteration, they are 308 microfractures corresponding to the spacing seen in microscopic scale. For this assemblage of microdiscontinuities, some of them are identified in the preparation. The preparation shows just a number of microfracturation that allows defining the fractal dimension. The fractal dimension associated to this operation is $d_f = 1.12$.

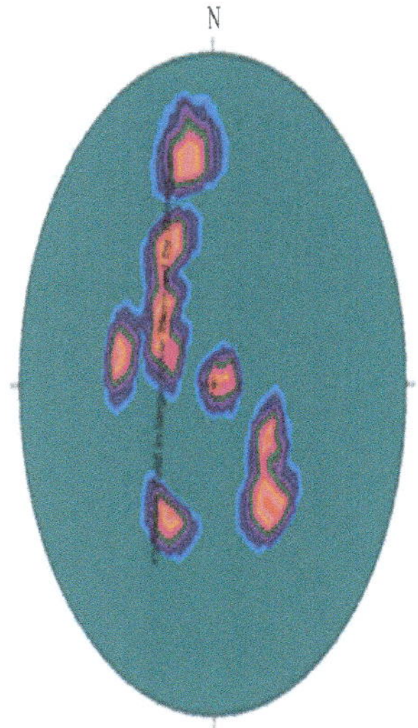

(a)

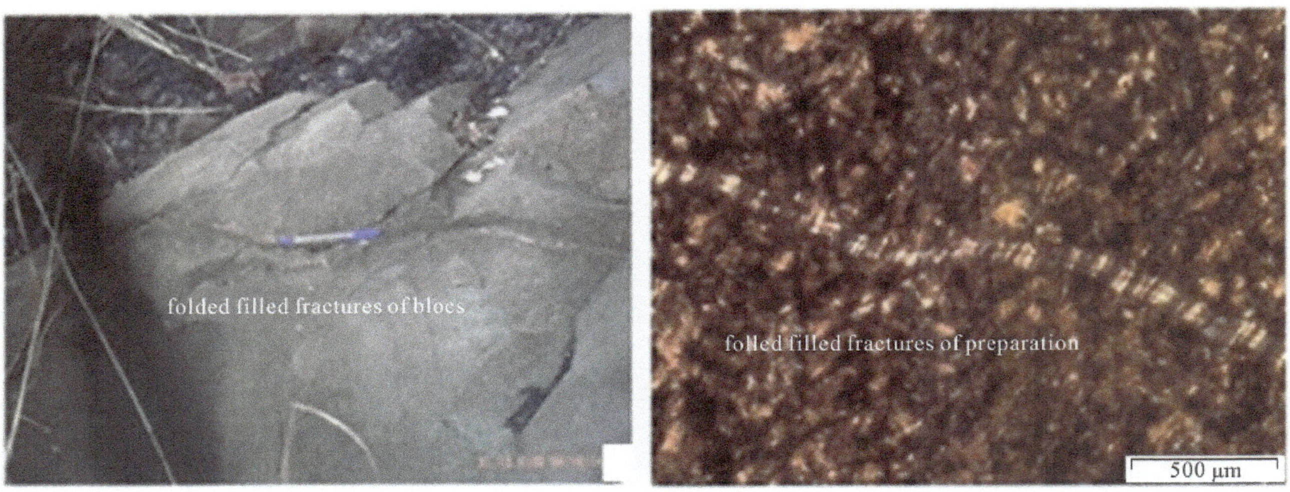

(b)

Figure 9. Fractal geometry of undulate structures.

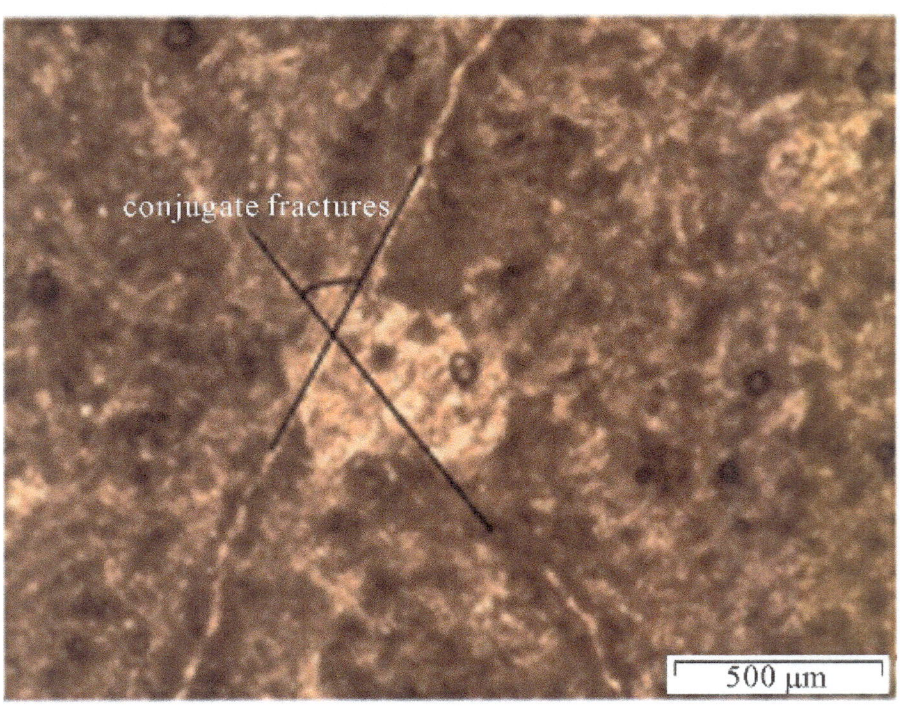

(a)

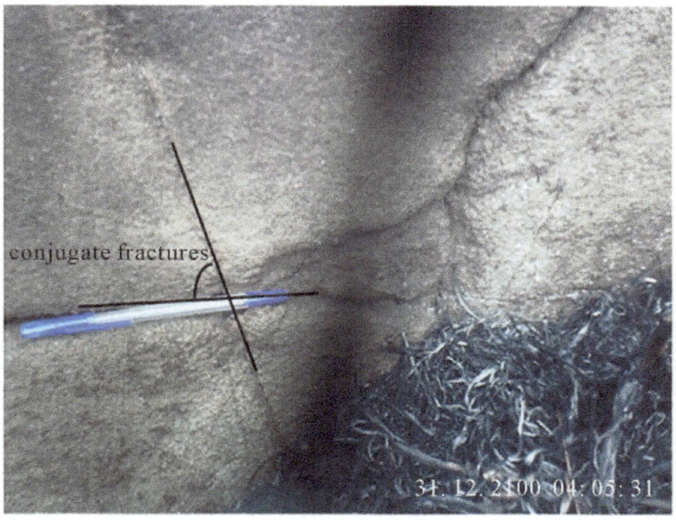

(b)

Figure 10. Fractal geometry in conjugate fractures ((a) outcrops scale; (b) preparation scale).

(a)

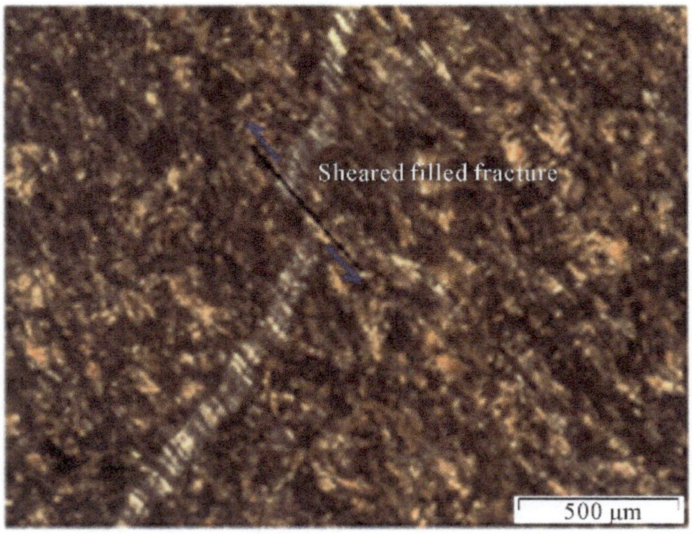

(b)

Figure 11. Microscopic shear of filled microdiscontinuity ((a) outcrops scale; (b) preparation scale).

(a)

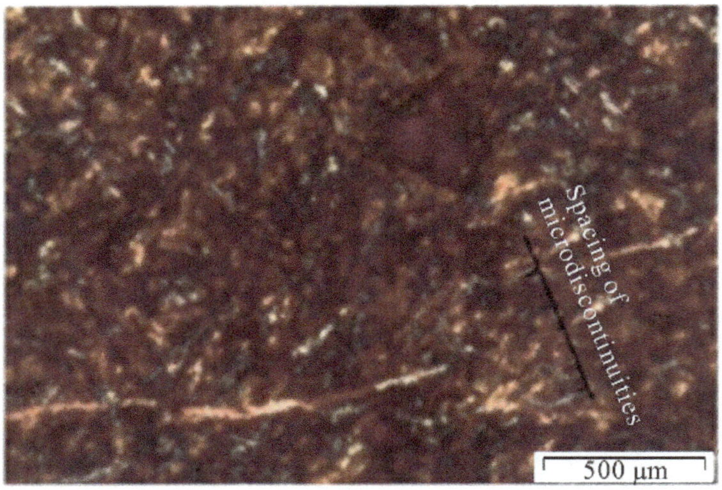

(b)

Figure 12. Fractal analysis of spacing of discontinuities ((a) outcrops scale; (b) preparation scale).

Aperture of discontinuities shows also the morphology of fractal objects (**Figure 13**). For the aperture of discontinuities, we consider that initial stage corresponds to an aperture of 6.4 centimeters in the rock mass. After fragmentation, it appears in the rock matrix some microscopic discontinuities with 41.1 micrometers as aperture. The fractal dimension associate to this system is $d_f = 2.09$.

Another system presenting a fractal aspect is the asperities of fractures (**Figure 13**). These asperities present some length of 1.03 centimeters as escarpment for macrofractures and some of 41.1 micrometers for microscopic discontinuities. From macroscopic scale to microscopic one, the dimension fractal is $d_f = 1.486$.

The same phenomena are observed in macroscopic and microscopic scales of igneous rocks (basalts of Mako) Discontinuities appears with same characteristics but they are continuous system. So they are affected by the same geological history. The consequence of this story appears in macroscopic and microscopic scale.

CONCLUSIONS

The results suggest that in the Mako area, there is a genetic link between the fractures from one location to another. In fact, the development of the fracturation can be synchronized with the development of folds. These results have important impacts on the geometry and the growth propagation of the development of fractures during the folding phases; where the fracture sets are interpreted as tectonic in origin and are sometimes synchronous with the post-eburnean folding of the basaltic rocks and are not preor post-folding fractures.

For these rock masses, the studies of mechanical parameters (Young moduli and uniaxial compressive strength) show that values decrease from laboratory to field. Decreasing of the values is the consequence of scale effect, which is caused by the heterogeneity of the materials. The heterogeneity of the material depends on the discontinuities, which are present in the rock masses, the filling products and their natures. When all factors of heterogeneity are taken into account in the samples, it deals with ERV. In this paper, we also define a parameter which is characteristic of the scale effect and called scale coefficient (ech). This scale coefficient increases with the presence of anisotropy in the rock masses. The more complex the anisotropy of the rock mass is, the higher the ech is. At least, the scale coefficient is the function of scale effect which depends on fractures and their fillings.

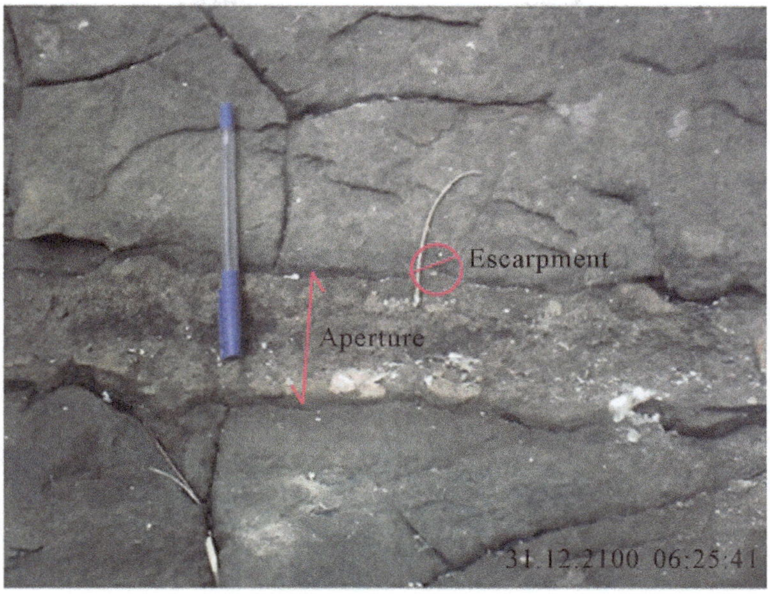

(a)

(b)

Figure 13. Fractal analysis of aperture and escarpment of discontinuities ((a) outcrops scale; (b) preparation scale).

The fractal geometry is represented also on rocks. It appears on the spacing and the aperture of discontinuities. Escarpments of asperities also show this configuration. All these fractal characteristics are studied on filled fractures. Their dimensions vary from an element to another. The process of iterations is more pronounced for aperture of discontinuities in the filled fractures than for escarpment and at least spacing of fractures.

REFERENCES

1. E. Bonnet, O. Bour, N. Odling, I. Main, B. Berkowitz, P. Davy and P. Cowie, "Scaling of Fracture Systems in Geological Media," Reviews of Geophysics, Vol. 39, No. 3, 2001, pp. 347-383. http://dx.doi.org/10.1029/1999RG000074
2. H. Théveniaut, P. M. Ndiaye, F. Buscail, R. Couëffé, C. Delor, T. Fullgraf and J. C. Goujou, "Notice Explicative de la Carte Géologique du Sénégal Oriental à 1/500000," Ministère des Mines, de l'Industrie, de l'Agro-Industrie et des PME, Direction des Mines et de la Géologie, Dakar, 2010, 122 p. [Paper reference:3]
3. P. Ledru, J. Pons, J. P. Milési, J. L. Feybesse, A. Dommanget, V. Johan, M. Diallo and C. Vinchon, "Tectonique Transcurrente et Evolution Polycyclique dans le Birimien, Protérozoïque Inférieur du Sénégal-Mali (Afrique de l'Ouest)," Comptes Rendus de l'Académie des Sciences, Série II, Vol. 308, 1989, pp. 117-122.
4. P. Ledru, J. Pons, J. P. Milési, J. L. Feybesse and V. Johan, "Transcurrent Tectonics and Polycyclic Evolution

in the Lower Proterozoic of Senegal-Mali," Precambrian Research, Vol. 50, No. 3-4, 1991, pp. 337-354. http://dx.doi.org/10.1016/0301-9268(91)90028-9

5. M. Diène, M. Gueye, D. P. Diallo and A. Dia, "Structural Evolution of a Precambrian Segment: Example of the Paleoproterozoic Formations of the Mako Belt (Eastern Senegal, West Africa)," International Journal of Geosciences, Vol. 3, No. 10, 2012, pp. 153-165.

6. J. Pons, P. Barbey, D. Dupuis and J. M. Leger, "Mechanisms of Pluton Emplacement and Structural Evolution of a 2.1 Ga Juvenile Continental Crust: The Birimian of Southwestern Niger," Precambrian Research, Vol. 70, No. 3-4, 1995, pp. 281-301.http://dx.doi.org/10.1016/0301-9268(94)00048-V

7. M. Gueye, S. Siegesmund, K. Wemmer, S. Pawlig, M. Drobe, N. Nolte and P. Layer, "New Evidences for an Early Birimian Evolution in the West African Craton: An Example from the Kédougou-Kénieba Inlier, Southeast Senegal" South African Journal of Geology, Vol. 110, No. 4, 2007, pp. 511-534. http://dx.doi.org/10.2113/gssajg.110.4.511

8. J. P. Milési, J. L. Feybesse, P. Ledru, A. Dommanget, M. F. Ouedraogo, E. Marcoux, A. Prost, C. Vinchon, J. P. Sylvain, V. Johan, M. Tegyey, J. Y. Calvez and P. Lagny, "Les Minéralisations Aurifères de l'Afrique de l'Ouest: Leurs Relations avec l'Evolution Lithostructurale au Protérozoïque Inférieur," Chronique de la Recherche Minière, Vol. 497, No. 987, 1989, pp. 3-98.

9. J. P. Bassot and A. Dommanget, "Complexe Volcanoplutonique Calco-Alcalin de la Rivière Daléma (Est Sénégal): Discussion de sa Signification Géodynamique dans le Cadre de l'Orogenèse Eburnéen (Protérozoique Inférieur)," Journal of African Earth Sciences, 1986, 14 p.

10. P. M. Ndiaye, "Evolution au Protérozoïque Inférieur de la Région Est—Saraya, (Supergroupe de Mako, Sénégal oriental). Tourmalisation, Altérations Hydrothermales et Minéralisations," Thèse d'Etat, Université Cheikh Anta Diop de Dakar, 1994, 372 p.

11. J. P. Bassot, "Etude Géologique du Sénégal Oriental et de ses Confins Guinéo Maliens," Mém. B.R.G.M. No. 40, 1996, 332 p.

12. D. Sarr, M. Fall, P. M. Ngom and M. Ndiaye "Mechanical Behavior of Pillow Lavas in Mako Supergroup: Case of South Mako Hill," International Journal of Geosciences, Vol. 2, No. 4, 2011, pp. 640-647. http://dx.doi.org/10.4236/ijg.2011.24065

13. D. Sarr, M. Fall, P. M. Ngom, M. Ndiaye, C. H. Kane and M. Ba, "Geomechanical Characterization of Sandstones Cliffs of Segou (Senegal, West Africa) in the Madina Kouta Basin," International Journal of Geosciences, Vol. 3, No. 1, 2012, pp. 166-174.http://dx.doi.org/10.4236/ijg.2012.31018 [Paper reference:1]

14. B. Mandelbrot, "Les Objets Fractals," Flammarium, Paris, 1975.

15. B. Mandelbrot. "Multifractal Measures, Especially for Geophysicist," Pageoph, Vol. 131, No. 1/2, 1989, pp. 5-42. http://dx.doi.org/10.1007/BF00874478

16. Carton "Les Fractales: Une Géométrie de la Nature," 2003.

17. H. Legrain, "Etude de l'Influence de la Rugosité sur l'Ecoulement des Fluides dans les Fissures Rocheuses," Thèse de Sciences Appliquées, Académie Universitaire Wallonie Bruxelles, 2006, 162 p.

18. N. H. Tran, "A Dissertation Submitted to the University of New South Wales in Partial Fulfillment of the Requirements for the Degree of Doctor Philosophy in Petroleum Engineering," University of New South Wales, Sydney, 2004, 267 p.

19. K. Thuro, R. J. Plinninger, S. Zäh and S. Schütz, "Scale Effect in Rock Strength Properties. Part 1: Unconfined Compressive Test and Brazilian Test. Rock Mechanics—A Challenge for Society," Särkkä et Eloranta, Eds. Swets et Zeitinger Lisse, 2001.

20. R. Simon, "Développement d'une Approche pour Estimer la Résistance des Roches dures à l'Echelle du bloc Unitaire," Rapport R-594 de l'IRSST du Québec, 2005, 59 p.

21. M. Pierce, M. Gaida, D. DeGagne, "Estimation of Rock Block Strength," ROCKENG09 Proceedings of the 3rd CANUS Rock Mechanics Symposium, Toronto, May 2009.

22. C. C. Barton, "Fractal Analysis of Scaling and Spatial Clustering of Fractures," In: C. C. Barton and P. R. LaPointe, Eds., Fractals in the Earth Sciences, Plenum Press, 1995, pp. 141-178.

23. D. Ellsworth and C. Fairhurst, "Engineering Research Opportunities in the Subsurface: Geo-Hydrology and Geo-Mechanics," The American Rock Mechanics Association Foundation. 600 Woodland Terrace, Alexandria, VA 22302. A Report to the National Science Foundation Award No. CMS-0234136, 2003.

24. W. A. Illman, "Strong Field Evidence of Directional Permeability Scale Effect in Fractured Rock," Journal of Hydrology, Vol. 319, No. 1-4, 2006, pp. 227-236.

25. L. Zhang and H. H. Einstein, "Using RQD Estimating the Deformation Modulus of Rock Masses," International Journal of Rock Mechanics and Mining Sciences, Vol. 41, No. 2, 2004, pp. 337-341. http://dx.doi.org/10.1016/S1365-1609(03)00100-X

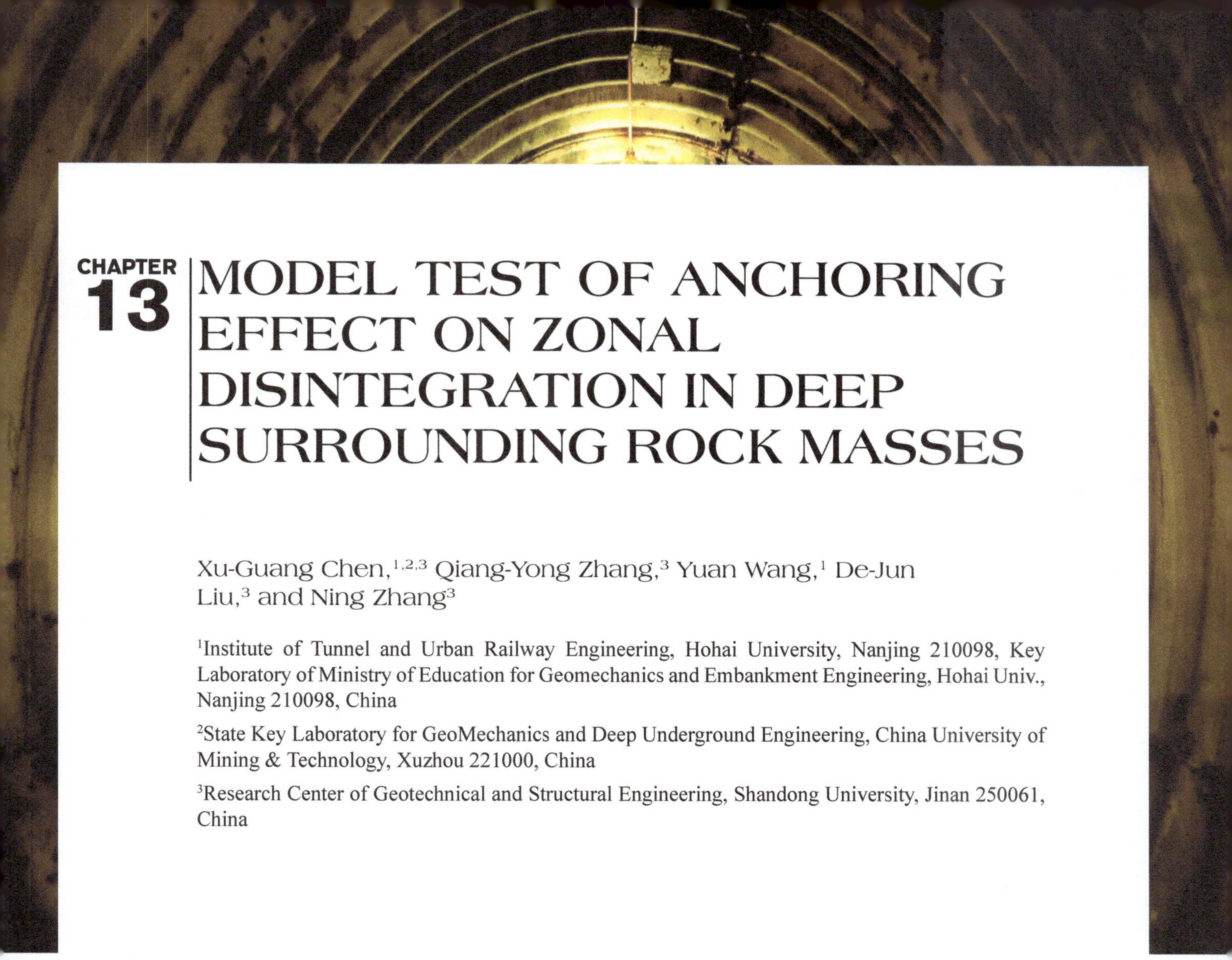

CHAPTER 13 MODEL TEST OF ANCHORING EFFECT ON ZONAL DISINTEGRATION IN DEEP SURROUNDING ROCK MASSES

Xu-Guang Chen,[1,2,3] Qiang-Yong Zhang,[3] Yuan Wang,[1] De-Jun Liu,[3] and Ning Zhang[3]

[1]Institute of Tunnel and Urban Railway Engineering, Hohai University, Nanjing 210098, Key Laboratory of Ministry of Education for Geomechanics and Embankment Engineering, Hohai Univ., Nanjing 210098, China

[2]State Key Laboratory for GeoMechanics and Deep Underground Engineering, China University of Mining & Technology, Xuzhou 221000, China

[3]Research Center of Geotechnical and Structural Engineering, Shandong University, Jinan 250061, China

ABSTRACT

The deep rock masses show a different mechanical behavior compared with the shallow rock masses. They are classified into alternating fractured and intact zones during the excavation, which is known as zonal disintegration. Such phenomenon is a great disaster and will induce the different excavation and anchoring methodology. In this study, a 3D geomechanics model test was conducted to research the anchoring effect of zonal disintegration. The model was constructed with anchoring in a half and nonanchoring in the other half, to compare with each other. The optical extensometer and optical sensor were adopted to measure the displacement and strain changing law in the model test. The displacement laws of the deep surrounding rocks were obtained and found to be nonmonotonic versus the distance to the periphery. Zonal disintegration occurs in the area without anchoring and did not occur in the model under anchoring condition. By contrasting the phenomenon, the anchor effect of restraining zonal disintegration was revealed. And the formation condition of zonal disintegration was decided. In the procedure of tunnel excavation, the anchor strain was found to be alternation in tension and compression. It indicates that anchor will show the nonmonotonic law during suppressing the zonal disintegration.

INTRODUCTION

Shallow-buried resources have been decreasing with the rapid progress in global economy. Thus, the exploitation of deeply buried resources has drawn interest from a number of countries. South Africa, Russia, India, and China have recently conducted a series of exploitations of deep mines with embedded depths of more than 1,000 m. In China, the Jinchuan nickel mine, Tongling copper mine, and Dingji coal mine are more than 1,000 m deep, whereas the Jinping II Hydropower Station is 2,600 m deep. The Kolar gold mine in India is 2,400 m deep, and the deepest gold mine worldwide residing in South Africa is 3,700 m deep. This series of new phenomena in underground engineering with increasing embedded depth has caused the emergence of a failure phenomenon called zonal disintegration. Shemyakin et al. [1] defined zonal disintegration as the alternated regions of fractured and relatively intact rock masses appearing around or in front of the working slope during the

Citation: Xu-Guang Chen, Qiang-Yong Zhang, Yuan Wang, De-Jun Liu, and Ning Zhang, "Model Test of Anchoring Effect on Zonal Disintegration in Deep Surrounding Rock Masses," The Scientific World Journal, vol. 2013, Article ID 935148, 16 pages, 2013. doi:10.1155/2013/935148.

excavation of tunnels in the deep rock mass. This phenomenon has not been observed in shallow rock engineering. Moreover, zonal disintegration presents a serious hazard to the stability of surrounding rocks (Qian) [2].

Zonal disintegration in many deep tunnels has been monitored using the physical probe method. Adams and Jager [3] were the first to observe such phenomenon by using a bore periscope at an embedded depth of 2,000 m to 3,000 m in the Witwatersrand gold mine in South Africa. He reported that zonal disintegration occurred when the tunnel was excavated either by drilling and blasting or the mechanized method. However, explosion was eliminated as the result of zonal disintegration after Shemyakin [1, 4–6] explored zonal disintegration in Taimyrskii deep mine in Russia by using a resistivity meter.

Zonal disintegration phenomenon differs from the engineering response in shallow rock excavation and as such cannot be explained perfectly under the framework of traditional rock theory. In accordance with the concepts of traditional continuum mechanics, the enclosing rock mass around a deep tunnel is divided into fractured, plastic, and elastic regions from the periphery of the tunnel to infinity. Zonal disintegration, a characteristic of deep rock masses, has been the focus of recent investigations.

A number of experts have used various methods to explain zonal disintegration. Sellers and Klerck [7] indicated that the discontinued surface could be one of the derivations of zonal disintegration. Malan and Spottiswoode [8] analyzed the relationship between the shock bump and the zonal disintegration of a top plate in the surrounding rocks of a mining field. Zhou et al. [9] investigated the dynamic excavation of a deep tunnel to determine the residual strength and the forming time of fractured zones. Gu et al. [10] conducted a compression test on cylinder specimen and regarded axial stress as an important factor for zonal disintegration. Other studies on zonal disintegration have applied different techniques such as a series of compression tests Pan [11, 12], nonequilibrium thermodynamics (Metlov et al.) [13], Hamiltonian time-domain variation (Li et al.) [14], and the non-Euclidean model (Guzev and Paroshin) [15]. In addition, some elastic-plastic theories have been adopted to analyze the forming mechanism of zonal disintegration (Wang et al. [16, 17]; He et al., [18]; Zhou et al. [19–24]; Reva and Tropp, [25]; Tan et al., [26]; Wu et al., [27]; Odintsev, [28]). A zonal disintegration phenomenon is shown in Figure 1.

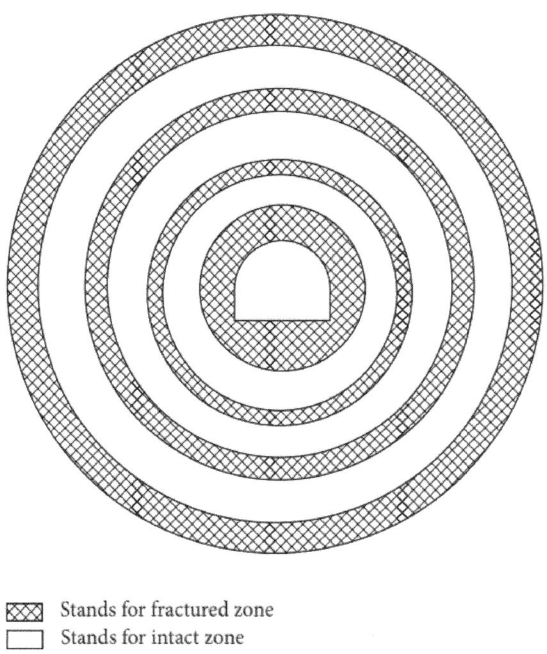

Figure 1: Sketch of the zonal disintegration phenomena in deep tunnel.

Zonal disintegration is a unique failure phenomenon posing a large-scale disaster during excavation of deep rock masses (Laptev and Potekhin) [29]. It threatens the stability of deep tunnel and will cause large collapse of rock mass which induces a great loss. It is of great importance to know the anchoring effect on zonal disintegration and the mechanical behavior under anchoring condition in deep rock masses, for the stability of deep tunnel. To the authors' knowledge, anchoring effect on zonal disintegration phenomenon in deep rock masses is not investigated previously.

In this paper, the Huainan coal mine in which zonal disintegration occurs in China was taken as the engineering background. The model tests on zonal disintegration were carried on in the condition of anchoring and without anchoring, in separate. The model was built using an independently developed barites-iron-sand cementation analogical (BISA) material. Through the analogical model test, the damage pattern with and without anchoring was observed. The nonlinear deformation changing laws were clarified by using a precise optical apparatus. Based on this, the anchoring effect and forming condition of zonal disintegration in deep rock masses is revealed.

SIMILARITY THEORY AND ANALOGICAL MATERIAL

The geomechanical model test is an important scientific research method. Similar to prototype engineering, the model was designed based on the similarity principle. An optical measuring apparatus was used in the geomechanical model test. The stress and displacement changing rules of the model and strain of the anchor were monitored to determine the deformation laws of prototype engineering. The model test exhibits an advantage in studying the failure mechanism of underground cavities over in situ observation, which relies on auditory-visual perception and is time-consuming.

The geomechanical model test is an effective reduced scale method used for investigating special engineering problems based on the similarity principle. The changing laws of stress, strain, and displacement can be monitored by designing the model similar to that of prototype engineering. Following the similarity principle, the data observed from the model test can be used to reveal the stress distribution laws and the mechanism in prototype engineering, thereby solving actual problems.

Similarity Theory

The geomechanical model test requires a suitable similar material that can reflect the mechanical behavior of a rock type. The similar material and its prototype must comply with the similarity principle. The theory requires several similarity coefficients defined as ratios of prototype parameters to model parameters to be constant (Fumagalli) [28, 30]:

$$C_\sigma = C_\gamma C_L, \quad (1a)$$

$$C_\delta = C_\varepsilon C_L, \quad (2b)$$

$$C_\sigma = C_\varepsilon C_E, \quad (3b)$$

$$C_\varepsilon = 1, \qquad C_f = 1, \quad (4b)$$

$$C_\phi = 1, \qquad C_\mu = 1, \quad (5b)$$

where C_σ, C_ε, C_E, C_L, C_γ, C_f, C_ϕ, and C_μ indicate the similarity ratios of stress, strain, MOE, geometry, volume weight, COF, internal friction angle, and passion ratio, respectively.

The analogical material should have the properties of high volume-weight, low deformation module, and changeable inner friction angle. No crude material can fulfill all these demands, and thus, the similar material should be assembled artificially. According to the similarity theory, the mechanical parameters of the model can be readily obtained through the prototype.

Proportion of the Similar Material

There are several Institutes researching on the similar material, such as ISMES (Institute of Experimental Models and Structures) in Italy, LNEC (National Laboratory for Civil Engineering) in Portugal, and Tsinghua University in China [30, 31]. Their work shows that whether the model test can reflect the prototype engineering's mechanical response depends on the chosen materials. The suitable material should reflect the mechanical behavior of prototype engineering. The proportion of each component is important for model simulation.

The barite powder, iron powder, and quartz sand are selected to form the aggregate, whereas the alcoholic solution of rosin is used as the mucilage glue (Figure 2). The proportion of the aggregates and the concentration of the alcohol solution of rosin decide the mechanical behavior of the BISA material. The barite-iron-sand (BISA) material was developed through hundreds of groups of proportioning tests. The specimens of similar material were built by pouring the material into a mould and compressing it (Figure 3). The material exhibits the following advantages: stability in performance, widely variable mechanical parameters, low price, high volume-weight, easy processing, and no toxicity or side effects. The BISA material, which can be used for modeling a tunnel or underground powerhouse, has obtained a patent in China. The material can be used to simulate all kinds of rocks, including hard and soft rocks. The proportion of the material composition for surrounding rocks in the Dingji coal mine was determined via the physical-mechanical parameters test on the analogical material. The mechanical parameters of the material were tested in the proportion of each component in the material (Figure 3).

Figure 2: Proportion of the analogical material. The barites powder, iron powder and quarts sand are mixed together to make the aggregate; the solution of rosin and alcohol make the glue.

Figure 3: Mechanical test on analogical material. (a) is the specimen mould. (b) is the specimens for UCS. (c) is the specimens for direct shear test. (d) is the Brazilian test. (e) is the uniaxial test. (f) is the direct shear test.

The laws between the mechanical parameters of material and components proportion were derived by mechanical testing on hundreds of specimens.

The medium sandstone in the bed stratum from the Dingji coal mine and the Huainan mining area was processed into a specimen. The physical-mechanical parameters of equivalent anchors were tested. The mechanical parameters of the medium sandstone consist of the following: unconstrained compressive strength (UCS) of 88.55 MPa, tensile strength of 14.01 Mpa, and deformation modulus of elasticity (Edef) of 12.97 GPa. The similarity ratio of volume weight for the analogical material is set to 1 : 1, whereas the similarity ratio of geostress is set to 1 : 50. Thus, according to the similarity principle, the mechanical parameters of the prototype and similar material are as follows (Table 1).

Table 1: Physical-mechanical parameters of the prototype rock and model material.

Material type	Volume weight/KN·m^{-3}	Edef/MPa	Cohesion/MPa	φ	UCS/MPa	TS/MPa	NUXY
Prototype	2.62	12970	10	43	88.55	14.01	0.268
Model	2.62	51.88	0.4	43	3.54	0.56	0.268

According to the curves between the mechanical parameters and the material proportion, the proportion of each component for medium sandstone is as follows (Table 2).

Table 2: Proportion of the analogical material.

I : B : S	Portion of the gypsum	Concentration of the solution	Portion of the solution
1 : 1.1 : 0.42	2.5%	7.5%	5.0%

Equivalent Anchors

The parameters of anchor adopted in engineering are: ϕ 20@800 × 800 mm, L = 2.2 m. According to the similar principle, the parameters of model anchor can be got from the prototype anchor (Table 3).

Table 3: Physical-mechanical parameters of prototype and equivalent anchor.

Anchor	Edef/GPa	TS/MPa	Yield strength/MPa	length/cm
Prototype	210	510	345	220
Model	4.2	10.2	6.90	4.4

After the mechanical test on the serious metal materials (Figure 4), the aluminum wire is selected as the equivalent anchor.

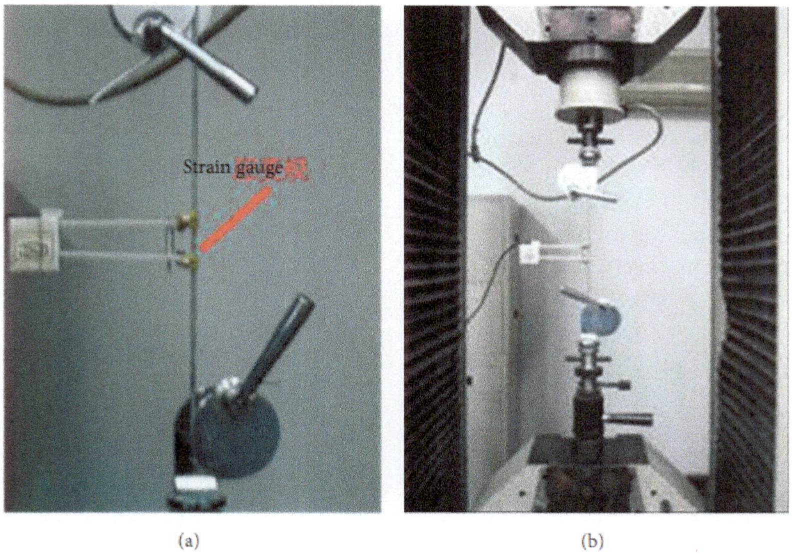

Figure 4: Mechanical test on equivalent anchor.

DEVELOPMENT OF THE MODEL TEST SYSTEM

Development of the Triaxial Model Test System

In order to simulate the 3 D geostress state of tunnel precisely, the high geostress-triaxial loading model test system was developed independently. Figure 5 is the design sketch and the photo of the system.

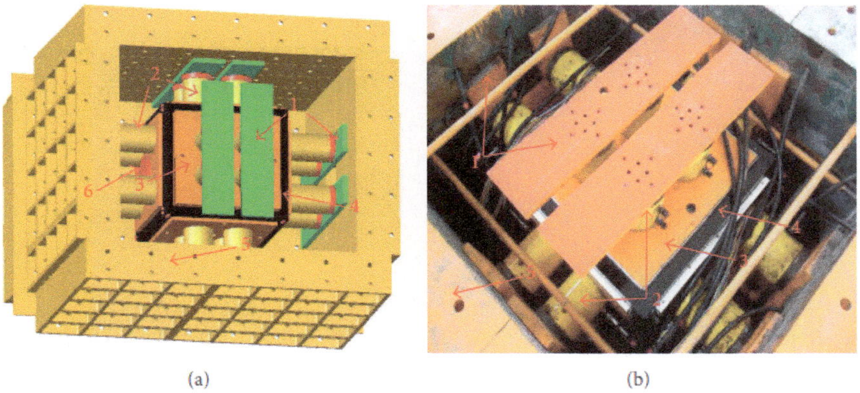

Figure 5: Sketch and photos of the true 3D model test system. The model was set inside the frame, and the platen was set into the frame at 0.5 cm depth. Then, each direction of the model has the reserved displacement of 9 cm. Notes: (1) junction plate, (2) hydraulic jacks, (3) loading platen, (4) oriented frame, and (5) combination reaction frame for loading (6) excavation guiding platen.

The model test system is comprised of the high geostress loading system, the computerized numerical control (CNC) hydraulic system, and the support reaction apparatus. The high geostress loading system consists of jacks and loading platens, which are used to apply high geostress. It is set inside the support reaction apparatus. One end of the jacks was disposed with loading platen, while the other was fixed to the support reaction apparatus with bolts. There are 6 platens in all, and one contacts each surface of the model, which can apply loading at 3 directions. The dimension of each platen is 0.6 × 0.6 × 0.04 m, which is equal to the size of model surface. There arrange 4 hydraulic jets on each platen, and the capacity of each jack is 40 tons. The loading capacity of each platen is 1600 KN. The high geostress loading system is connected with the CNC hydraulic system. The CNC hydraulic system is used to regulate and control loading value. The support reaction apparatus is composed of box cast steel components, which can be assembled flexibly. It is solid enough to undertake support reaction to 5000 KN high.

During the triaxial compression on 3D model, two adjacent loading platens will interrupt each other because of the volume contraction. So, an oriented frame was developed to solve the problem. The frame consists of 12 steel poles which are connected with each other. Each pole was 5 cm thick and 70 cm long. The model was set inside the frame, and the platen was set into the frame at 0.5 cm depth. Then, each direction of the model has the reserved displacement of 9 cm. A circle platen whose size is similar to the tunnel was fixed into the two platens in front and back of the model. And the platens will be fixed during the loading on the model and taken off when excavation of the tunnel (Figure 5). So, the excavation in 3D model can be solved.

The advantages of the system are shown as follows.(1) High geostress can be applied to the model in all 3 directions independently and synchronously. And the geostress can be remained stable for long term.(2)The problem of two adjacent loading platens that interrupted each other which was caused by the triaxial compression was solved. And the excavation in 3D model was realized.(3)The loading capacity is large enough to 2000 KN, which can simulate the cavern deep to 5000 m. And the loading precision reaches up to 0.5%.

Development of the High Precision Optical Measuring System

A micrometer is necessary because the multi-point displacement meters are too huge for the model. Thus, a micrograting ruler multipoint displacement measuring System (GRDS) was developed to monitor the displacement in the rock mass of the model (Figure 6). The Moiré fringe is used to enlarge the displacement in the model. The GRDS is composed of rack, grating rulers, steel wires, balance weight, fixed end, signal translating system (STS), and data acquisition system (DAS). Several fixed ends were buried at the measuring spots inside the model. The fixed ends were connected to the grating ruler outside the model with the steel wire. The steel wire was made up of special sterepsinema without any axial deformation, and thus, it is very flexible and can be bent arbitrarily. The steel wire was wrapped with Teflon pipe to eliminate friction between the steel wire and the model. The GRDS transformed the displacement of the model to the grating ruler through the mechanical method. Then, the real displacement was transferred to the optical signal by using an optical ruler. The STS was connected to the grating ruler with the guide line, and the optical signal was transformed to digital signal and then transferred to the DAS. The displacement data of the model was displayed and stored instantaneously. The high precision (1 μm) GRDS can be used to measure the displacement in any direction as it can be bent at any angle.

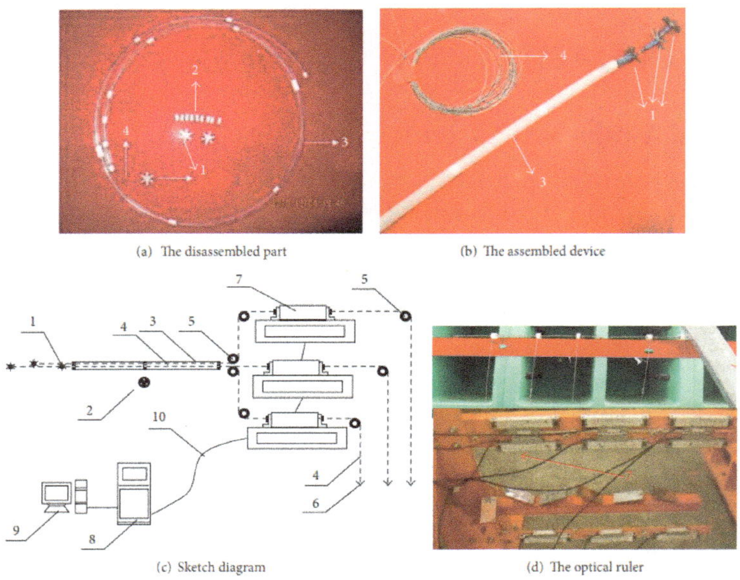

Figure 6: Grating ruler multipoints displacement measuring system (GRDS). (1) Fixed end, (2) guding frame, (3) soft pipe, (4) flexible steel wire, (5) displacement transfer roller, (6) balance weight, (7) optical ruler, (8) STS, (9) DAS, (10) cable.

The fiber bragg grating (FBG) was stuck to the analogical material blocks to make the grating strain sensor. And it can be used to measure the strain in each direction inside the model (Figure 7).

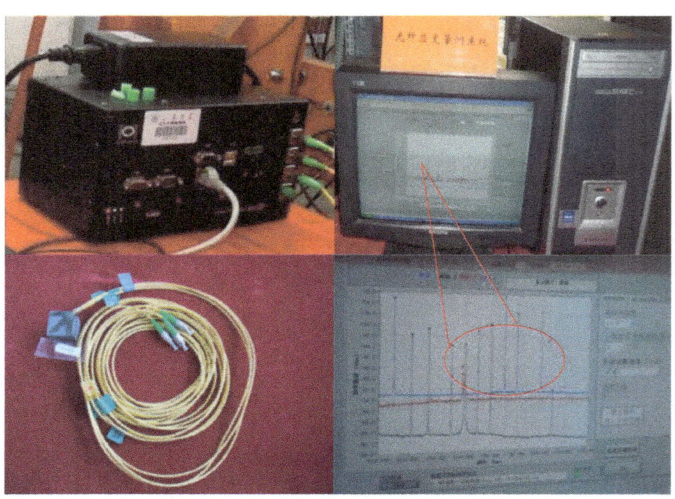

Figure 7: Grating strain sensor measuring system.

The FBG can measure the strain changing outside through the centre wavelength mobile. The relationship between the centre wavelength λ_B and the effective refraction index n_{eff} of grating and the period of grating Λ is

$$\lambda_B = 2n_{\text{eff}}\Lambda, \tag{2}$$

where λ_B is centre wavelength of the Bragg grating; Λ is the period of grating; n_{eff} is the effective refraction index.

When the fiber is stretched, Λ and n_{eff} will change, and then the centre wavelength λ_B will drift. The wavelength gets larger when the fiber is stretched, while it gets smaller when the fiber is compressed. The linear relationship will be satisfied:

$$\Delta\lambda_B = \lambda_B (1 - P_e)\, \varepsilon = K_e \varepsilon, \tag{3}$$

where, $\Delta\lambda_B$ is the variation of wavelength; P_e is the effective optical coefficient (0.22, generally); ε is the axial strain of fiber; K_e is the sensitivity of strain measuring.

When the grating generates strain in the stress condition, the pitch of the grating will change to $\Delta\Lambda$, which will make the wavelength change to $\Delta\lambda$, so the strain can be got:

$$\varepsilon = \frac{\Delta l}{l} = \frac{\Delta\lambda}{\lambda}. \tag{4}$$

CONSTRUCTION OF THE MODEL

To study the anchoring effect on zonal disintegration, a model was built and divided to two halves to make compassion: one half is anchored and the other half is nonanchoring.

Simulation Range of the Model

The model size is limited by the reasonable size of the steel frame. Within the frame, if the similarity constant for the geometry C_L is smaller than 1 : 50, then the model tunnel will be too small to be excavated. If the C_L is very large, then the relevant monitoring devices will be too difficult to install and the monitoring data will be inaccurate. Considering these prior factors, 100 is taken as the optimal similarity coefficient C_L.

The simulation range of the prototype is 30 m × 30 m × 30 m. According to the geometry similar scale 1/50, the dimension of the model is determined to 0.6 m × 0.6 m × 0.6 m and the model tunnel is 100 mm × 77.6 mm.

Flow of Model Construction

The model was made delaminating. Each layer is 10 cm high, so 7 times are needed in all to finish the entire model. The measuring components were set up when the material reached the designed height (see Figure 8).

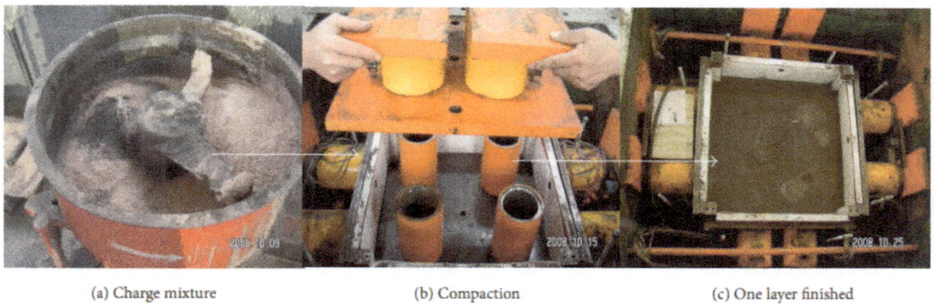

Figure 8: Procedure of the model building.

BURYING OF THE MEASURING COMPONENTS

The measuring components were disposed in the model, including the optical multipoint displacement meter and the grating Bragg optical strain sensor, which were used to monitor the displacement and strain changing in the surrounding rocks, respectively. Figures 9–10 show the section sketch and burying procedure of the measuring components, in separate.

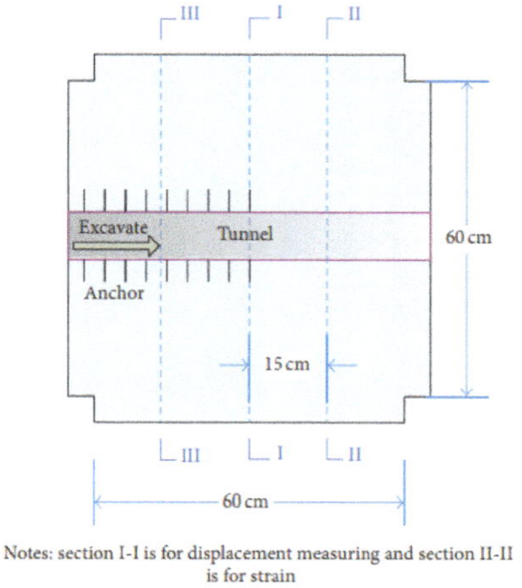

Figure 9: Layout sketch of monitoring sections in the model. The model is divided to two parts: anchored area and the area without anchoring. The measuring sections were set to 3, which divides the model to 4 parts evenly. The II measuring section was chosen to be the middle part

of the model, while the I and III sections are quarter and three-quarters of the model. Then, the measuring section can be used to monitor the deformation inside the model.

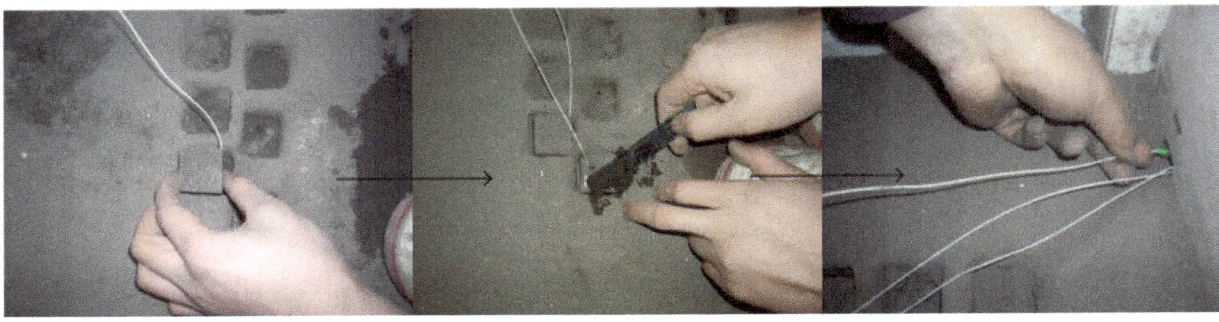

(a) Burying of optical Bragg fiber strain sensor

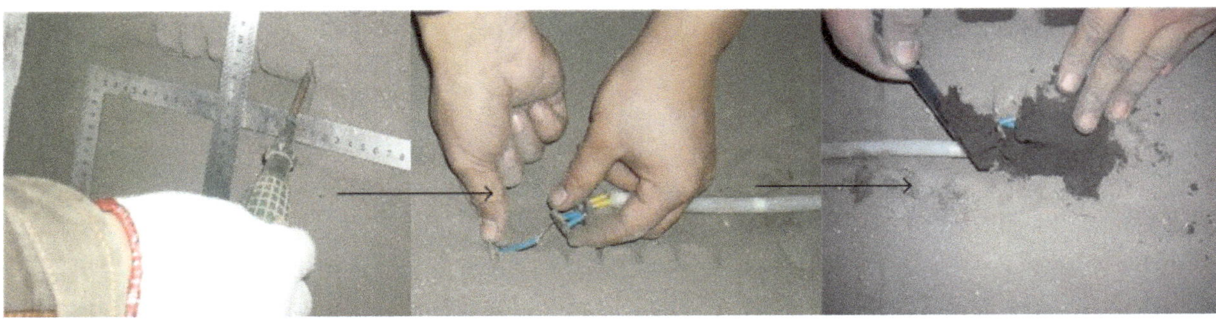

(b) Burying of GRDS

Figure 10: Burying procedure of the optical measurement sensor. The sensor is Fiber Optical Bragg grating sensor, where the precision is 10^{-3} με. The spacing between those sensors is 1 cm.

Distribution and Burying of the Anchor

The anchored area was applied with anchors, while the other half was not. Because the anchor surrounding the tunnel is too intensive, an equivalent principle of pulling resistance is adopted to dispose the anchor. A thick anchor is replaced by 4 thin anchors. Considering of the quincuncial disposal of anchor, it is not equal between the two adjacent sections. There are 8 anchors in section A and 7 anchors in section B, in separate. The two sections are arranged alternately (Figure 11). The grouting material is mixed by the high thickness alcohol solution of rosin, whose stickiness and fluidity are both well to meet the grouting demand.

(1) Arrangement of anchors. Figure 11 is the layout sketch of the equivalent anchor. As the figure shows, the equivalent anchor is arranged according to the quincunx. There are 8 anchors in section A and 7 anchors in section B. The inter-row spacings are both 1.6 m. The two sections are arranged alternately along the axial direction of tunnel.

(2) Burying of the anchors. The embedded method and grouting burying method are adopted together. The embedded method is adopted for the anchor of the middle of model. The grouting burying method is adopted for the entrance of the the tunnel.

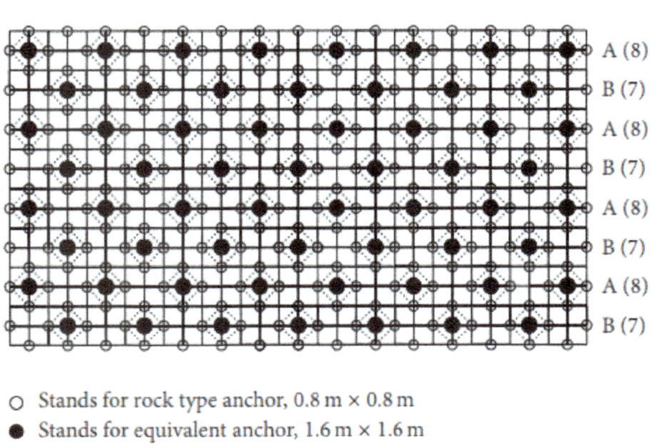

(a) Distribution sketch along the axial direction

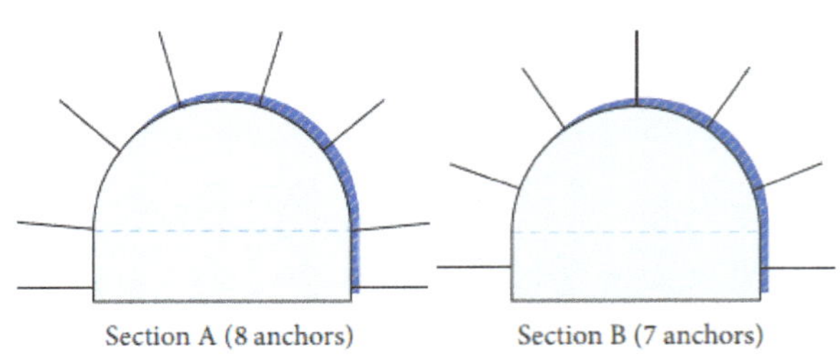

(b) Section sketch

Figure 11: Layout sketch of the equivalent anchor.

Figure 12 is the burying procedure of the embedded anchor, where (a) is for the side wall and (b) is for the arc crown. After burying, the material is backfilled and tamped, till the completion of the entire model.

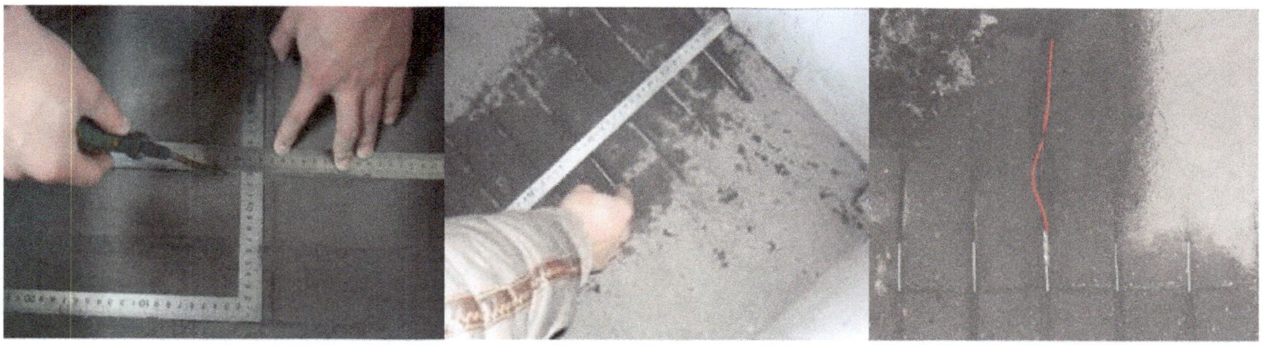

(a) Anchor in the side walls

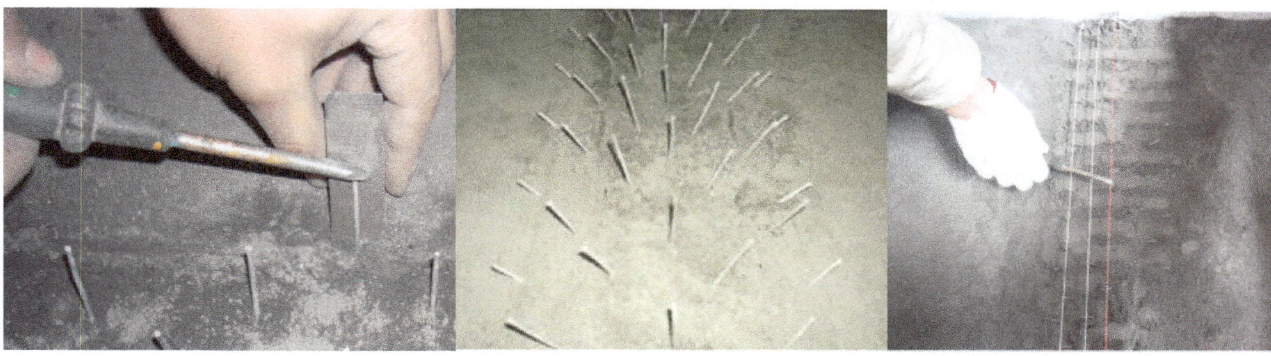

(b) Anchor in the arc crown

Figure 12: Burying procedure of analogical anchor.

The anchor burying of two sections near the cavern was carried during the procedure of tunnel excavation. The details are shown in Figures 14(b)–14(d).

Tunnel Excavation and Model Testing

The deep tunnel is in the 3D geostress state. In order to simulate the geostress fields accurately, the model was loading in true 3D state (Figure 13). During the excavation, the displacement and strain changing laws were monitored and recorded (Figure 13).

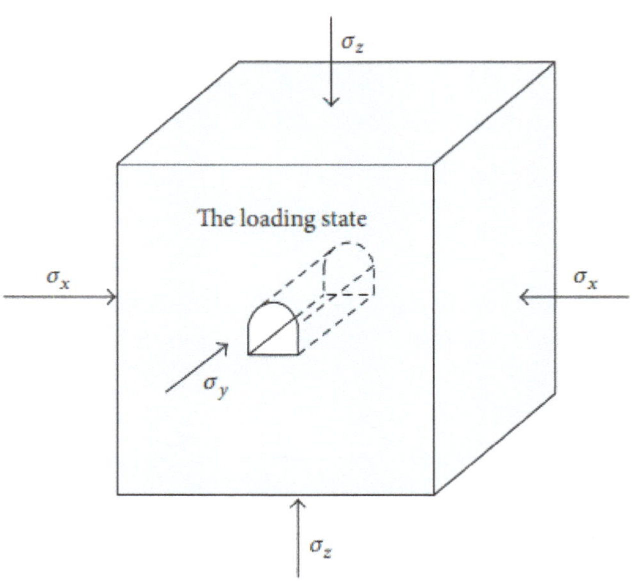

(a)

(b)

Figure 13: Loading state and test procedure of the model.

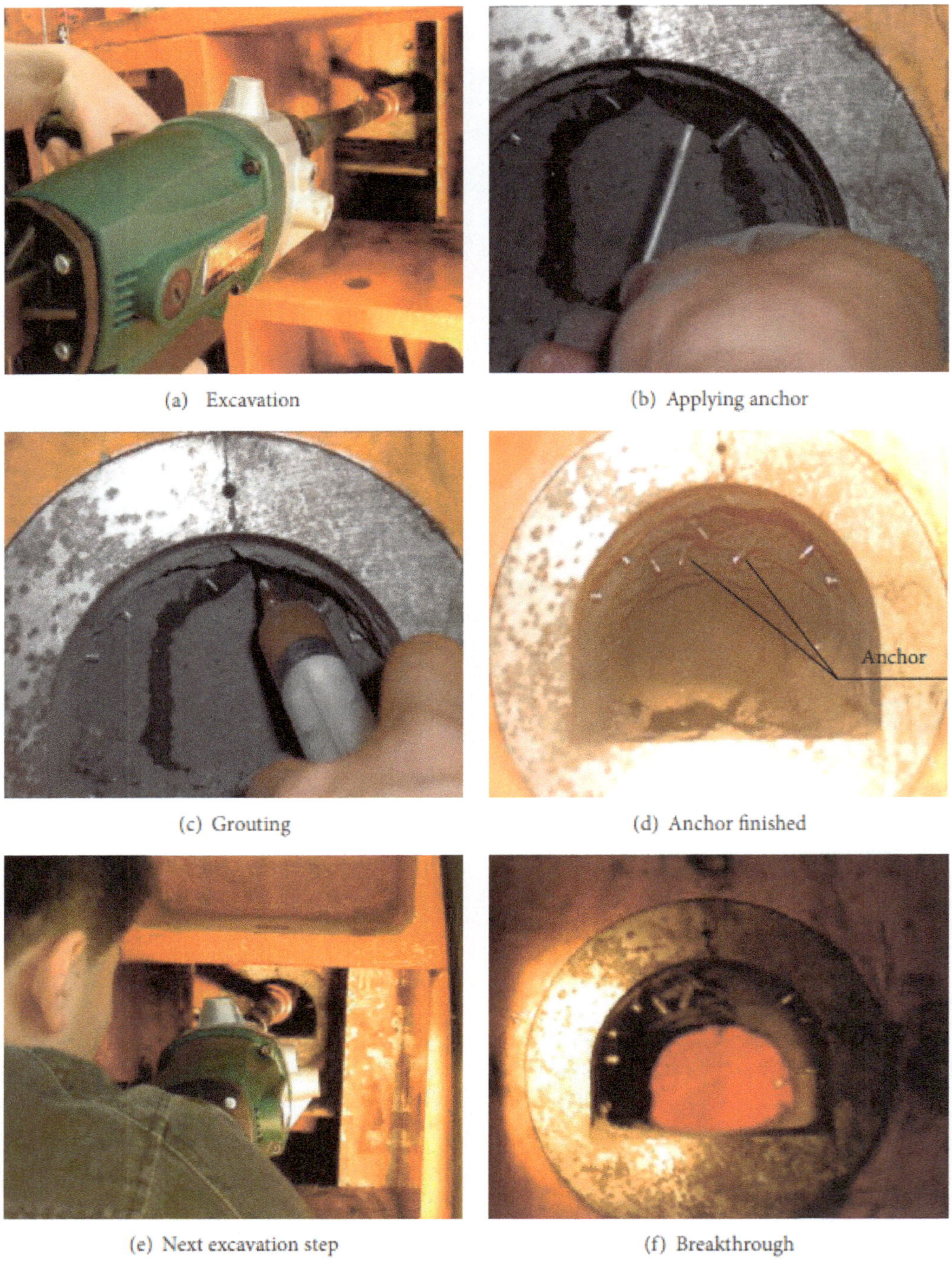

(a) Excavation (b) Applying anchor

(c) Grouting (d) Anchor finished

(e) Next excavation step (f) Breakthrough

Figure 14: Excavation procedure of model tunnels.

The self-weight stress is γh . The loading which is perpendicular to the tunnel axial is $1.5\gamma h$ (coefficient of horizontal pressure is 1.5). Where γ is the volume weight, h is the embedded depth of the tunnel. The loading value is listed in Table 4.

Table 4: The loading value of the model.

Loading direction	Self-weight stress	Perpendicular to the axial	Parallel to the axial
Loading value/Mpa	0.5	0.75	0.75
Loading direction	Self-weight stress	Perpendicular to the axial	Parallel to the axial

The excavation procedure of the model is shown as follows: (Figure 14).

RESULTS OF MODEL TEST AND DISCUSSION

Displacement from the Grating Extensometer

Figure 15 shows the drawing sketch of grating extensometer. The displacement of surrounding rocks is labeled and connected using the smooth curve after excavation. Then, the changing laws of displacement can be got. The displacement changing sketch is placed together to compare. It is known from the sketch of displacement surrounding the tunnel. (1)The displacement shows a very different changing law between the anchored model and nonanchoring model. In the anchored model, the displacement decreases monotonously as the distance to the tunnel wall increases. It is similar to the shallow embedded tunnel. While in the nonanchoring model, the displacement presents the undulate changing status, wherein the wave crest and the trough are arranged alternately. It completely differs from the shallow embedded tunnel. (2)Compared with the dissembled model, the area of larger displacement is the severely damaged region. In the anchored model, the displacements of no. 1 spot near the periphery are larger than the other area. It indicates that this area is the damage zone in the traditional perception. This is in accordance with the phenomenon of periphery damage seriously during the excavation. The displacement in arc crown is larger than the side walls. This is corresponding to the phenomenon of serious damage and collapse in arc crown. In the model without anchoring, the wave crest region with a larger displacement is the fractured zone, whereas the trough region with a smaller displacement is the intact zone. The additional displacement in the wave crest area, which is caused by the circular fracture, increases the total displacement.

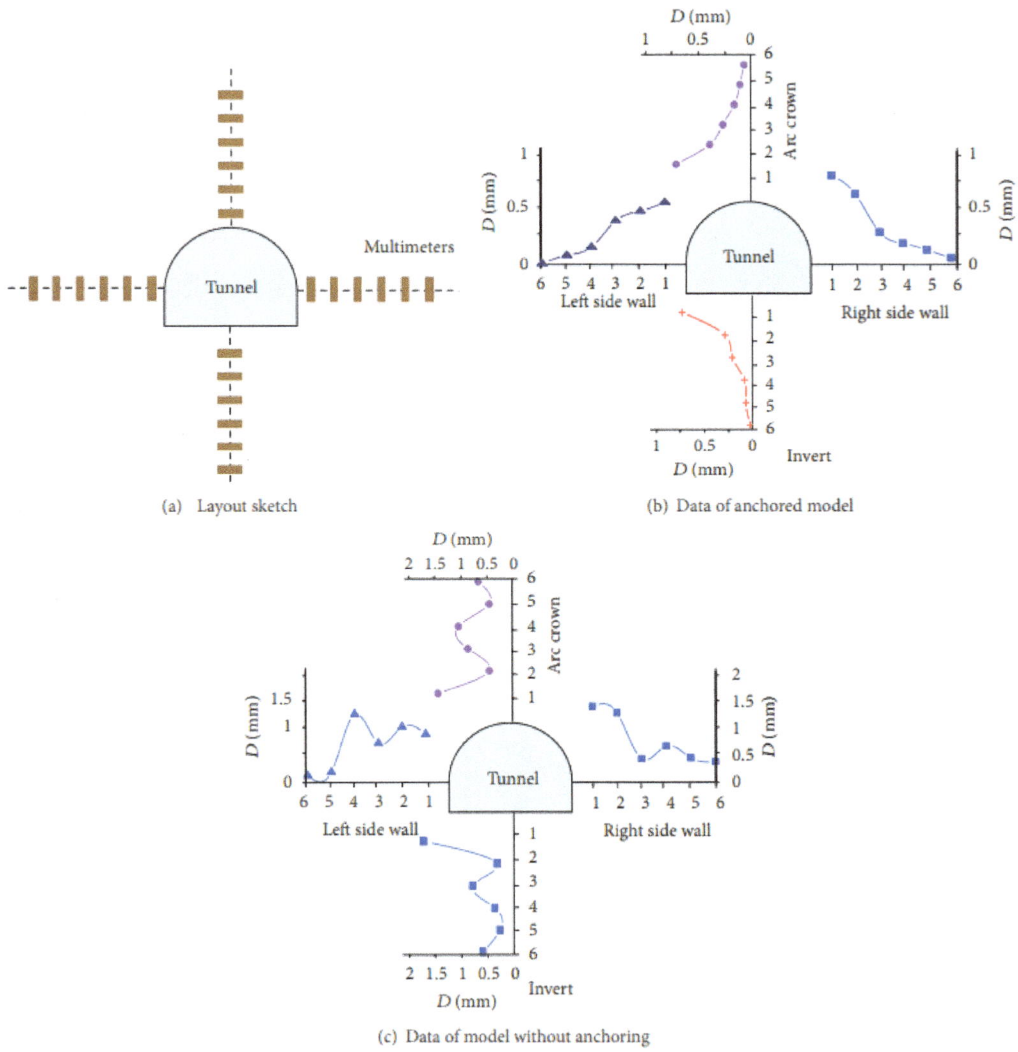

(a) Layout sketch

(b) Data of anchored model

(c) Data of model without anchoring

Figure 15: Displacement surrounding the model tunnel.

Strains from FBG Strain Sensor

The strain results show that both the radial and tangential strains were negative before the excavation. This indicates that the surrounding rocks were in compressive state. When it is excavated to the strain monitoring section, the radial strain ε_r turns to positive, which indicates that the tensile strain appear in the radial direction. The phenomenon was in accordance with the elastic-plastic strain field analysis of the ideal circular tunnel.

Figure 16 is the diagram sketch of FBG strain sensor. The radial strain is labeled in the surrounding rocks nearby the tunnel after the tunnel excavation and measurement.

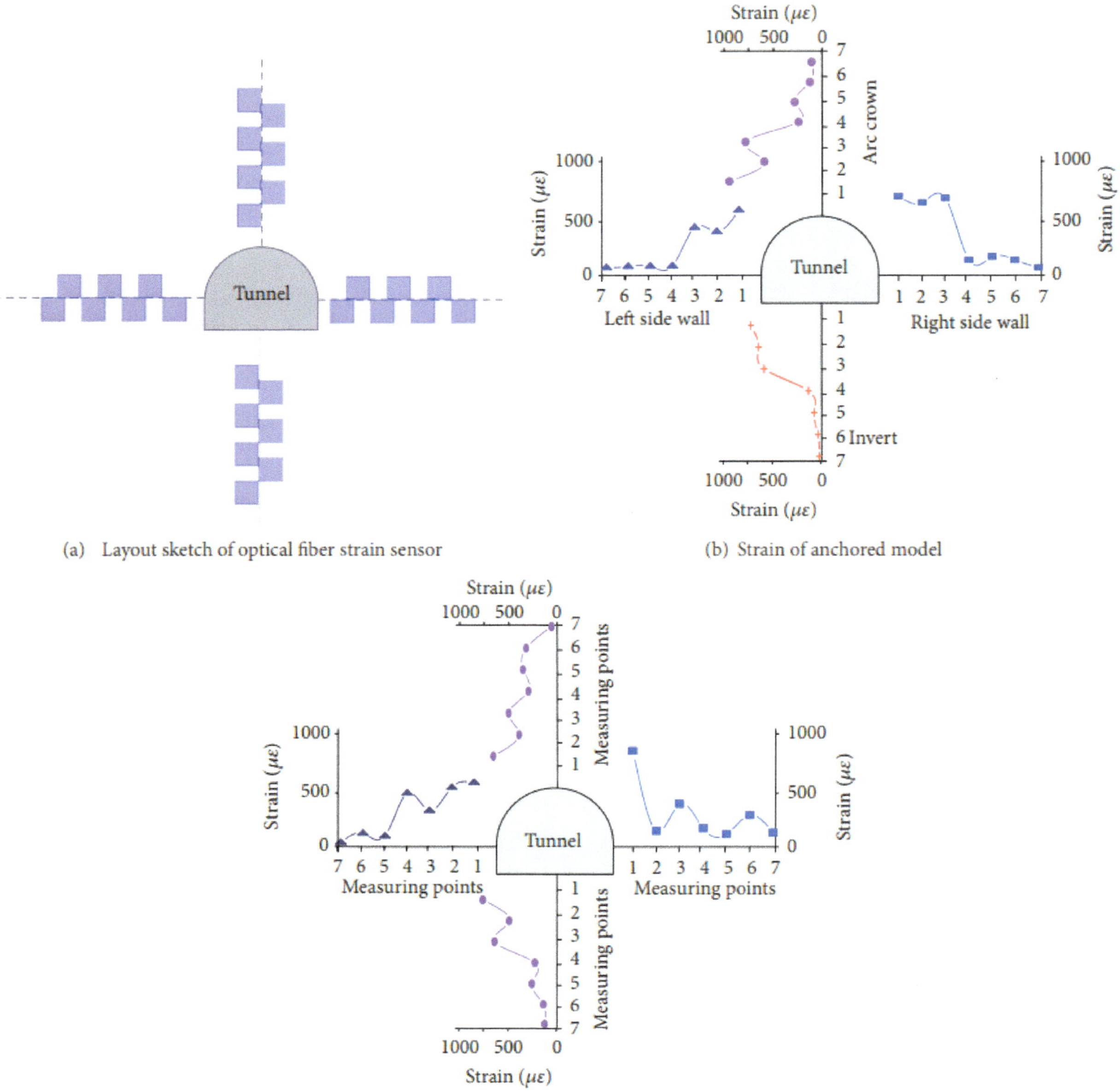

(a) Layout sketch of optical fiber strain sensor

(b) Strain of anchored model

(c) Strain of model without anchoring

Figure 16: Radial strain around tunnel measured by optical fiber strain sensor.

As is shown in Figure 16, the radial strain in the surrounding rocks both in the anchored model and nonanchoring model shows the same changing law: it presents the fluctuate distribution with the distance increasing from the periphery. The wave crest and the trough distribute alternately. This law is totally different from that in the shallow embedded tunnel. The phenomenon indicates that under the high geostress loading, the radial tensile strain is the highest in a certain area of the surrounding rocks. This area is almost circular and concentrated in the tunnel, and identified as the elastoplastic zone of the surrounding rock.

Zonal disintegration phenomenon does not occur in the side walls and arc crowns of anchored model. It is because the reinforcement effect improves the capacity of fracture-resistance of surrounding rocks. There is no reinforcing effect in the nonanchoring area, such as the invert of anchored half model and the half model without anchoring. Zonal disintegration phenomenon occurs in these areas.

Anchor Strain in the Middle Section of the Model

Figure 17 is the changing laws of anchor strain in the middle section of the model measured by the strain gages.(1)The anchor strain is minus at the beginning of tunnel excavation. It indicates that the anchors are in compressive state in high loading condition. It turns to positive at the excavation of step 6 which is near the middle of the model (there are 14 excavation steps in all). It indicates that anchors turn into tensile state and work.(2)Most anchors (except nos. 1, 6, and 11) turn to maximum state at excavation of 7 or 8. This is because the anchor works at the largest effect when excavated in the middle of the model, which is the section of anchor strain measurement.(3)A mount of anchors (nos. 1, 4, 5, 8, 10, 13, and 14) shows the phenomenon of positive and negative alternation during the excavation process. It indicates the tension and compression alternating of anchor, which is first observed in the geomechanical model test.

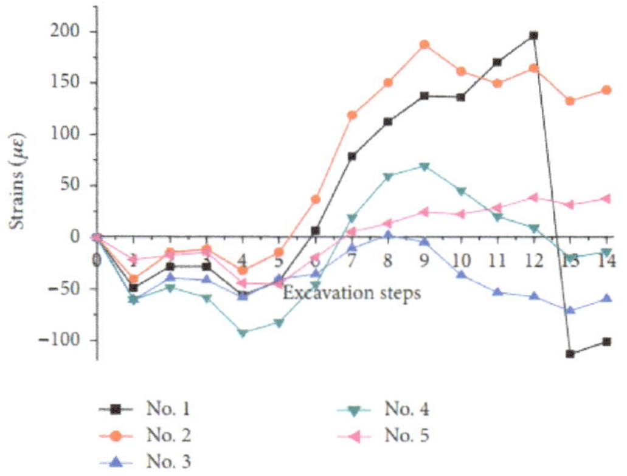

(a) Nos. 1–5 anchor strain changing with excavation steps

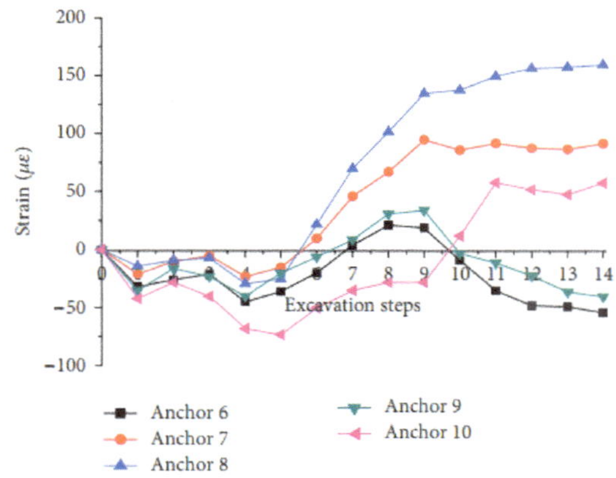

(b) Nos. 6–10 anchor strain changing with excavation steps

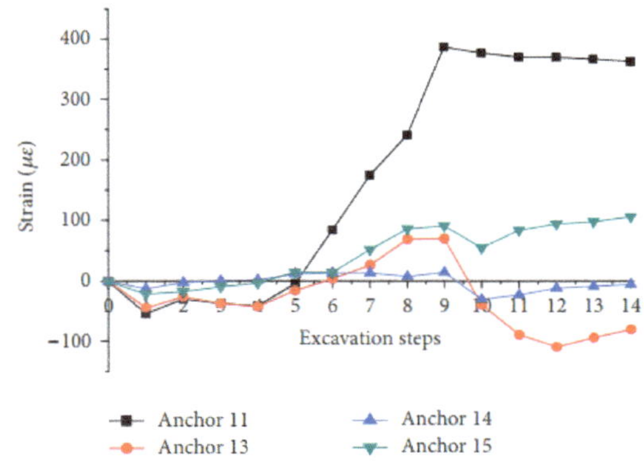

(c) Nos. 11–15 anchor strain changing with excavation steps

Figure 17: Anchor strain changing with the excavation steps. Note: No. 12 anchor was damaged during the test and the data could not be recorded.

Fang [32] used to observe the similar special phenomenon of tension and compression alternating in the deep tunnel of Jinchuan nickel mine, 1150 m deep. He believes that it is the self-organized phenomenon and the inherent characteristic of deep tunnel.

Fracture Shape inside the Model

Figure 18 is the photos of dissembled model which show the damage pattern.

(a) The non-anchoring half model

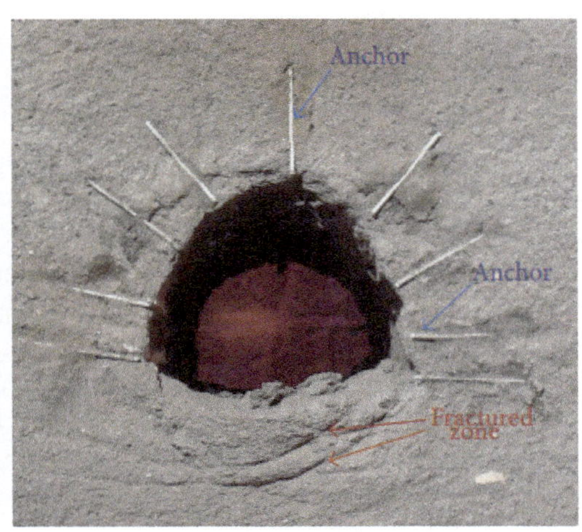

(b) The anchored half model

Figure 18: Failure distribution of the surrounding rock mass after tunnel excavation.

The zonal disintegration phenomenon of alternating fractured zone and intact zone occurs in nonanchoring model. There are 4 fractured zones and 4 intact zones arranging in space which are in accordance with the fractured shape of the in-situ observation in DINGJI coal mine. There are 2 obvious fracture lines in the bottom of the anchored model. Zonal disintegration also occurs in the invert of the model. But there is not any fracture line occurs in the arc crown and side walls; that is, zonal disintegration phenomenon did not occur in the anchored part of model.

After comparing with the model of anchored and nonanchoring, it indicates that the anchoring effect reinforces the anchoring area of model to be a unity. This impact makes zonal disintegration difficult to occur.

Analysis of the Anchoring Effect on Zonal Disintegration

Under high geostress conditions, the surrounding rocks near the cavern wall yield to plasticity. The principal stress field in the plastic and the elastic zones is as follows (Figure 19). As is shown, the tangential stress is the maximum principal stress. At the location of $r=R_p$, the tangential stress is the summit value.

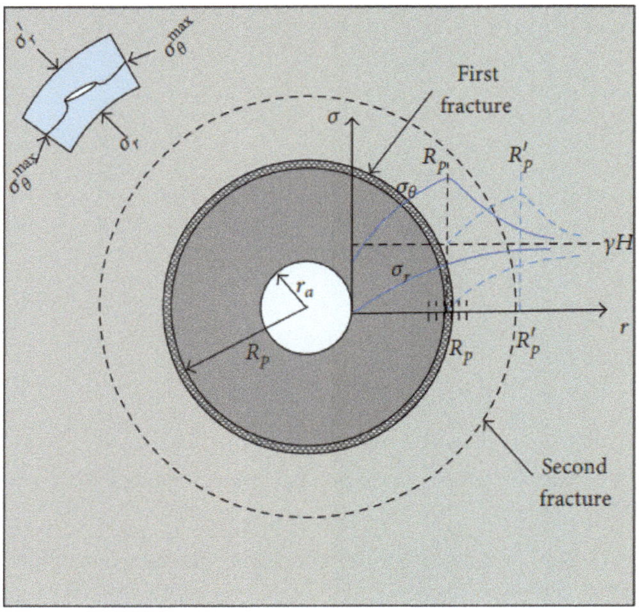

Figure 19: Elastoplastic geostress fields in surrounding rock mass and forming process of zonal disintegration.

According to Griffith's criterion, a fracture occurs when the UCS of rocks under pressure reaches the threshold value. Fairhurst and Cook (1966) [33] indicated that microcracks would initiate and extend in the direction of the maximum principal stress when compressive stress reaches Griffith's strength σ_s. This finding explains the longitudinal splitting of the rock specimen and the slabbing of the surrounding rocks in the rectangular openings, which also holds true for the circular cavern. Given a unit rock in the system of the polar coordinates (Figure 19), the second circular fracture extends towards the direction of the maximum principal stress (i.e., the tangential direction) when stress fulfills its relationship with the mechanical parameters of the surrounding rocks. The fracture is expected to transfix and form the circular fracture (i.e., the first fracture of the zonal disintegration) when compressive loading is sufficiently large. Fracture formation causes geostress redistribution in the surrounding rocks, inducing the formation of other ultimate equilibrium plastic zones. The second circular fracture occurs when the summit tangential stress is sufficiently large. Zonal disintegration occurs during the process cycle (Figure 19).

In the condition of high axial geostress, the surrounding rock mass has the trend of zonal disintegration. When model is anchored, it is reinforced and the threshold value of fracture is enlarged. So, at the same geostress, the zonal disintegration did not occur. The model test indicates the trend of zonal disintegration phenomenon existing under a condition of high axial geostress. And the reinforcement of anchor suppresses the zonal disintegration in the anchored area. During the anchor working, the trend of zonal disintegration which induces it shows the tension and compression alternation.

CONCLUSION

Under high axial geostress, zonal disintegration phenomena occur in the nonanchored area, while it did not occur in the fully anchored area. It indicates that anchoring suppresses the zonal disintegration obviously.(2)The radial strain of surrounding rocks displays the fluctuation state where wave crest and rough arrange in interval, both in anchored and nonanchoring area. It indicates that the deep rock masses have the trend of zonal disintegration under the high axial geostress. (3)The tension and compression alternation phenomenon of anchor is observed during the tunnel excavation procedure. This special phenomenon is different from the shallow buried tunnel. It indicates that the trend of zonal disintegration exists in the deep tunnel. The reinforcement of anchor suppresses the occurrence of zonal disintegration. During the work of anchor, it shows the alternation of tension and compression. It indicates that the mechanical behavior of the surrounding rocks is nonmonotonic.

The anchor suppresses the growth of zonal disintegration, so the optimization of parameters and disposal of anchor are of great importance to the stability of deep tunnel. The further study is needed to conduct in the future.

ACKNOWLEDGMENTS

This work was supported by the National Natural Science Foundations of China (Grant nos. 51209074 and 41172268), China Postdoctoral Science Foundation (Grant nos. 2012M511189and2013T60494),andtheFundamentalResearch Funds for the Central Universities (Grant no. 2012B02714), supported by State Key Laboratory for Geomechanics and Deep Underground Engineering, China University of Mining & Technology, under Grant SKLGDUEK1206, the Open Research Fund of State Key Laboratory of Geomechanics and Geotechnical Engineering, and the Institute of Rock and Soil Mechanics, Chinese Academy of Sciences, under Grant no. Z012008, and funded by CRSRI Open Research Program CKWV2012306/KY, by the Key laboratory of coal-based CO_2 capture, and geological storage Open Research Program 2012KF08. The authors are deeply grateful for these supports.

REFERENCES

1. E. I. Shemyakin, G. L. Fisenko, M. V. Kurlenya et al., "Zonal disintegration of rocks around underground workings, part II: rock fracture simulated in equivalent materials," Soviet Mining Science, vol. 22, no. 4, pp. 223–232, 1986.
2. Q. H. Qian, "The characteristic scientific phenomena of engineering response to deep rock mass and the implication of deepness," Journal of East China Institute of Technology, vol. 27, no. 1, pp. 1–5, 2004.
3. G. R. Adams and A. J. Jager, "Etroscopic observations of rock fracturing ahead of the stope faces in deep-

level gold mines," Journal of the South Africa Institute of Mining and Metallurgy, vol. 21, no. 2, pp. 115–127, 1980.

4. E. I. Shemyakin, G. L. Fisenko, M. V. Kurlenya et al., "Zonal disintegration of rocks around underground workings, part 1: data of in situ observations," Soviet Mining Science, vol. 22, no. 3, pp. 157–168, 1986. · ·
5. E. I. Shemyakin, G. L. Fisenko, M. V. Kurlenya et al., "Zonal disintegration of rocks around underground mines, part III: theoretical concepts," Soviet Mining Science, vol. 23, no. 1, pp. 1–6, 1987.
6. E. I. Shemyakin, M. V. Kurlenya, V. N. Oparin et al., "Zonal disintegration of rocks around underground workings. IV. Practical applications," Soviet Mining Science, vol. 25, no. 4, pp. 297–302, 1989.
7. E. J. Sellers and P. Klerck, "Modelling of the effect of discontinuities on the extent of the fracture zone surrounding deep tunnels," Tunnelling and Underground Space Technology, vol. 15, no. 4, pp. 463–469, 2000.
8. D. F. Malan and S. M. Spottiswoode, "Time-dependent fracture zone behavior and seismicity surrounding deep level stopping operations," in Rockbursts and Seismicity in Mines, S. J. Gibowicz and S. Lasocki, Eds., pp. 173–177, A. A. Balkema, Rotterdam, The Netherlands, 1997.
9. X. Zhou, Q. Qian, and B. Zhang, "Zonal disintegration mechanism of deep crack-weakened rock masses under dynamic unloading," Acta Mechanica Solida Sinica, vol. 22, no. 3, pp. 240–250, 2009.
10. J. Gu, L. Gu, A. Chen, J. Xu, and W. Chen, "Model test study on mechanism of layered fracture within surrounding rock of tunnels in deep stratum," Chinese Journal of Rock Mechanics and Engineering, vol. 27, no. 3, pp. 433–438, 2008. ·
11. Y. Pan, Y. Li, X. Tang, and Z. Zhang, "Study on zonal desintegration of rock," Chinese Journal of Rock Mechanics and Engineering, vol. 26, supplement 1, pp. 3335–3341, 2007. ·
12. X. Tang, Y. S. Pan, and M. T. Zhang, "Mechanism analysis of zonal disintegration in deep level tunnel,"Journal of Geological Hazards and Environment Preservation, vol. 17, no. 4, pp. 80–84, 2006.
13. L. S. Metlov, A. F. Morozov, and M. P. Zborshchik, "Physical foundations of mechanism of zonal rock failure in the vicinity of mine working," Journal of Mining Science, vol. 38, no. 2, pp. 150–155, 2002.
14. S. Li, Q. Qian, D. Zhang, and S. Li, "Analysis of dynamic and fractured phenomena for excavation process of deep tunnel," Chinese Journal of Rock Mechanics and Engineering, vol. 28, no. 10, pp. 2104–2112, 2009. ·
15. M. A. Guzev and A. A. Paroshin, "Non-Euclidean model of the zonal disintegration of rocks around an underground working," Journal of Applied Mechanics and Technical Physics, vol. 42, no. 1, pp. 131–139, 2001.
16. M. Wang, H. Song, D. Zheng, and S. Chen, "On mechanism of zonal disintegration within rock mass around deep tunnel and definition of 'deep rock engineering'," Chinese Journal of Rock Mechanics and Engineering, vol. 25, no. 9, pp. 1771–1776, 2006. ·
17. M.-Y. Wang, P.-X. Fan, and Z. K. Guo, "Elastoplastic model for discontinuous shear deformation of deep rock mass," Journal of Central South University of Technology, vol. 18, no. 3, pp. 866–873, 2011.
18. Y. He, B. Jiang, L. Han, P. Shao, and H. Zhang, "Study of intermittent zonal fracturing of surrounding rock in deep roadways," Journal of China University of Mining and Technology, vol. 37, no. 3, pp. 300–304, 2008. ·
19. X. Zhou and Q. Qian, "Zonal fracturing mechanism in deep tunnel," Chinese Journal of Rock Mechanics and Engineering, vol. 26, no. 5, pp. 877–885, 2007. ·
20. X. Zhou, Q. Qian, and H. Yang, "Effect of loading rate on fracture characteristics of rock," Journal of Central South University of Technology, vol. 17, no. 1, pp. 150–155, 2010.
21. X. P. Zhou, H. F. Song, and Q. H. Qian, "Zonal disintegration of deep crack-weakened rock masses: a non-Euclidean model," Theoretical and Applied Fracture Mechanics, vol. 55, no. 3, pp. 227–236, 2011. · ·
22. Q. Qian, X. Zhou, and E. Xia, "Effects of the axial in situ stresses on the zonal disintegration phenomenon in the surrounding rock masses around a deep circular tunnel," Journal of Mining Science, vol. 48, no. 2, pp. 276–285, 2012.
23. X. P. Zhou, G. Chen, and Q. H. Qian, "Zonal disintegration mechanism of cross-anisotropic rock masses around a deep circular tunnel," Theoretical and Applied Fracture Mechanics, vol. 57, no. 1, pp. 49–54, 2012.
24. X. P. Zhou and J. Bi, "Zonal disintegration mechanism of cross-anisotropic rock mass around a deep circular tunnel under dynamic unloading," Theoretical and Applied Fracture Mechanics, vol. 60, no. 1, pp. 15–22, 2012.
25. V. N. Reva and E. A. Tropp, "Elastoplastic model of the zonal disintegration of the neighborhood of an underground working," in Physics and Mechanics of Rock Fracture as Applied to Prediction of Dynamic Phenomena (Collected Scientific Papers), pp. 125–130, Mine Surveying Institute, Saint Petersburg, Russia, 1995.
26. Y. L. Tan, J. G. Ning, and H. T. Li, "In situ explorations on zonal disintegration of roof strata in deep coalmines," International Journal of Rock Mechanics and Mining Sciences, vol. 49, pp. 113–124, 2012.
27. H. Wu, Z. Guo, Q. Fang, Y. Zhang, and J. Liu, "Mechanism of zonal disintegration phenomenon in enclosing rock mass around deep tunnels," Journal of Central South University of Technology, vol. 16, no. 2, pp. 303–311, 2009.
28. V. N. Odintsev, "Mechanism of the zonal disintegration of a rock mass in the vicinity of deep-level workings," Journal of Mining Science, vol. 30, no. 4, pp. 334–343, 1994.
29. B. V. Laptev and R. P. Potekhin, "Burst triggering by zonal disintegration of evaporites," Soviet Mining Science, vol. 24, no. 3, pp. 238–241, 1988.
30. E. Fumagalli, Statical and Geomechanical Models, Springer, New York, NY, USA, 1973.

31. E. Fumagalli, "Geomechanical models of dam foundation," in Proceedings of the International Colloquium on Physical and Geomechanical Models, Bergamo, Italy, March 1979.

32. Z. Fang, "Support principles for roadway in soft rock and its controlling measures," in Soft Rock Tunnel Support in Chinese Mines: Theory and Practice, H. E. Manchao, Ed., pp. 64–70, Coal Industry Publishing Press, Beijing, China, 1996 (Chinese).

33. C. Fairhurst and N. G. W. Cook, "The phenomenon of rock splitting parallel to the direction of maximum compression in the neighborhood of a surface," in Proceedings of the 1st Congress of the International Society for Rock Mechanics, pp. 687–692, Lisbon, Portugal, September 1966.

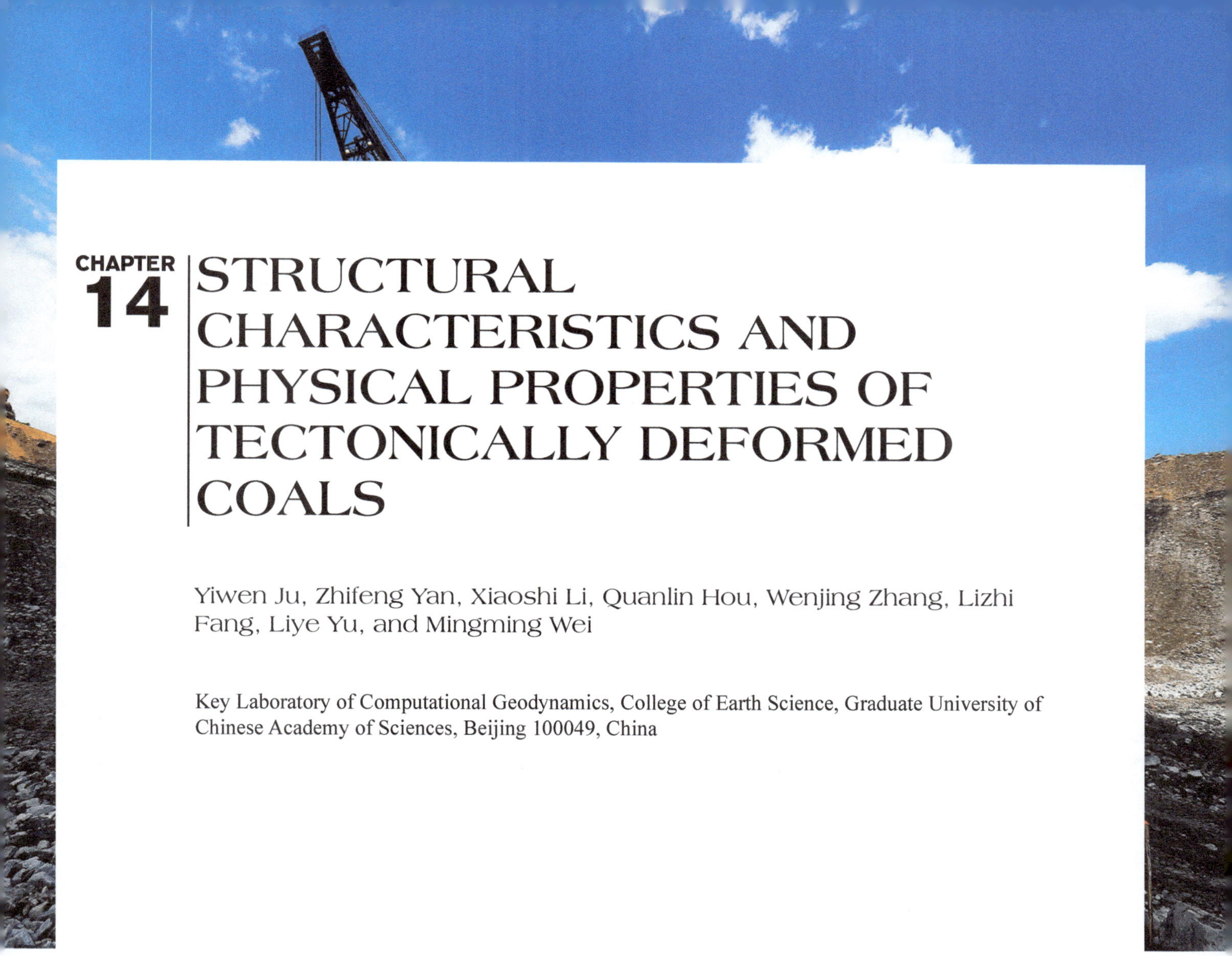

CHAPTER 14

STRUCTURAL CHARACTERISTICS AND PHYSICAL PROPERTIES OF TECTONICALLY DEFORMED COALS

Yiwen Ju, Zhifeng Yan, Xiaoshi Li, Quanlin Hou, Wenjing Zhang, Lizhi Fang, Liye Yu, and Mingming Wei

Key Laboratory of Computational Geodynamics, College of Earth Science, Graduate University of Chinese Academy of Sciences, Beijing 100049, China

ABSTRACT

Different mechanisms of deformation could make different influence on inner structure and physical properties of tectonically deformed coal (TDC) reservoirs. This paper discusses the relationship between macromolecular structure and physical properties of the Huaibei-Huainan coal mine areas in southern North China. The macromolecular structure and pore characteristics are systematically investigated by using techniques such as X-ray diffraction (XRD), high-resolution transmission electron microscopy (HRTEM), electron paramagnetic resonance (EPR), nuclear magnetic resonance (NMR), and low-temperature nitrogen adsorption method. The results suggest that under the directional stress, basic structural units (BSU) arrangement is closer, and the orientation becomes stronger from brittle deformed coal to ductile deformed coal. Structural deformation directly influences the macromolecular structure of coal, which results in changes of pore structure. The nanoscale pores of the cataclastic coal structure caused by the brittle deformation are mainly mesopores, and the proportion of mesopores volume in ductile deformed coal diminishes rapidly. So the exploration and development potential of coalbed gas are good in reservoirs such as schistose structure coal, mortar structure coal and cataclastic structure coal. It also holds promise for a certain degree of brittle deformation and wrinkle structure coal of low ductile deformation or later superimposed by brittle deformation.

INTRODUCTION

^{13}C Tectonically deformed coal (TDC) is a kind of coal in which, under mono- or multiphase tectonic stress fields, its primary texture and structure is significantly destroyed. Tectonic deformation could influence coal macromolecular or deform structure and enhance coalification to a certain degree [1–6]. Through structural shearing, graphitization can be facilitated [7, 8]. In fact, tectonic deformation can not only further alter coal's molecular structure at different degrees but also change the physical properties of reservoirs (such as porosity and permeability). Based on measurement of mercury intrusion, researchers compared micropore features of TDCs collected from different coalfields [9–12]. These experiments have indicated that the pore structure of TDCs showed greater medium pores and mesopore volumes than normal undeformed coals, but there was rare difference in micropores between

Citation: Yiwen Ju, Zhifeng Yan, Xiaoshi Li, et al., "Structural Characteristics and Physical Properties of Tectonically Deformed Coals," Journal of Geological Research, vol. 2012, Article ID 852945, 14 pages, 2012. doi:10.1155/2012/852945.

primary structure coals and TDCs. This indicated that structural deformation did not obviously influent micropore structure (less than 10 nm in diameter). Meanwhile, the nanoscale pore structure of TDCs is the main adsorption space for coalbed gas, and the macromolecular structure is an important factor for restricting the adsorption capacity in coal. Therefore, the problem of macromolecular structure and nanoscale pores in TDCs are worthy research topics [13–15], and the behavior of gas adsorption/desorption and permeability of micro-pore in coal have strong relationship with the occurrence and distribution of coalbed gas [16–24].

The deformation of coals causes changes of structure and chemical composition of coal because of differences in the property, mode, and intensity of stress and the deformation environment. Moreover, the physical properties of TDCs also vary significantly [5, 22]. Therefore, comprehensive studies on stress action, strain environment, physical and chemical structures of TDCs, and physical properties of deformed coals will provide the theoretical basis for exploration and evaluation of coalbed gas resources and aid in understanding the prevention and control of coal and gas outbursts. These studies will also provide insight into gas emission, which is studied in the coalbed gas research field. In this study, the author would like to reveal the structures and the physical properties of tectonically deformed and metamorphic coals.

GEOLOGIC BACKGROUND AND COAL STRUCTURE IN HUAINAN AND HUAIBEI MINE AREAS

The Paleozoic coal-bearing basins, especially in eastern China, underwent complexly multiphase structural evolution by compressing, shearing, and extending. This resulted in structural reworking of the primary structure of coal at different degrees, forming the different types of TDCs. The existing data from China have shown that tectonic deformation is an important factor in restricting the exploration and development of coalbed gas; it is also a direct cause of gas outbursts of coal mines.

In the Huaibei and Huainan coalfields, situated in the southern North China (Figure 1), coal measurement revealed Permo-Carboniferous systems which were formed in the Paleozoic and structurally influenced in the Mesozoic. The main coal-bearing areas were found in fault depression basins, especially in the syncline. The upwelling areas between depression basins have undergone destruction and denudation at different degrees. The different types of faults and folds in the distribution area of coal have given rise to obvious breakages and deformations in the coal seams, leading to changes of the coal structure. At the end of this process, different types of TDCs, such as cataclastic structure coal, mortar structure coal, schistose structure coal, mealy structure coal, and even wrinkle structure coal and mylonitic structure coal, were formed [5].

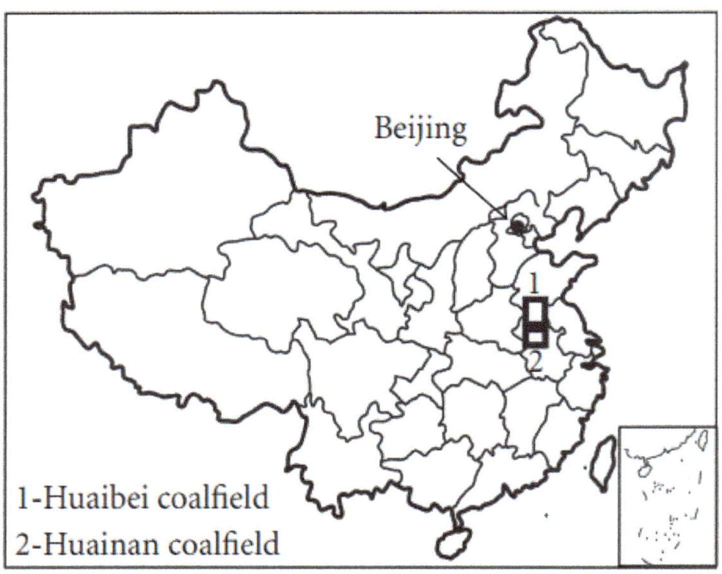

Figure 1: Huaibei-Huainan coalfield in the southern North China.

THE SAMPLE AND METHODS

The research areas possess six mine wells of the Suzhou mine area and the Linhuan mine area in the Huaibei coal mine region. Three of them are contained by the Xinji mine region, Zhangji mine region and Panji mine region, which represent different tectonic units. The sample sets were collected from the Huaibei and Huainan coalfields (Figure 1). These samples belong to different kinds of TDCs, suffering from multiphase tectonic deformation and magmatic events.

In this study, two significant aspects were taken into account in the selection of the samples. The first aspect is the evolutional degree of coal, namely, reflectance ($R_{o,max}$) of samples ranging from 0.76% to 3.59%, including low, middle, and high metamorphic coal. The second aspect is the characteristics of TDCs formed by different mechanisms. The samples of brittle deformation series, ductile deformation series, and brittle-ductile series were selected separately.

Because of different sedimentation and tectonic environments, coal often contains different maceral compositions and various kinds of mineral components, which greatly affect the results of coal structural testing. Therefore, coal samples must be prepared before testing and analysis to extract vitrinite components and take off inorganic (mineral) components.

On the basis of in situ investigation, detailed microstructural observations of these samples using optical and scanning electron microscopy (SEM) were described elsewhere. $R_{o,max}$ measurements were carried out on polished sections under halogen lamp and oil immersion conditions by using a Leitz Orthoplan/MPV-SP microscope photometer.

Using X-ray diffraction (XRD), high-resolution transmission electron microscopy (HRTEM), electron paramagnetic resonance (EPR), and nuclear magnetic resonance (NMR), systematic research has been conducted on the characteristics of macromolecular structure and chemical structure of different kinds of TDCs.

The instrument of X-ray diffraction testing is a D/Max-IIIB X-ray diffractometer which is made by Rigaku Inc. Of Japan; the instrument of the HRTEM is a JEM-2010 HRTEM made in Japan by JEOL Inc.; the instrument of EPR testing is the E-109 electron paramagnetic resonance spectrometer made by Varian Inc. of America; the instrument of NMR testing is the Infinity 400 nuclear magnetic resonance spectrometer made by Varian Inc. of America. The experimental conditions of different instruments are discussed in detail by Ju et al. [5, 6].

The pore characteristics of TDCs have been determined by use of the low-temperature nitrogen adsorption method and scan electron microscopy (SEM). The permeability has been analyzed through experiments of in-lab permeability measurements of large diameter, in-site samples from TDCs.

The instrument used for low-temperature liquid nitrogen adsorption experiments is the ASAP-2010 made in America by Micromeritics Instrument Inc. This instrument is used for measuring the micropore specific surface area and distribution of pore diameter. In the gas-water phase permeability experiment, the whole core flow system, made by Terra Tek company, was utilized as the main instrument. This system was primarily used to simulate crustal pressure and rock permeability under natural oil reservoir pressure. The highest simulated confined pressure and fluid pressure were 70 MPa and 65 MPa, respectively. The measurement system was composed of the pressure system, the constant temperature system, the control system, the core clamper and separator equipment.

RESULTS

Forming Environments and Structure-Genetic Classification of Tectonically Deformed Coals

Forming Environments of Tectonically Deformed Coals

The Huaibei and Huainan coal mine areas underwent complexly multiphase structural and magma-thermal activities. Thus, under different metamorphic and deformed environments, various kinds of TDCs were formed in these mine areas [11, 20–22]. By combining deformational characteristics of macrocosms and microcosms of different kinds of TDCs, we recognized three kinds of metamorphic and deformed environments where TDCs were formed in the Huaibei and Huainan coal mine areas. The first type is the metamorphic and deformed environment of the low-rank coals ($R_{o,max}$ lower than 1.30%), which is based on the plutonic metamorphism; the second type is the metamorphic and deformed environments of the middle-rank coals ($R_{o,max}$ from 1.30 to 2.00%), which resulted from plutonic metamorphism and a superimposed magmatic thermal and dynamic metamorphism. The third type of metamorphic and deformed environment of the high-rank coals ($R_{o,max}$ greater than 2.00%) resulted from plutonic metamorphism and a superimposed, stronger magmatic thermal and dynamic metamorphism.

Structural-Genetic Classification of Tectonically Deformed Coals

Based on the results of the in situ measurement of coal mines and microstructural observation, a structural-genetic classification system has been proposed that can be applied not only to the exploitation of coalbed gas but also to the prevention and control of coal and gas outbursts. In terms of the scale of hand specimens or samples of coal core from boreholes, TDCs formed by different deformational mechanisms have been divided into three deformation series and ten classes (Table 1). The brittle deformation series includes cataclastic structure coal, mortar structure coal, granulitic structure coal, mealy structure coal, schistose structure coal, and thin-layer structure coal. The ductile deformation series includes wrinkle structure coal, mylonitic structure coal, and ductile structure coal. The brittle-ductile series includes scaly structure coal. Compared to the previous classifications [17, 19, 25, 26] of brittle deformational coals, schistose structure coal and thin-layer structure coal have been added into this series. This is because these coals are quiet different from the cataclastic structure coals in structural and physical properties. Similarly, ductile structure coal has been divided into the ductile deformation series. For ductile deformational coals, wrinkle structure coal was considered to be formed in a compressing and shearing process, mylonitic structure coal is formed by strong ductile shearing, while ductile structure coal is formed during the creeping process of higher temperatures. Moreover, a brittle-ductile transition type of coal has been proposed; this may include scaly structure coal.

Table 1: The structural-genetic classification of TDCs.

Deformation series	Type of tectonically deformed coals	Texture and structure	Structural fracture and wrinkle	Breaking degree	Micro characteristics	Deformational mechanism and environment
Brittle deformation series	Cataclastic structure coal	Band texture could be seen; layer structure was preserved	Multidirectional fracture cuts; no obvious displacement	More hardness; crumbing is difficult	Tensile fracture, shear fracture, and compressional fracture	Multi directional compression and extension; extension was the predominant role
	Mortar structure coal	Little band texture was seen; lentoid structure was formed	Multidirectional fracture cuts; obvious displacement in the mortar	Broken into detritus from 1 to 5 cm; angular shape		
	Granulitic structure coal	Primary structure not present; disorder of players	Multi directional cross fracture; particles were rotated	Broken into detritus, 1 cm		
	Mealy structure coal	Primary structure not present; powdered particles formed	No obvious directions for the particles	Be pinched into powered particles		Strong compressional fracture zone or scaly structure coal be reconstructed later
	Schistose structure coal	Band texture could be seen; layer structure was preserved	Single directional fracture; little displacement on the surface	Broken into detritus from 1 to 5 cm, shard shape		Single directional compression and extension or shearing strain environment
	Thin-layer structure coal	Little band texture was seen; layer structure was not obvious	Single directional fracture; obvious displacement on the surface	Broken into detritus, 1 cm		
Brittle-ductile series	Scaly structure coal	Primary structure not present; scaly structure was formed	Multi directional cross-fracture; coal was cut and wrinkled; the particles could be rotated	Broken into detritus, 0.5 cm; shard shape	Shear fracture, cleavages	Strong shearing or cleaving strain environment
Ductile deformation series	Wrinkle structure coal	Primary structure not present; wrinkle structure was formed	Coal was wrinkled	Broken into detritus, 0.5 cm, shard shape	Wrinkle, foliated coal, S-C structure, eyed structure, optical anisotropy structure and wavy extinction	Strong shearing or long and low stress deformation
	Mylonitic structure coal	Primary structure not present; mylonitic structure was formed	The particles had directional arrangement; flow structure was formed	Was pinched into powered particles		
	Ductile structure coal	Primary structure not present; lump and lentoid structures were formed	No obvious physical deformation	More hardness; crumbing was difficult		Long and low stress creep; ductile flow at high temperatures

Structural Evolution and Deformational Mechanisms of Tectonically Deformed Coals

Temperature is the main factor that causes the change of coal chemical structure; it also facilitates metamorphism and improves coal rank [27, 28]. The temperature and stress together lead to change in coal macromolecular structure and its chemical components [4, 5]. Under thermal and stress conditions, how does the structural evolution of a deformational mechanism, over different kinds of TDCs, take place?

Thermal action and directional stress are the important factors in changing coal macromolecular structures and destroying bond forces. The vitrinite reflectance anisotropy (VRA) characteristic of deformed coals was well known and was interpreted from the ordering degree promoted by the stress [5, 29, 30]. Meanwhile, $R_{o,max}$ optical characteristics are shown in different kinds of TDCs (Table 2). Because of thermal values and stress, the value of *Ro* , m a x, the size of stacking (L_c) and extension (L_a) of basic structural units (BSU), carbon aromaticity (f_a) and radical density (N_g) increase. Accordingly, layers of the microcrystal (*N*) and the number of benzene rings of each layer also increase, whereas the spacing among monolayers of aromatics (d_{002}), L_a/L_c and the aliphatic spacing (d_r) continually decrease. In addition, the values of L_a/L_c (formed under the condition of stress) are obviously lower than those formed under conditions of thermal action. Therefore, the whole coalification process is the course of condensation of aromatic rings in which pores change continually [31]. Furthermore, when the structural deformation enhances, the faint orientation and bad ordering domain of schistose structure coal formed by brittle deformation extend to a local orientation, a strong orientation, and an integer orientation orderliness (Table 2 and Figure 2). It is obvious that stress has affected the chemical structure of coal. However, in the past, researchers had not convinced that the effect of tectonic stress is not the same for different kinds of ductile deformation coals. The macromolecular structure of wrinkle structure coal shows smaller changes, while the macromolecular structures of mylonitic structure coal and ductile structure coal change significantly.

Table 2: The characteristics of structural parameter of tectonically deformed coals.

Type of tectonically deformed coals	Coal sample's number	$R_{o,max}$ (%)		NMR	EPR	X-ray diffraction							HRTEM
		$R_{o,max}$	$R_{o,bi}$	f_a^* (%)	$N_g \times 10^{18}$ spins/g	d_{002}	L_c (nm)	L_a (nm)	d_r	N	n	L_a/L_c	lattice images
Cataclastic structure coal	QN03	1.23	0.33		75.0365	0.3667	1.3755	1.8575	0.568	3.75	51.9	1.35	
Mortar structure coal	TY02	1.51	0.32		44.8972	0.3662	1.4903	1.8546	0.535	4.07	51.8	1.24	
Mortar structure coal	XY02	1.99	0.44		45.5163	0.3699	1.2997	1.8444	0.561	3.51	51.2	1.42	
Granulitic structure coal	XY01	1.26	0.31		33.8709	0.3665	1.4379	1.9163	0.540	3.92	55.3	1.33	
Mealy structure coal	ZXZ06	1.33	0.28		65.7937	0.3561	1.5138	2.0003	0.528	4.25	60.3	1.32	
Schistose structure coal	ZJ01	0.95	0.23	0.791	52.4347	0.3828	0.9011	1.7254	0.613	2.35	44.8	1.91	Dispersedly isolated points, with weak orientation
Thin-layer structure coal	XJ03	1.01	0.23		44.1508	0.3716	0.928	1.8524	0.592	2.50	51.7	2.00	
Scaly structure coal	LL01	1.45	0.43	0.795	64.0869	0.3659	1.3645	1.8997	0.535	3.73	54.3	1.39	
Wrinkle structure coal	BJ05	1.08	0.24	0.778	36.1217	0.3716	1.0827	1.7684	0.57	2.91	47.1	1.63	
Mylonitic structure coal	QN09	0.96	0.27		46.3902	0.3665	1.3782	1.8187	0.536	3.76	49.8	1.32	Gracile worm shape, with local orientation
Mylonitic structure coal	ZXZ04	1.82	0.41	0.809	136.9129	0.3684	1.4733	1.9754	0.558	4.00	53.0	1.27	
Ductile structure coal	XY06	2.21	0.53		77.4869	0.3601	1.6195	2.1802	0.528	4.50	71.6	1.35	
Schistose structure coal	LH05	1.75	0.38	0.802	61.2011	0.3617	1.5694	2.0895	0.538	4.34	65.8	1.33	
Schistose structure coal	TT10	1.94	0.40		41.9478	0.3619	1.566	1.9791	0.534	4.33	59	1.26	
Cataclastic structure coal	HZ01	2.97	0.58	0.873	107.7825	0.3537	2.0291	2.4107	0.485	5.73	87.5	1.19	
Mylonitic structure coal	HZ07	3.59	0.71	0.925	84.3871	0.3497	2.5172	2.4489	0.467	7.20	90.3	0.97	Striations patch group, with strong orientation

* f_a: carbon aromaticity; N: layers of the microcrystal; n: numbers of benzene ring; L_c: size of stacking of basic structural units; L_a: extension of basic structural units; d_{002}: the spacing among monolayers of aromatics; d_r: the aliphatic spacing; N_g: radical density.
Broad horizontal lines in the table distinguish TDCs formed in different metamorphic and deformational environments; narrow horizontal lines distinguish tectonically deformed coals of different deformational series. Parts of Table 3 are the same.

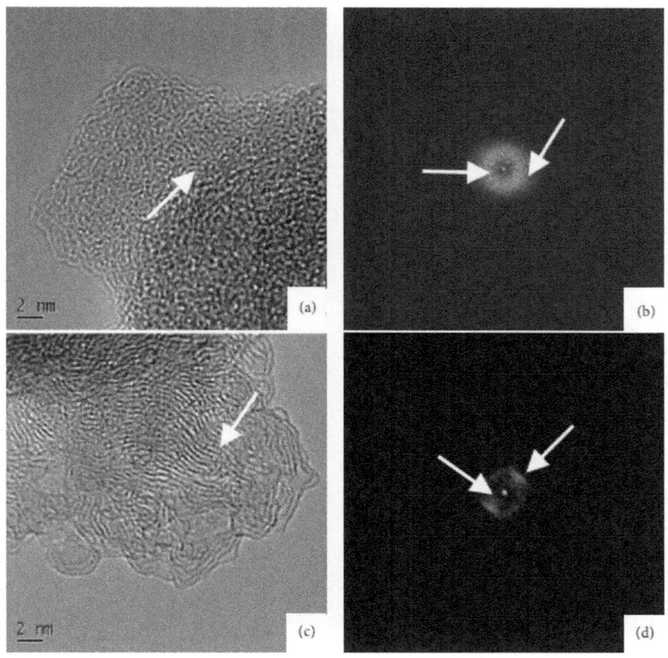

Figure 2: HRTEM images of TDCs. (a) Sample ZJ01 of schistose structure coal, in the lattice image; BSU is dispersive and isolated, its diameter is smaller, and BSU has weak orientation, Zhangji Coal Mine, Huainan Coalfield; (b) Sample ZJ01 of schistose structure coal, (002) the (inner ring) diffraction ring lightness is stronger, (101) outer ring looms, Zhangji Coal Mine, Huainan Coalfield; (c) Sample HZ07 of mylonitic structure coal, in the image; BSU is striation group shape, its diameter is large, and BSU has stronger orientation, HRTEM, Haizi Coal Mine, Huaibei Coalfield; (d) Sample HZ07 of mylonitic structure coal, (002) diffraction ring lightness is dispersive, (101) can be seen clearly, and symmetry light spot appears, HRTEM, Haizi Coal Mine, Huaibei Coalfield.

Physical Properties of Tectonically Deformed Coals and Their Relation with Structure

Porosity of Tectonically Deformed Coals

Coal contains a large numbers of pores [10, 13, 15], even more in TDCs. The nanoscale pore diameters of different kinds of TDCs have been grouped into four types: mesopores (15 ~ 100 nm), micropores (5~ 15 nm), submicropores (2.5~5 nm), and ultramicropores (<2.5 nm) [22]. This was done according to the average pore diameter of the natural distribution at three points: 15 nm, 5 nm, and 2.5 nm, and through combining the adsorbed gas and diffused features.

In this study, the parameters of pore structure were tested and the experimental results were shown in Table 3. Research was performed on the nanoscale pore structure in metamorphic-deformed environment of low-rank coal and the different deformational series of structural coal. Our results have demonstrated that the volume of mesopores reduces rapidly while the volume of micropores and pores whose diameters are lower than micropores increase in the metamorphic deformed environment of low-rank coal. Also, submicropores and ultramicropores can be found, the specific surface area of mesopores reduces greatly while and submicropores increase rapidly. The change of pore parameters in nonhomogeneous coal structure is similar to the weak brittle deformation coal. For different types of tectonic coals formed in metamorphic-deformed environments of middle- and high-rank coal, pore parameter change is basically consistent with the metamorphic-deformed environment of low-rank coal.

But, there are differences in the changes among different kinds of tectonic coals. To sum up, the conditions of temperature and confining pressure can play an important role in the evolution of nanoscale pore characteristics, but structural stress is the predominant factor influencing the formation of the characteristics of nanoscale pores.

Table 3: The characteristics of pore structural parameter of TDCs.

Type of tectonically deformed coals	Coal sample's number	Pore volume					Pore specific surface					N_2 adsorption volume (cm^3/g)
		V_t (cm^3/g)	V_4/V_t	V_5/V_t	V_6/V_t	V_7/V_t	S_t (m^2/g)	S_4/S_t	S_5/S_t	S_6/S_t	S_7/S_t	
Cataclastic structure coal	QN03	0.00046	95.652	4.348			0.484	83.33	16.667			0.3105
Mortar structure coal	TY02	0.00079	78.481	11.392	7.595	2.532	0.224	30.357	21.429	27.678	20.536	0.9606
Mortar structure coal	XY02	0.00055	92.727	5.455	0	1.818	0.072	59.722	16.667	8.333	15.278	0.3566
Granulitic structure coal	XY01	0.00229	81.223	13.100	5.240	0.437	0.491	43.789	28.31	26.28	1.221	1.9401
Schistose structure coal	ZJ01	0.00066	86.364	7.576	6.060		0.123	46.341	15.447	38.212		3.762
Thin-layer structure coal	XJ03	0.00491	70.061	18.534	11.202	0.203	1.377	25.853	28.467	44.080	1.600	3.7620
Scaly structure coal	LL01	0.00215	82.791	10.698	6.511		0.464	41.164	22.845	35.991		1.9104
Wrinkle structure coal	BJ05	0.00407	76.167	17.700	6.143		1.008	37.401	33.929	28.670		3.1771
Mylonitic structure coal	QN09	0.00930	67.229	19.277	13.253	0.251	2.636	24.772	28.376	45.524	1.328	6.1405
Mylonitic structure coal	ZXZ04	0.01871	74.025	19.401	6.146	0.428	4.575	33.880	34.972	27.607	3.541	12.9174
Ductile structure coal	XY06	0.00078	85.900	5.128	7.692	0.282	0.158	35.443	11.392	48.101	5.064	0.5320
Schistose structure coal	LH05	0.00093	94.624	4.301	1.075		0.101	72.277	12.871	14.852		0.6216
Schistose structure coal	TT10	0.00090	96.667	3.333			0.081	85.185	13.580	1.2350		0.5834
Cataclastic structure coal	HZ01	0.00049	81.663	8.163	10.204		0.105	43.810	10.476	45.714		0.4836
Mylonitic structure coal	HZ07	0.00882	28.458	14.286	54.875	2.381	6.365	5.027	10.008	79.214	5.751	6.9903

V_4: 15 ~ 100 nm; V_5: 5 ~ 15 nm; V_6: 2.5 ~ 5 nm; V_7: <2.5 nm; V_t: Total pore volume; S_4: 15 ~ 100 nm; S_5: 5 ~ 15 nm; S_6: 2.5 ~ 5 nm; S_7: <2.5 nm; S_t: Total specific surface area.

Permeability of Tectonically Deformed Coals

Often, the TDCs are believed to be low-permeability coals without much value for exploitation [32–34]. In fact, this is not always true. The porosity and permeability of TDCs vary according to changes in their tectonic deformation [19, 22, 35]. In this study, the parameters of permeability in different types of TDCs were tested and the experimental results were shown in Table 4. Those samples were drilled in the coal mine samples collected from the Huaibei and Huainan coalfields. The gas-water phase permeability of TDCs is mainly determined by their type and their sample drilling style. It can be seen clearly from Table 4 that the gas-water phase permeability of brittle deformation increases sequentially. The increase begins with cataclastic structure coal and moves to mortar structure coal, then to schistose structure coal, which is perpendicular to the fracture, and, finally, to the schistose structure coal, which is parallel to the fracture.

Table 4: The experiment results of gas-water phase permeability for TDCs.

Deformation series	Type of tectonically deformed coal	Coal sample's number*	Apparent fracture porosity (%)	K_0** of CH_4 ($10^{-3}\,\mu m^2$)	K_0 of He ($10^{-3}\,\mu m^2$)	K_w ($10^{-3}\,\mu m^2$)
Brittle deformation	Cataclastic structure coal	HZ01	4.1	0.162	0.180	0.099
	Cataclastic structure coal	QN03	1.5	0.019	0.020	0.017
	Cataclastic structure coal	TY10	2.4	0.178	0.205	0.075
	Mortar structure coal	TY02	2.6	0.183	0.211	0.087
	Mortar structure coal	XY02	3.1	0.471	0.495	0.138
	Schistose structure coal	ZJ01 (vertical sampling)	2.9	0.534	0.535	0.155
	Schistose structure coal	ZJ01 (lateral sampling)	2.9	1.630	2.120	0.359
Ductile deformation	Wrinkle structure coal	BJ05	2.7	0.362	0.378	0.124
	Ductile structural coal	XY06	2.0	0.089	0.098	0.056

*The direction and degree of fractures affect the drilling coal samples. We made vertical drilling in samples when there is the cross of two sets of fractures (e.g., schistose structure coal, mortar structure coal). And lateral drilling when there is only one set of fracture (Cataclastic structure coal).

**K_0: absolute permeability; K_w: water phase permeability.

On the other hand, the value of gas-water phase permeability of coal in ductile deformation is less than that in brittle deformation. Moreover, the permeability of TDCs in which ductile deformation is superimposed by later brittle deformation has been significantly increased. Thus, TDCs suffering from certain degrees of deformation have a higher permeability than those of primitive structure coals or very weak TDCs [36].

DISCUSSION

Structural Evolution and Deformational Mechanisms of Tectonically Deformed Coals

Main Geological Factors Affecting Deformation and Metamorphism of Coals

The relationship between coal structure and metamorphism has been discussed over a century [37]. It is thought that temperature, pressure, and time are the general geologic factors that control coalification. Pressure factors of coalification include static pressure (confining pressure) and dynamic pressure (tectonic stress). Previous studies focused on the influence of temperature and confining pressure on coal structure. As a result, the existed studies have indicated that temperature is the main factor causing the changes in the chemical structure of coal, facilitating metamorphism and improving coal rank [37, 38]; confining pressure benefits physical coalification but suppresses chemical coalification. Tectonic stress has seldom involved the metamorphism of coals [37, 38]. From BSU testing and the extraction of dissoluble organic molecules of tectonic coals, Cao et al. [4, 39] and Cao et al. [40] concluded that temperature and stress lead to changes in coal macromolecular structure and chemical components. The changes in coal macromolecular structure parameters are mainly caused by compression or shearing. Both brittle and ductile deformation can affect the macromolecular structure of the coal, which leads to dynamic metamorphism and elevation in the coal rank, to a certain degree. The influence of macromolecular structure in ductile deformation coals is stronger than that in brittle deformation coals. The reason is that carbon aromaticity and the degree of ring condensation have been obviously increased in the former. In this research, by using the XRD, HRTEM,

EPR, and NMR, L_c, L_a, f_a, and N_g increase, while L_a/L_c, d_{002}, and d_r decrease. The results showed that the cracking and removal of straight and side-chain compounds of organic matter in coal caused reduction in molecular weights and directly changed the chemical composition of macromolecules in coal, especially in low and medium coalification stages. This shows "dynamic hydrocarbon generation". The change of structural parameters promotes the BSU rearrangement, as well as ordering enlargement and orientation growth. The result is an increase of the aromatic condensed ring system. The stress polycondensation is usually represented by VRA increasing. Therefore, structural stress promotes physical coalification and chemical coalification.

The Types of Tectonic Stress Affecting Deformation and Metamorphism of Coals

There has been speculation as to how stress affects the optical properties of coal and whether deformed coal changes chemically when its physical structure changes. Based on the present studies, the influence model of the tectonic stress on the dynamic evolution of organic matter can be concluded as the followings: friction heat view, strain energy view, and mechanochemistry chemical view.

Stone and Cook [41] suggested that anisotropic stress may influent coalification by strain being imparted to vitrinite during coalification. Teichmüller and Teichmüller [42], Bustin [43] and Suchy et al. [44] suggested that local increases in coal rank and local graphitization resulted from frictional heating. Levine and Davis [45] assumed that the applied stress field caused increased bireflectance by the preferred nucleation and growth of favorably oriented lamellae. Based on XRD, TEM, and micro-Raman spectroscopy evaluations, Ju et al. [5, 6] concluded that the coal structure generated under shearing was changed through elongation and contraction of pores between the aromatic lamellae, thus enhancing growth of aromatic stacks. Observation of increased reflectance or local coalification can be classified either by a thermal mechanism [42–44] or mechanical strain mechanism [41].

In the brittle deformational coals, structural influence of different TDCs is not the same. As the deformation becomes stronger, coal macromolecular structures change from a cataclastic structure coal, mortar structure coal, or granulitic structure coal to mealy structure coal. It is similar when going from schistose structure coal to thin-layer structure coal. Under the effect of stress, the structural fractures quickly spread along weak belts of bond forces in coal; the coal body breaks, and the blocks and particles glide, rotate, and displace; friction heat is then produced, which promotes local coal metamorphism. From the observation of XRD, HRTEM and the highest bireflectance in shear zones, stress-chemistry and frictional heating of shearing strain areas might play a more important role in increasing brittle deformation which is because the rate of strain is high enough to destroy bond forces on the stress direction, generate thermal heat, and locally alter coal structure under stress.

Levine and Davis [45] postulated a mechanical-chemical mechanism in which stress reoriented the lamellae and facilitated growth. Bustin et al. [8] and Ross and Bustin [46], on the other hand, thought strain energy transformation played an significant role in graphitization from coal under the shear stress regime. A previous study of ductile deformation showed that the size of aromatic stacks and reflectance increase in deformed coals [4, 6, 30]. It is obvious that for distinct kinds of ductile deformation coals, the effect of tectonic stress is not the same. There are three main types of changes in macromolecular structures for TDCs: dislocation glide between the aromatic layers of coal cores, shear glide of the aromatic cores, and destruction/recombination of the coal molecular structure bonding energy. The changes of macromolecular structure in ductile deformation become greater than that in brittle deformation. Superimposition of these types of deformations in coal macromolecular structure is, therefore, the basic process of coal structure shearing and also the essential process of strong ductile deformation of coal. These findings are consistent with the hypothesis of aromatic lamellae growing and reorientating proposed by Levine and Davis [45] and the strain energy transformation presented by Bustin et al. [8] and Ross and Bustin [46]. Since deformation was the result of strain effects in coal, the chemical coalification derived from deformation should be attributable to the strain. During deformation of the coal seam, strain might play significant roles which depend upon the strain rate and the geological setting. In strong ductile shearing, reorientation of aromatic lamellae responsible for higher bireflectance or biaxial vitrinite reflectance indicatrices may occur without brittle failure under lower strain rates. In this case, coal would deform ductilely and the aromatic lamellae would become oriented with the new stress regime [4, 5]. A progressive strain may tend to flatten porosity and shorten the distance between lamellae. If the rearrangement of lamellae was slow enough, all of the chemical structural units might conform to the new stress setting with no chemical change. If the coal was pulverized to a fine grain size by ductile shearing, grain flow would parallel the stress direction. During grain flow, strain would be translated to particle surface thus causing thin areas of strain. Changes in the physical and chemical structures of coal might occur on the particle surface, where local changes in coalification might be produced. Strain rearrangement of aromatic lamellae on the particle surface would lead to local coalification [4]. The degree to which strain rearrangement translates to the particle interior may have a profound influence on the gross change in coal structure. Greater strain on the particle surface not only causes reorientation of aromatic lamellae but also may result in some chemical change.

The fact that the highest bireflectance occurs in shear zones, similar to strain area, has been observed from the anthracite specimens deformed at high temperature and pressure [8]. Thus, it may be supposed that strain energy might play a more important role than frictional heating in increasing coal deformation and metamorphism because the rate of strain is insufficient to generate temperatures high enough to thermally alter coal.

Physical Properties of Tectonically Deformed Coals and Their Relation with Structure

Porosity of Tectonically Deformed Coals and Their Relation with Structure

Coal is one of the few highly porous media found in nature. Porosity, pore diameter distribution, and pore specific surface area are the main physical properties of coals. These physical properties are important in coalbed methane exploitation and evaluation of coal and gas outburst tendency [4, 16–22, 47]. The pore volumes and specific surface areas of coals have been widely studied [9, 13, 15, 48–56]. Under a scanning electron microscope and high-resolution transmission electron microscopy, pore shape, pore diameter, the molecular structure and relationship of these features with thermal deformation mechanisms have been identified and discussed [31, 57–59]. Duber and Rouzaud [31] found that the pore structure change determined the carbon microcosmic structure and the bireflectance of coal. The size and shape of the pores depended on the processes of coalification, temperature, and tectonic stress. Coal was considered to be formed with different size and shape of pores and aromatic clusters around the pore wall. Coalification was considered to be a process of aromatic ring concentration with pores changing size and shape. Studies of pore structure are therefore scientifically significant.

Tectonic deformation can not only alter coal's macromolecular structure to distinct degrees but can also change pore structure. Based on measurements of mercury intrusion, researchers compared micropore features of TDCs collected from different coalfields [9–11]. These experiments showed that TDCs possessed with greater medium pore volume than normal coal. No difference in mesopores and micropores was found, however. This suggested that tectonic deformation did not significantly influence micropore structure (less than 10 nm in diameter). Ju et al. [5, 6] suggested that tectonic deformation can not only change the micropore structure (>10 nm) of coal but also influence coal's nanoscale pore structure (<10 nm), which is the main adsorption space for coalbed methane.

From the results listed in Tables 3 and 4, the parameters of pore structure, such as pore diameter structure, pore volume and specific surface area, are seriously influenced by changes in thermal conditions. In contrast, the structural deformation of coal results from variability in pore porosity. Brittle and ductile deformation might change the structural parameters of coal pores, but the structural parameters of different kinds of TDC pores are distinct. It can be seen that the volume and specific surface area of nanoscale pores (<10 nm) in TDCs are more than those in normal coals. This finding indicates that TDCs have a higher potential of gas accumulating and pooling. On the other hand, Zhang et al. [10] and Ju et al. [23] have showed that the volumes and specific surface areas of macro- and medium pores of TDCs have a similar variation. The macro- and medium pores of TDCs reveal stronger laminar flow.

Thermal and stress action can facilitate coalification and change the macromolecular structure and nanoscale pore structure of TDCs [28, 39], but manner and degree of changes are different. Under thermal conditions, the meso- and micromolecules can be arranged to recomposition and condensation by means of physical resultant force. The increase of ordering degree is not quite obvious in orientation, leading to the aromatic layers remaining thin. This commonly enhances the thermal stability of the aromatic structure. Because the aromatic layers are disordered, a more concentrated arrangement results in the formation of mesopores and micropores, forming one or several channels that are connected by pores. Under stressful conditions, the amount of coal macromolecules increases continuously. Additionally, the bordering aromatic rings continue to polymerize and form superpositions. For the local directional and surrounding nondirectional aromatic layers to assemble together, the dislocation of the aromatic cores and slippage of aromatic layers must occur. The larger pores are produced between the aromatic cores and aromatic layers. The pore of TDCs are mainly micropores and submicropores, while ultramicropores are also developed with worse connection. In the local direction process of aromatic structure units, special thin neck, bottle-shaped pores have developed. These are mainly built up by pores between the directional and undirectional structure units.

Permeability of Tectonically Deformed Coals and Their Role in CBM Exploitation

This experimental result shows that the gas-water phase permeability of TDCs is mainly determined by type. The permeability of TDCs could be very high. The permeability increases sequentially, beginning with cataclastic structure coal, to mortar structure coal, to schistose structure coal, which is vertical to the fracture, and, finally, to the schistose structure coal that is parallel to the fracture. TDC with early low ductile deformation superposed by brittle deformation, gas-water phase permeability in the range of $0.1–5.0\times10^{-3}\,\mu m^2$, which is in the middle or higher than the primary structure coals and TDCs with weak deformation (Table 4, Figure 3). If the structural fractures are filled with calcite veins or the other mineral matter (Figure 3), gas-water phase permeability would decrease. This means that traditional theory, which states that the permeability in the area where TDCs are common is low, is inaccurate. The permeability of TDCs obtained from the experiment of samples with in situ large diameter in the indoor test can be an important parameter that reflects the real permeability of the coalbed.

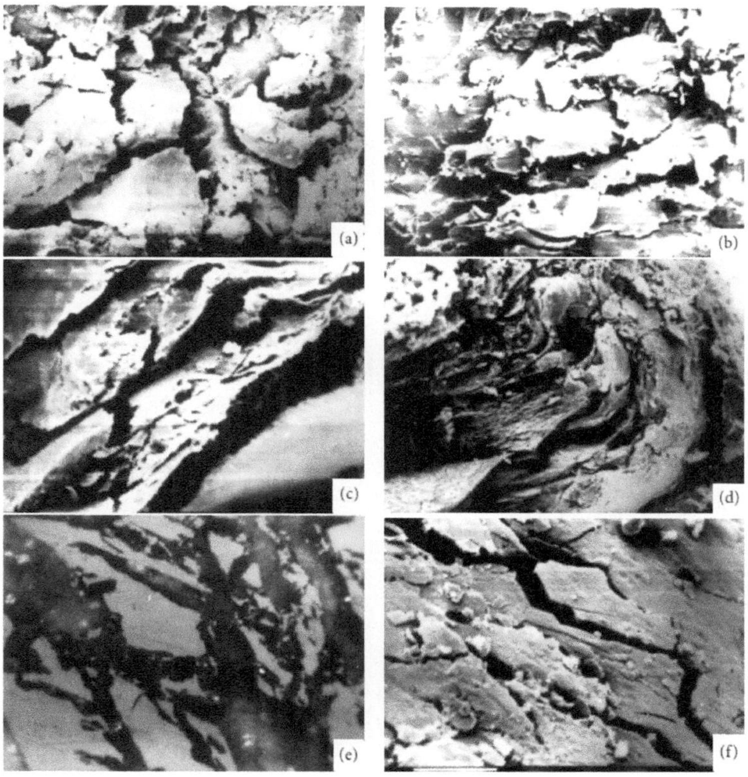

Figure 3: SEM and optical microscopy images of TDCs (a) cataclastic structure coal, multidirectional fracture cutting, and no obvious displacement in the blocks, SEM; (b) mortar structure coal, multidirectional fracture cutting, and obvious displacement in the blocks, SEM; (c) schistose structure coal, single directional fracture, and little displacement on its surface, SEM; (d) wrinkle structure coal; coal was wrinkled, and later superimposed by brittle fractures, SEM; (e) cataclastic structure coal with the structural fractures filled with calcite veins, optical microscopy; (f) cataclastic structure coal with the structural fractures expanded the cleats, SEM.

When we study the in situ permeability of the coalbed, we should get representative samples of each type of coal in order to obtain integrative permeability. Because of the heterogenous characteristics of the coal seams, especially in the area developing TDCs, we may predict a permeability change of the TDCs according to the types present and the coal gas mining style in different places and seams.

As the stress increases, the coals are further crushed into cataclastic-angular and/or cataclastic-granular pieces. Original cleats [60, 61] that are completely covered with deformation may wrench and tightly compress the fractures, even grinding the coals. Locally intensive deformation may wrench and tightly compress the fractures, even grinding coal into powder, which may plug sheared fractures. However, a combination of the original cleats and tectonic-induced fractures with deformation style (brittle or ductile), size, continuity, and connectivity of the sheared fractures contribute significantly to the overall permeability and are likely to play a major role in the flow of methane through structurally deformed coal in both diffusion at the micropore level and laminar flow at the cleat level [35].

In general, structurally deformed coals are formed through a brittle deformation mechanism and are called cataclastic coals and schistose coals. In such coals which have a hierarchy of open, continuous, and connecting fractures and cleats, the effective block size is not defined at present by the cleat spacing [60, 61], but somewhere between cleats and microfractures. A large number of brittle fractures in structurally deformed coals will make methane migration faster than that in normal coals. This tendency is extremely important for the development of CBM projects and may not only preserve significant gases but also increase the inherent permeability.

Nevertheless, not all structurally deformed coal possesses the properties of fast desorption and laminar flow. Ductile deformation coals, such as mylonitic coals, which are locally and ductilely generated in an intensively compressive and shearing environment, always display tightly compressed and broken fractures to be pushed into one another with worse connectivity for methane migration.

The various types of TDCs formed by the structural deformation of coal seams, especially the coals such as schistose, mortar and cataclastic structure coals subjected to a certain degree of brittle fractures, have good exploration and development potential for coalbed gas. Low ductile deformation superimposed by later brittle deformation also has this potential. On the other hand, changes in coal seam thickness from structural stress and destruction of coal structures caused by the rheology of coal seams are the major factor resulting in the gas outbursts of the coal mines [62, 63]. Therefore, according to the intensity, type and distribution of the rheology of coal seams, the zone of pooling coalbed gas and gas outbursts could be predicted. TDCs of strong brittle-ductile and ductile deformation, such as scale structure coal and mylonitic structure coal, are the main problems in preventing and controlling coal and gas outbursts and gas emissions. These problems need to be resolved urgently.

CONCLUSIONS

Taking Huaibei and Huainan as typical examples of coal mine areas in the southern North China, the structural characteristics and the physical properties of TDCs have been studied by the authors. Some conclusions are drawn as follows.

(1) Tectonically deformed coals (TDCs) formed by different deformational mechanisms have been divided into three deformation series and ten classes. In the metamorphic-deformed environments, from brittle deformed coal to ductile deformed coal, the changes in characteristics of the macromolecular structures and chemical compositions are that as the increase in structural deformation becomes stronger, from the brittle deformation coal to ductile deformation coal, the ratio of width at the half height of the aromatic carbon and aliphatic carbon peaks (H_{fa}/H_{fal}) was increased. As carbon aromaticity is raised further, carbon aliphaticity reduced obviously and different compositions of macromolecular structure appeared as a jump and wave pattern except that in wrinkle structure coal. Under the directional stress, BSU arrangement is closer, and the orientation becomes stronger from brittle deformed coal to ductile deformed coal. Under the effect of oriented stress, the orientation of the macromolecular structure becomes locally stronger, and the ordering degree of the arrangement of the BSU is dramatically enhanced.

(2) The characteristics of structural evolution of different kinds of tectonically deformed coals have been further investigated. Structural deformation directly influences the macromolecular structure of coal, which results in dynamic metamorphism. In the brittle deformational coals, frictional heating and stress chemistry of shearing strain areas might play a more important role in increasing brittle deformation because the rate of strain is big enough to generate high temperature, locally altering coal structure under stress. In ductile deformation coals, it may be assumed that strain areas might play a more significant role than frictional heating in increasing coal deformation and metamorphism because the rate of strain is insufficient to generate temperatures high enough to thermally alter coal.

(3) In comparison with the previous results, the characteristics of macromolecular structures and micro- and nanoscale pore structures of different kinds of TDCs have been analyzed in detail. In particular, advanced research on the formation mechanism of TDCs has been carried out. The nanoscale pore of the cataclastic structure coals caused by the brittle deformation is mainly mesopores; the rest are micropores and submicropores. As the action of tectonic deformation increases, the volume of micropores and the pores whose diameters are even smaller, increases dramatically, and submicropores and ultramicropores can be found. In coals that have experienced ductile deformation, the proportion of mesopores volume in wrinkle structure coal diminishes rapidly, and an increase in the proportion of micropores and submicropores is observed. The proportion of mesopores volume of mylonitic structure coal decreases continually as tectonic stress increases, while the volume of micropores and submicropores increases rapidly, and ultramicropores can also be found. Ductile structure coal has a characteristic similar with the weak brittle deformation coals, but submicropores and ultramicropores are present. The main factor that produces a change in pore structure is strongly tectonic deformation. It is proposed that brittle and ductile deformation can change the pore distribution, which then change spacing and chemical structure.

(4) The permeability of different kinds of TDCs has been investigated which could be very high. The permeability increases sequentially, beginning with cataclastic structure coal, to mortar structure coal, to schistose structure coal, which is perpendicular to the fracture, and, finally, to the schistose structure coal, which is parallel to the fracture. The permeability of TDCs where weak ductile deformation was superimposed by later brittle deformation is better than primary structural coal. According to the results of the coalbed gas testing well, the permeability of TDCs obtained from the experiment of in situ major diameter samples in the indoor test can be an important parameter that reflects the real permeability of the coalbed.

ACKNOWLEDGMENTS

This study was financially supported by the National Basic Research Program of China (also called 973 Program) (Grant no. 2009CB219601), the National Natural Science Foundation of China (Grant no. 40772135; 41030422; 40972131; 40172058), and Strategic Priority Research Program of the Chinese Academy of Sciences (Grant no. XDA05030100).

REFERENCES

1. M. Teichmüller, "Organic material and very low-grade metamorphism," in Temperature Metamorphism, M. Frey, Ed., pp. 114–161, Blackie Academic & Son, London, UK, 1987.
2. M. Nishioka, "The associated molecular nature of bituminous coal," Fuel, vol. 71, no. 8, pp. 941–948, 1992.
3. P. Fowler and R. A. Gayer, "The association between tectonic deformation, inorganic composition and coal rank in the bituminous coals from the South Wales coalfield, United Kingdom," International Journal of Coal Geology, vol. 42, no. 1, pp. 1–31, 1999.
4. Y. X. Cao, G. D. Mitchell, A. Davis, et al., "Deformation metamorphism of bituminous and anthracite coals from China," International Journal of Coal Geology, vol. 43, no. 1–4, pp. 227–242, 2000.
5. Y. W. Ju, B. Jiang, Q. L. Hou, and G. L. Wang, "Relationship between nano-scale deformation of coal structure and metamorphic-deformed environments," Chinese Science Bulletin, vol. 50, no. 16, pp. 1784–1795, 2005.
6. Y. W. Ju, B. Jiang, Q. L. Hou, et al., "C13 NMR spectra of tectonic coals and the effects of stress on structural components," Science in China, Series D, vol. 48, no. 9, pp. 1418–1437, 2005. ·
7. K. R. Wilks, M. Mastalerz, R. M. Bustin, et al., "The role of shear strain in the graphitization of a high-volatile bituminous and an anthracitic coal," International Journal of Coal Geology, vol. 22, no. 3-4, pp. 247–277, 1993.
8. R. M. Bustin, J. V. Ross, and J. N. Rouzaud, "Mechanisms of graphite formation from kerogen: experimental

evidence," International Journal of Coal Geology, vol. 28, no. 1, pp. 1–36, 1995.

9. Y. A. Wang and S. J. Yang, "Some characteristics of coal seam with hazard of outburst," Journal of China Coal Society, vol. 1, pp. 47–53, 1980 (Chinese).

10. J. Zhang, B. Yu, and J. X. Tang, "Study on pore structures of coal seam Blowing out gas," Coal Geology of China, vol. 8, no. 2, pp. 71–74, 1996 (Chinese).

11. D. X. Yao and J. Lu, "Study on pority of coal from the First Mine, Xiejiaji, Huainan," Coal Geology of China, vol. 8, no. 6, pp. 31–33, 1996 (Chinese).

12. T. Wang and W. T. Huang, "Porous structure, characteristics of a coal seam's soft slice in Xinhua coal mine, Jiangxi," Coal Geology of China, vol. 6, no. 4, pp. 57–59, 1994 (Chinese).

13. H. Gan, S. P. Nandi, and P. L. Walker, "Nature of the porosity in American coals," Fuel, vol. 51, no. 4, pp. 272–277, 1972.

14. S. J. Gregg and K. S. W. Sing, Adsorption Surface Area and Porosity, Academic Press, London, UK, 2nd edition, 1982.

15. G. Gürdal and M. N. Yalçin, "Pore volume and surface area of the carboniferous coal from the Zonguldak basin (NW Turkey) and their variations with rank and maceral composition," International Journal of Coal Geology, vol. 48, no. 1-2, pp. 133–144, 2001. ·

16. H. Evans and K. M. Brown, "Coal structures in outbursts of coal and firedamp conditions," Mining Engineering, no. 148, pp. 171–179, 1973.

17. C. F. Yuan, "Tectonically deformed coal and coal and gas outbursts," Gas Geology Started, pp. 45–52, 1985 (Chinese).

18. Y. X. Cao, Tectonically deformed coal: its physicochemical properties and mining hazards, Ph.D. thesis, Peking University, Beijing, China, 1999.

19. H. Y. Li, "Major and minor structural features of a bedding shear zone along a coal seam and related gas outburst, Pingdingshan coalfield, Northern China," International Journal of Coal Geology, vol. 47, no. 2, pp. 101–113, 2001.

20. B. Jiang, Y. Qin, B. H. Fan, et al., "Physical property of coal reservoir and exploration prospects for coal bed methane in Huaibei area," Journal of China University of Mining and Technology, vol. 30, no. 5, pp. 433–437, 2001 (Chinese).

21. Y. W. Ju and G. L. Wang, "Rheology of coal seams and their relation with gas outbursts: a case study of the Haizi coal mine, Huaibei coalfield," Geological Review, vol. 48, no. 1, pp. 96–105, 2002 (Chinese).

22. Y. W. Ju, Characteristics of structural evolution and physical properties of reservoirs of tectonically deformed coals and theirs action mechanism, Ph.D. thesis, China University of Mining and Technology, Xuzhou, China, 2003.

23. Y. W. Ju, G. L. Wang, B. Jiang, and Q. L. Hou, "Microcosmic analysis of ductile shearing zones of coal seams of brittle deformation domain in superficial lithosphere," Science in China, Series D, vol. 47, no. 5, pp. 393–404, 2004.

24. L. J. Xu, C. L. Liu, and X. F. Xian, Fractal Structure and Properties of Coal in Outburst Zone, Chongqing Univerdity Press, Chongqing, China, 2004.

25. S. Q. Chen, "Characteristics of tectonic deformation Permian coal in Hubei, Hunan, Guangdong and Guangxi provinces and analysis of its origin," Journal of China Coal Society, vol. 4, pp. 1–9, 1989 (Chinese).

26. D. Y. Cao, S. R. Zhang, and D. Y. Ren, "Theinfluence of structural deformation on coalification: a case study of carboniferous coal measures in the northern foothills of the Dabie orogenic belt," Geological Review, vol. 48, no. 3, pp. 313–317, 2002 (Chinese).

27. E. Stach, M.-T. Mackowsky, and M. Teichmüller, Stach's Textbook of Coal Petrology, Gebrüder Borntraeger, Berlin, Germany, 3rd edition, 1982.

28. Y. Qin, Micropetrology and Structural Evolution of High-Rank Coals in P. R. China, China University of Mining & Technology Press, Xuzhou, China, 1994.

29. A. Oberlin and G. Terriere, "Graphitization studies of anthracites by high resolution electron microscopy," Carbon, vol. 13, no. 5, pp. 367–376, 1975.

30. B. Jiang and Y. Qin, Evolution Mechanism of Structures of Deformed Coal and Its Geological Significance, China of University of Mining and Technology Press, Xuzhou, China, 1998.

31. S. Duber and J. N. Rouzaud, "Calculation of relectance values for two models of texture of carbon materials," International Journal of Coal Geology, vol. 38, no. 3-4, pp. 333–348, 1999.·

32. M. Y. Sun, L. W. Yang, and X. W. Lu, "Geology possibility and technology feasibility to develop coal bed methane resources in China," Coal Science and Technology of China, vol. 29, no. 1, pp. 45–46, 2001.

33. L. W. Yang and M. Y. Sun, "Peculiarities of China CBM reservoirs and their dictation on CBM production technology," Natural Gas Industry, vol. 21, no. 6, pp. 17–19, 2001 (Chinese).

34. C. L. Lei and C. Y. Liu, "Reformed pattern of coal reservoirs in reformed coal basins and effect on favorable blocks of coalbed gas in the North China craton," Coal Geology and Exploration, vol. 32, no. 2, pp. 20–23, 2004 (Chinese).

35. H. Y. Li, Y. Ogawa, and S. Shimada, "Mechanism of methane flow through sheared coals and its role on methane recovery," Fuel, vol. 82, no. 10, pp. 1271–1279, 2003.·

36. X. H. Fu, Y. Qin, B. Jiang, W. F. Wang, and S. N. Zhou, "Research on permeability of multiphase medium of middle to high-rank coals," Journal of China University of Mining and Technology, vol. 13, no. 1, pp. 11–15, 2003.

37. J. C. Hower, "Observations on the role of the Bernice coal field (Sullivan County, Pennsylvania) anthracites in the development of coalification theories in the Appalachians," International Journal of Coal Geology, vol. 33, no. 2, pp. 95–102, 1997.

38. G. H. Taylor, M. Teichmüller, A. Davis, et al., Organic Petrology, Gebrüder Borntraeger, Berlin, Germany, 1998.

39. Y. X. Cao, A. Davis, R. X. Li, et al., "The influence of tectonic deformation on some geochemical properties of coals-a possible indicator of outburst potential," International Journal of Coal Geology, vol. 53, no. 2, pp. 69–79, 2003.

40. D. Y. Cao, X. M. Li, and S. R. Zhang, "Influence of tectonic stress on coalification: stress degradation mechanism and stress polycondensation mechanism," Science in China, Series D, vol. 50, no. 1, pp. 43–54, 2007.

41. I. J. Stone and A. C. Cook, "The influence of some tectonic structures upon vitrinite reflectance," Journal of Geology, vol. 87, no. 5, pp. 479–508, 1979.

42. M. Teichmüller and R. Teichmüller, "Geological causes of coalification," Coal Science, vol. 55, pp. 133–155, 1966.

43. R. M. Bustin, "Heating during thrust faulting in the rocky mountains: friction or fiction?" Tectonophysics, vol. 95, no. 3-4, pp. 309–328, 1983.

44. V. Suchy, M. Frey, and M. Wolf, "Vitrinite reflectance and shear-induced graphitization in orogenic belts: a case study from the Kandersteg area, Helvetic Alps, Switzerland," International Journal of Coal Geology, vol. 34, no. 1-2, pp. 1–20, 1997.

45. J. R. Levine and A. Davis, "The relationship of coal optical fabrics to Alleghanian tectonic deformation in the central Appalachian fold-and-thrust belt, Pennsylvania," Geological Society of America Bulletin, vol. 101, no. 10, pp. 1333–1347, 1989.

46. J. V. Ross and R. M. Bustin, "The role of strain energy in creep graphitization of anthracite," Nature, vol. 343, no. 6253, pp. 58–60, 1990.

47. B. B. Beamish and P. J. Crosdale, "Instantaneous outbursts in underground coal mines: an overview and association with coal type," International Journal of Coal Geology, vol. 35, no. 1–4, pp. 27–55, 1998.

48. R. A. Meyers, Coal Structure, Academic Press, New York, NY, USA, 1982.

49. S. Parkash and S. K. Chakrabartly, "Porosity of coal from Alberta planes," International Journal of Coal Geology, vol. 6, pp. 55–70, 1986.

50. P. L. Walker, S. K. Verma, J. Rivera-Utrilla, et al., "Densities, porosities and surface areas of coal macerals as measured by their interaction with gases, vapours and liquids," Fuel, vol. 67, no. 12, pp. 1615–1623, 1988.

51. Z. F. Lu, X. M. Zhang, L. W. Zhong, et al., "The pore features of lump coal and its influence factors,"Journal of China University of Mining & Technology, vol. 20, no. 3, pp. 45–50, 1991 (Chinese).

52. J. Wu, "Studies on characteristics of micropore and its relation with the migration and concentration of oil and gas," Science in China ,Series B, vol. 23, no. 7, pp. 77–84, 1993 (Chinese).

53. Y. Qin, Z. W. Xu, and J. Zhang, "Natural classification of the high-rank coal pore structure and its application," Journal of China Coal Society, vol. 20, no. 3, pp. 265–271, 1995 (Chinese).

54. C. R. Clarkson and R. M. Bustin, "Effect of pore structure and gas pressure upon the transport properties of coal: a laboratory and modeling study. 1. Isotherms and pore volume distributions," Fuel, vol. 78, no. 11, pp. 1333–1344, 1999.

55. K. Bratek, W. Bratek, I. Gerus-Piasecka, et al., "Properties and structure of different rank anthracites,"Fuel, vol. 81, no. 1, pp. 97–108, 2002.

56. A. M. Raichur and S. P. Vijayalakshmi, "The effect of nature of raw coal on the adhesion of bacteria to coal surface," Fuel, vol. 82, no. 2, pp. 225–231, 2003.

57. Q. Hao, "On morphological character and origin of micropores in coal," Journal of China Coal Society, vol. 4, pp. 52–54, 1987 (Chinese).

58. J. N. Rouzaud and A. Oberlin, "The characterization of coals and cokes by transmission electron microscopy," in Advanced Methodologies in Coal Characterization, Coal Science and Technology, C. Henri, Ed., vol. 15, pp. 311–355, Elsevier, New York, NY, USA, 1990.

59. F. H. Zhao and D. Y. Ren, "The application of high-resolution transmission electron microscopy to study the structures of coal macerals," Geological Review, vol. 41, no. 6, pp. 564–570, 1995 (Chinese).

60. C. Ö. Karacan and E. Okandan, "Fracture/cleat analysis of coals from Zonguldak Basin (northwestern Turkey) relative to the potential of coalbed methane production," International Journal of Coal Geology, vol. 44, no. 2, pp. 109–125, 2000.

61. X. B. Su, Y. L. Feng, J. F. Chen, et al., "The characteristics and origins of cleat in coal from Western North China," International Journal of Coal Geology, vol. 47, no. 1, pp. 51–62, 2001.

62. J. H. Hao, C. F. Yan, and Z. X. Zhang, "The tectonic coal and its effects on coal gas outburst," Journal of Jiaozho Institute of Technology, vol. 19, no. 6, pp. 403–406, 2000 (Chinese).

63. M. B. Díaz Aguado and C. González Nicieza, "Control and prevention of gas outbursts in coal mines, Riosa-Olloniego coalfield, Spain," International Journal of Coal Geology, vol. 69, no. 4, pp. 253–266, 2007. ·

CHAPTER 15

MULTISCALE EROSION SURFACES OF THE ORGANIC-RICH BARNETT SHALE, FORT WORTH BASIN, USA

Mohamed O. Abouelresh

Faculty of Petroleum and Mining Engineering, Suez University, Salah Naesem St., Etaka, Suez 43721, Egypt

ABSTRACT

The high frequency and diversity of erosion surfaces throughout the Barnett Shale give a unique view into the short-duration stratigraphic intervals that were previously much more difficult to detect in such fine-grained rocks. The erosion surfaces in Barnett Shale exhibit variable relief (5.08–61 mm) which commonly consists of shelly laminae, shale rip-up clasts, reworked mud intraclasts, phosphatic pellets, and/or diagenetic minerals (dolomite and pyrite) mostly with clay-rich mudstone groundmass. Several factors control this lithological variation, including the energy conditions, rate of relative sea-level fluctuation, rate of sedimentation, sediment influx, and the lithofacies type of the underlying as well as the overlying beds. The erosional features and their associated surfaces make them serve at least in part as boundaries between different genetic types of deposits but with different scales according to their dependence on base level and/or sediment supply. Accordingly, the studied erosion surfaces of the Barnett Shale can be grouped into three different scales of sequence stratigraphic surfaces: sequence-scale surfaces, parasequence-scale surfaces, and within trend-scale surfaces.

INTRODUCTION

The classic interpretation of organic-rich shale (≥0.5% total organic carbon) deposition emphasized continuous hemipelagic deposition in deep, quiet, low energy and stagnant basins, often with a stratified water column. However, Schieber [1] identified laterally continuous erosion surfaces in the Chattanooga Shale and he interpreted them as being the result of wave reworking and erosion of the sea floor. These surfaces are direct indications of major environmental events that may include nondeposition and/or erosion events.

The multiple erosion events, resultant surfaces, and erosion features are common within the coarser clastic rocks and are pivotal in classifying depositional sequences based upon relative sea-level fluctuations. On the other hand, in fine-grained sediments, these events are subtle [2], although the identification is important in elucidating and development of the sequence stratigraphic framework for such rocks.

Mud floored erosion in the geologic record is often associated with zones of intensely burrowed sediment;

Citation: Mohamed O. Abouelresh, "Multiscale Erosion Surfaces of the Organic-Rich Barnett Shale, Fort Worth Basin, USA," Journal of Geological Research, vol. 2013, Article ID 759395, 16 pages, 2013. doi:10.1155/2013/759395.

hiatus-concretions and prefossilized organic remains reveal complex cycles of exhumation and reburial associated with erosion (see [3–6]). Of particular significance are intervals of vertically mixed sediment, shells, and nodules associated with this erosion and/or reworking surfaces. Such sediment intervals yield a complete spectrum from in situ to extensively reworked shale clasts, thus allowing detailed reconstruction of the erosion event as an ongoing process; however, not all erosion surfaces are marked by shale clasts [7].

Noteworthy, the erosion surfaces and their resultant lags can provide an indirect measure of processes and palaeogeography that existed immediately preceding and during transgression, a time most commonly characterized by hiatus and erosional vacuity [8]. The erosion surfaces are relatively abundant throughout the Barnett Shale, punctuate the stacking patterns, and may obliterate the record of various strata [9]. However, erosion provides regional bounding surfaces for classifying this ~25 myr shale.

This work aims to identify the characters of erosion surfaces and their related features in organic-rich shales which occur on multiple scales from different erosional events. Moreover, this work also shows how these surfaces could be used as multiscale stratigraphic surfaces for classifying genetically related fine-grained rocks.

GEOLOGIC SETTING

The organic-rich Barnett Shale of Fort Worth Basin, Texas, is a world-class unconventional gas reservoir deposited over a period of 25 million years from 345 to 320 myr [10]. The Fort Worth Basin extends some 322–402 km from the Red River arch in the north to the Llano uplift in the south (Figure 1(a)). The Bend arch forms the western margin of the basin and the Ouachita thrust front forms the eastern margin. The basin covers approximately 24140 km². It is one of a series of foreland basins formed along the southern margin of the North American craton during the late Paleozoic Ouachita orogeny [11–13]. Paleozoic strata comprise almost the entire fill of the Fort Worth Basin. The basin fill thickens to the northeast, into the Oklahoma aulacogen, and thins to the south and west, toward the Llano uplift and the Bend arch, respectively. The Barnett Shale also thins to the west over the Bend arch where it interfingers with the Chappel Formation [14–17].

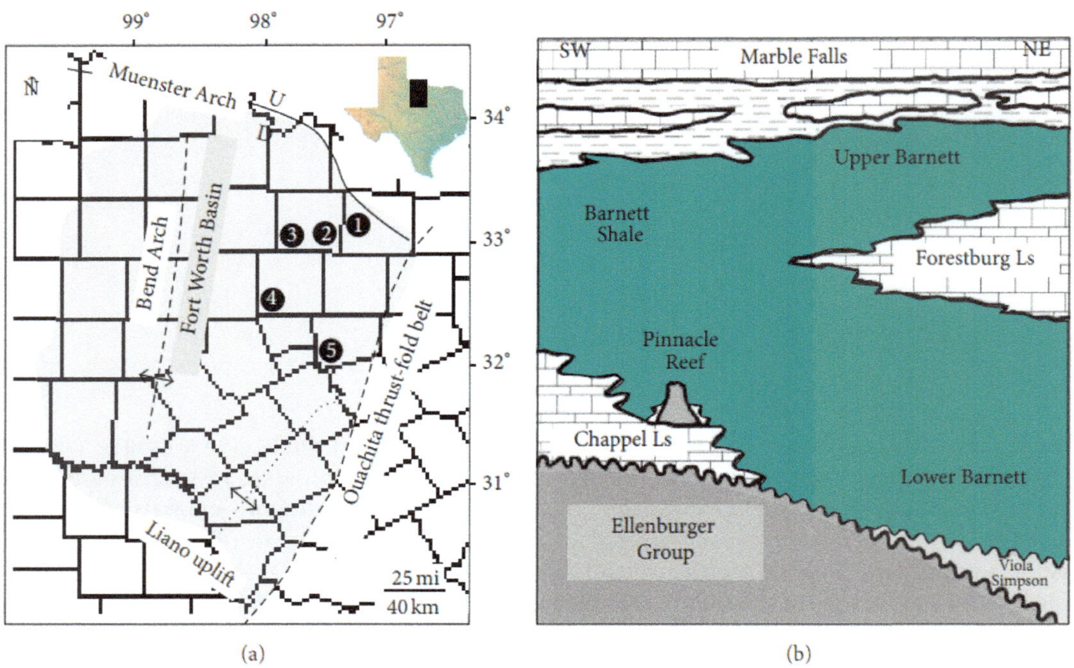

Figure 1: (a) Location map for the Fort Worth Basin showing the distribution of the Barnett Shale (shaded area) and structural and tectonic features; dots refer to the studied wells; 1: John Porter #3, 2: Sol Carpenter #7, 3: Adams SW #7, 4: Sugar Tree #3, and 5: Spencer Trussell 1-H (modified from [10]). (b) The schematic section shows an interpretation of the Mississippian stratigraphy (modified from [10]).

Following an extensive early Paleozoic transgression, erosion removed all of the Silurian and Devonian strata from the Fort Worth Basin [14]. The Barnett Shale was the first unit to be deposited when the seas returned (Figure 1(b)). The Barnett Shale was deposited on the karsted surface of the Ellenburger Group over a wide area [16]. Older stratigraphic studies suggest that most of the Barnett Shale accumulated either on a normal marine shelf [14] or in a relatively shallow, starved basin under euxinic conditions [16]. However, more recent work interprets the Barnett Shale in the Fort Worth Basin as deep-water slope-to-basin deposit [18, 19]. The relatively low energy and deep water environment, distant from a terrigenous sources, is inferred for the lower Barnett Shale. Whereas, depositional environment for the upper Barnett Shale is inferred to be shallower, proximal, well-oxygenated compared to the lower facies [9, 18].

Interestingly, the organic-rich Barnett Shale is characterized by a prolonged history of deposition and erosion due to frequent changes in relative sea level especially in the upper part; these fluctuations match with those proposed by Haq and Schutter [20] for the Late Mississippian. This type of fluctuation generated a variety of depositional, nondepositional, and erosional features, which have been placed within a sequence stratigraphic framework [9, 21].

DATA AND METHODOLOGY

The small grain size (<62 μm), subtle textural variations, lack of differences in erosion styles, and the gray to black color mean that the organic-rich Barnett Shale is difficult to characterize using conventional sedimentological techniques [22]. In this work, five continuous cores from the Barnett Shale (John Porter #3 (JP) in Denton County, Sol Carpenter H#7 (SC), Adams SW#7 (ASW) in Wise County, Sugar Tree #1 (ST) in Parker County and Spencer Trussell #1-H in Johnson County, Texas, USA) were studied.

Thirty three (33) thin sections collected from these cores represent a wide range of different erosion surfaces throughout the Barnett Shale. These thin sections were scanned using an Epson, 48-bit color, 4800 dpi scanner to give a higher resolution image than what can be obtained either from hand specimens or even under polarizing microscope. Core samples and thin sections observations were recorded and photographed in detail to shed light on the different characteristics of the erosional surfaces.

RESULTS

The detailed investigation of erosion surfaces throughout the Barnett Shale showed that these surfaces are part of zones vertical reliefs ranging from a few millimeters to several centimeters (Figure 2) as a result of differential erosion; considering that prior to compaction, this relief would have been up to 90% greater [23]. These zones consist of vertically mixed components within clay-rich mudstone groundmass. The most common components of the erosion surfaces include surface nature, shelly laminae, shale rip-up clasts, reworked intraclasts, phosphatic pellets, and diagenetic minerals (dolomite and pyrite). The following is a detailed description for each of these features which are also summarized in Table 1.

Table 1: Summary of the erosional features associated with the erosion surfaces.

Relief	Ranges between 5 mm and 60 mm with an average 19 mm.
Lithology	Siliceous noncalcareous to calcareous mudstone groundmass, silty laminated, and phosphatic-rich mudstone. Pyrite is common and rarely dolomitized.
Surface nature	
Lower	Sharp to irregular, scoured and down-cutting contact.
Upper	Sharp, gradational to irregular, rarely disturbed with burrowing.
Matrix components	
Shelly laminae	Very thin (1–15 mm), singular to multiple, fragmented shelly laminae. Highly compacted (grain to grain contact), horizontally alignment. The most common types of shell fragments are bivalves, bryozoans, brachiopods, filling branch mollusks, and echinoderms with mainly calcite filling.
Shale rip-up clasts	Mainly flakey-like shapes with straight outlines. Others are rounded to subrounded, with the same lithology of the underlying facies. Occurs as suspended clasts in the overlying facies.
Reworked concretions	In situ sandy-size to transported gravely-size concretions. Rounded, subrounded to irregular clasts. Slightly to highly compacted and subhorizontally oriented. Shaley to silty and phosphatic internal lithology. Pyritic effect is very common.
Phosphatic pellets	Very common, rounded to subrounded with multiple internal cores, sometimes broken, irregular, rarely elongated with their long axis parallel to bedding planes. They range in size from less than 1 mm up to 1 cm. The nuclei of the phosphatic intraclasts may include shell fragments, detrital quartz, and glauconite grains.
Pyrite	Common as framboidal shape and rarely as fine euhedral crystal. Not in all samples.

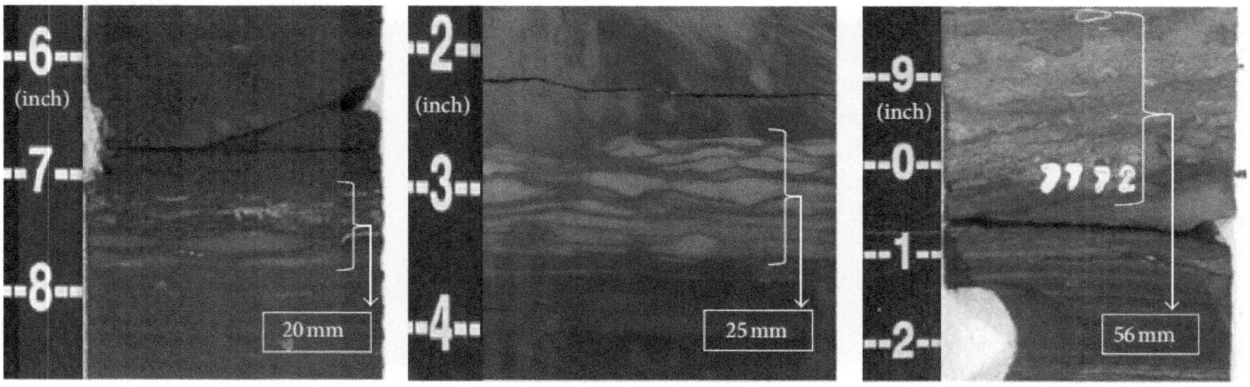

Figure 2: Core photos show different reliefs of erosion surfaces and their related features ranging from 20 mm to 56 mm.

Surface Nature

Description

The nature of erosion surfaces is frequent and identifiable on thin sections and core samples. The detailed investigation of the studied cores showed that these surfaces exhibit different shapes including (Table 1) sharp straight (though sharp irregular has been seen) (Figure 3(a)), scour, irregular (Figure 3(b)), and gradational upper contact (Figure 3(c)). In some cases the surface may down-cut into older deposits (Figure 3(d)). The lack of burrowing with most of lower sharp contacts is common.

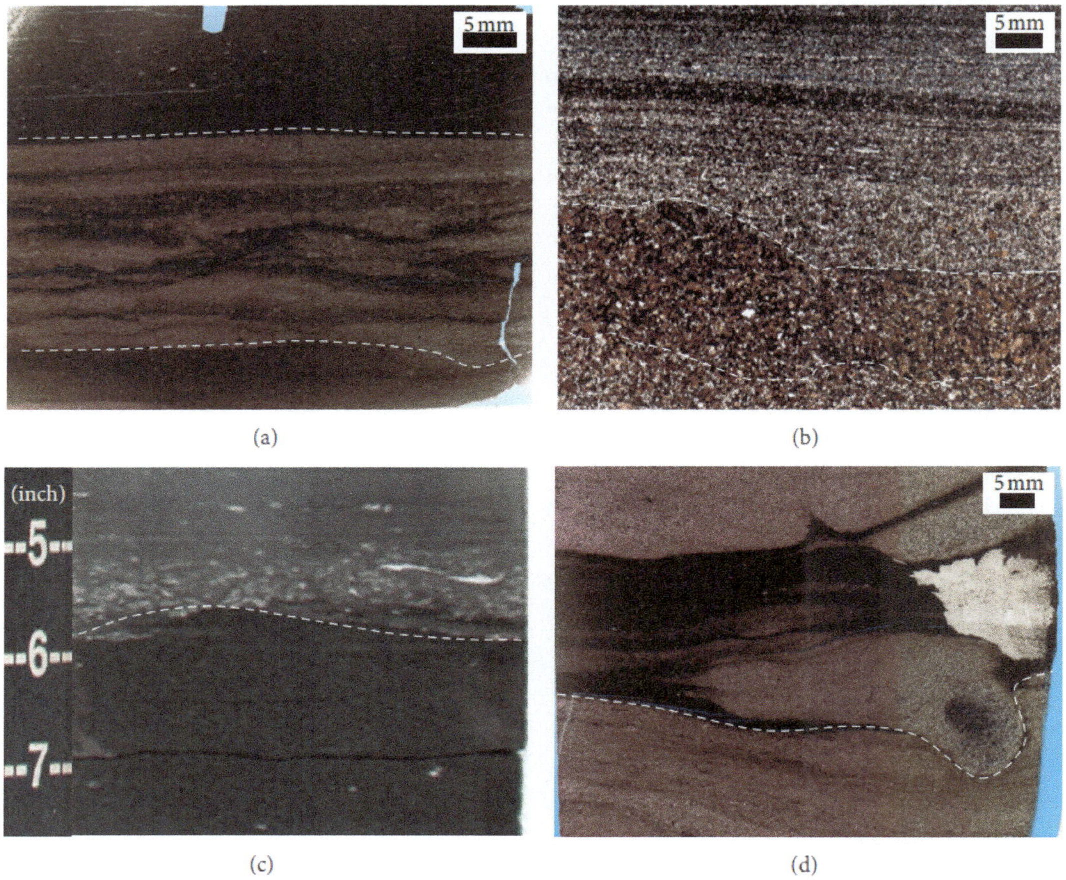

Figure 3: (a) Thin section photo shows upper sharp straight contact and lower scoured contact. (b) Thin section photo shows upper scoured contact and lower irregular contact. (c) Core photo shows lower sharp irregular contact and upper gradational contact. (d) Thin section photo shows lower contact that cut down into underlying facies.

Interpretation

The lithology changes marked by erosion surface could be associated with reworking and winnowing of the preexisting deposits as a result of prevailing tide, wave, or storm action [24]. However, these surfaces may represent a hiatus or periods of low sedimentation rate and/or sediment starved conditions. The sharp contact (Figure 3(c)) reflects an erosion event associated with a sudden sea-level fluctuation [1, 7] whereas the displacement of some surfaces into the underlying lithology is interpreted as an effect of soft sediment deformation processes (Figure 3(d)). On the contrary, the absence of soft sediment deformations below the sharp surface (Figure 3(a)) suggests the eroded shale was of firm consistency [1]. The sharp nature of the erosion surface suggests that erosion may have removed the mixed layer close to the sediment-water interface prior to deposition of the overlying unit. However, when the sharp contact marks the top of an erosion zone (Figure 3(a)), then sediment starvation was most likely and probably bottom currents that affected this location. Inversely, if the sharp nature is located at the base of the erosion zone (Figure 3(c)), then a sudden subsidence or a fall in sea level would be expected and the winnowed materials would be transported down-dip during a transgression. The organic-rich shale interval with sharp base is more erosional resistant due to the higher organic content [7]. The absence of burrowing along most of lower sharp surfaces suggests that these surfaces are not associated with a prolonged depositional hiatus and that erosion of the underlying sediments was immediately followed by, if not synchronous with, deposition of the overlying lithofacies blanket.

Shelly Laminae

Description

shelly laminae refer to dense concentrations of thin-walled, coalesced fragments, grain to grain contacts, and broken and disarticulated shells which are common in both upper and lower Barnett Shale [9, 21]. The laminae range to a maximum thickness of 15 mm (Table 1), with most between 1 and 3 mm. Surfaces are gently curved; bedding-parallel mantled the underlying sharp irregular to straight erosion surface and blanketed gradationally by siliceous noncalcareous to calcareous mudstone facies (Figures 4(a) and 4(b)). Shelly laminae of the Barnett Shale are composed mainly of bivalves, bryozoans, brachiopods, filling branch mollusks, and echinoderms with mainly calcite filling (Figure 4(c)).

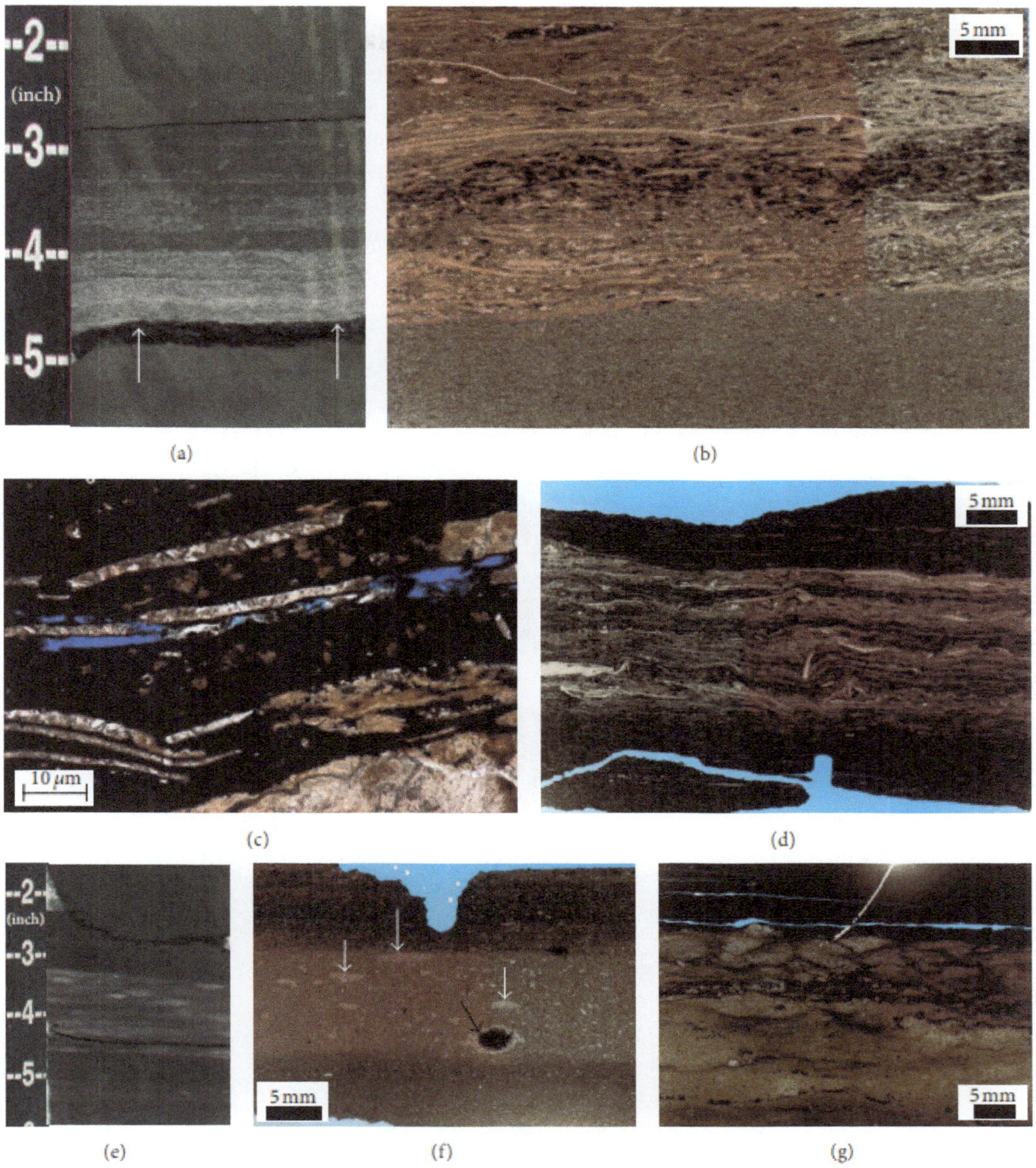

Figure 4: (a) Core photo shows highly compacted shelly laminae (arrows) with lower sharp contact and upper gradational contact. (b) Thin section photo shows lower sharp contact with overlying shelly laminae. (c) Thin section photo shows the shell fragments with calcite filling. (d) Thin section photo shows multiple shelly laminae. (e) Core photo shows shale rip-up clasts with lower and upper sharp contacts. (f) Thin section photo shows scattered shale rip-up clasts (white arrows), note the pyrite filling of some shale rip-up caslt (black arrow). (g) Thin section photo shows rounded to subrounded shale rip-up clasts with lower gradational contact and upper sharp contact.

Interpretation

Shelly laminae represents a condition of low sediment supply [25] due to the dense compacted shell fragments (Figure 4(a)); however, the rapid accumulation at the scale of such laminae should be considered [26] especially with the abundance of fossil material. The frequent occurrence of shelly laminae (Figure 4(d)) (regardless of their thickness) is interpreted as resulting either from thin compacted hiatal concentrations or multiple, varying duration episodes of erosion for the underlying deposits, that is, associated with a residual concentration of preexisting and broken fossils [26].

Shale Rip-Up Clasts

Description

Rip-up clasts are muddy chunks that swept up from the preexisting sediments and are also termed mudstone clasts, intrabasinal clasts, or clay chips. The shale rip-up clasts appear mostly without internal structures and occur as materials eroded from the underlying lithofacies (Figures 4(e) and 4(f)). These clasts exhibit different shapes and sizes: rounded to subrounded (Figure 4(g)) but flat to sharp edges; flaky-like (Figure 4(e)) is the most characteristic feature for the shale rip-up clasts in the examined samples.

Interpretation

The shale rip-up clasts occur due to the movement of coarser materials down-dip and erosion of the softer seafloor sediments. These clasts are transported and probably then suspended in basal currents and deposited when the current velocity reduced. However, shale rip-up clasts may be transported over considerable distances as evidenced by the rounded to subrounded and/or irregular surfaces (Figure 4(e)). Additionally, the flat and flaky-like shapes of shale rip-up clasts reflect compaction.

Reworked Concretions

Description

Reworked concretions are common features that are associated with the studied erosion surfaces. Typically these concretions are elongate, rounded, flat and/or irregular in shape (Figures 5(a), 5(b), and 5(c)), with a different internal grain size from the matrix ranging from a pebbly, sandy, and silty to a mud. (Figure 5(d)). Commonly concretions are light gray; however, black and dark brown concretions also occur. Interestingly, the concretions show more than one pyritized rim, indicating a multiphase origin (Figure 5(e)). The groundmass of the concretions varies between clay-rich mudstone (Figure 5(b)) and calcareous-rich mudstone (Figure 5(c)). In most cases, concretions that occur are associated with thin intervals of densely packed shells and/or shell debris (Figure 5(c)).

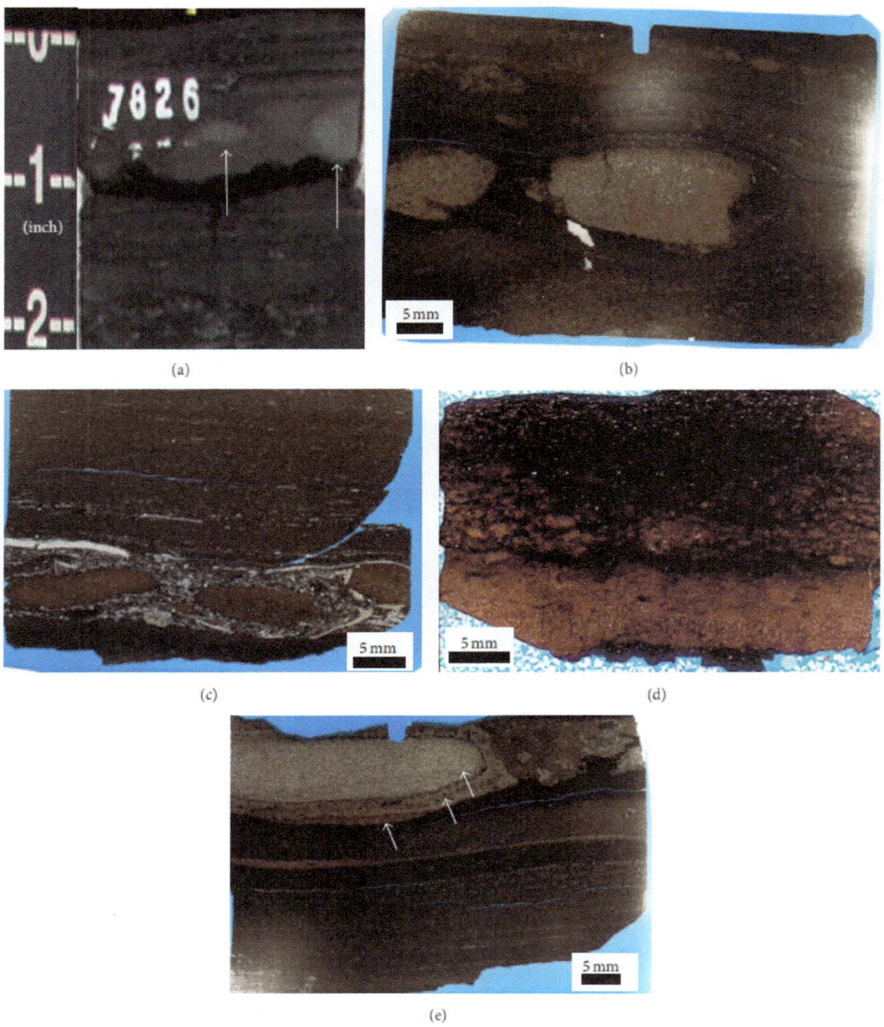

Figure 5: (a) Core photo shows reworked concretion (white arrows) associated with erosion surface with lower irregular contact and upper gradational contact. (b) Thin section photo shows large, transported subrounded concretions. (c) Thin section photo shows elongate, reworked concretions associated with irregular sharp erosional surface. (d) Thin section photo shows sandy-size reworked concretions. (e) Thin section photo shows reworked concretion with several core (white arrows) that rimmed with pyrite.

Interpretation

The roundness of concretions, different lithology from the host sediment, and the soft deformation structures in the underlying sediments suggest the reworking origin of these concretions. The occurrence of reworked concretions along erosional surfaces indicates formation at depths sufficiently shallow to be reworked by storms [23]. The dark brown color could be due to either phosphate and/or iron oxide impregnation; the latter might prevail due to the oxidation of pyrite in an oxic zone as a result of shallowing conditions. A storm event is one process that can rework concretions and is associated with a fining upward sequence afterstorm [27].

Description

The phosphatic pellets that are associated with erosion surfaces within the Barnett Shale are mostly brownish to black and range in size from few millimeters to 1 centimeter (Figure 6(a)). The pellets occur as rounded (Figure 6(b)) to subrounded, broken (Figure 6(c)), irregular, and rarely elongated with long axis parallel to bedding planes. The nuclei of these phosphatic grains may include shell fragments, detrital quartz, and glauconite grains.

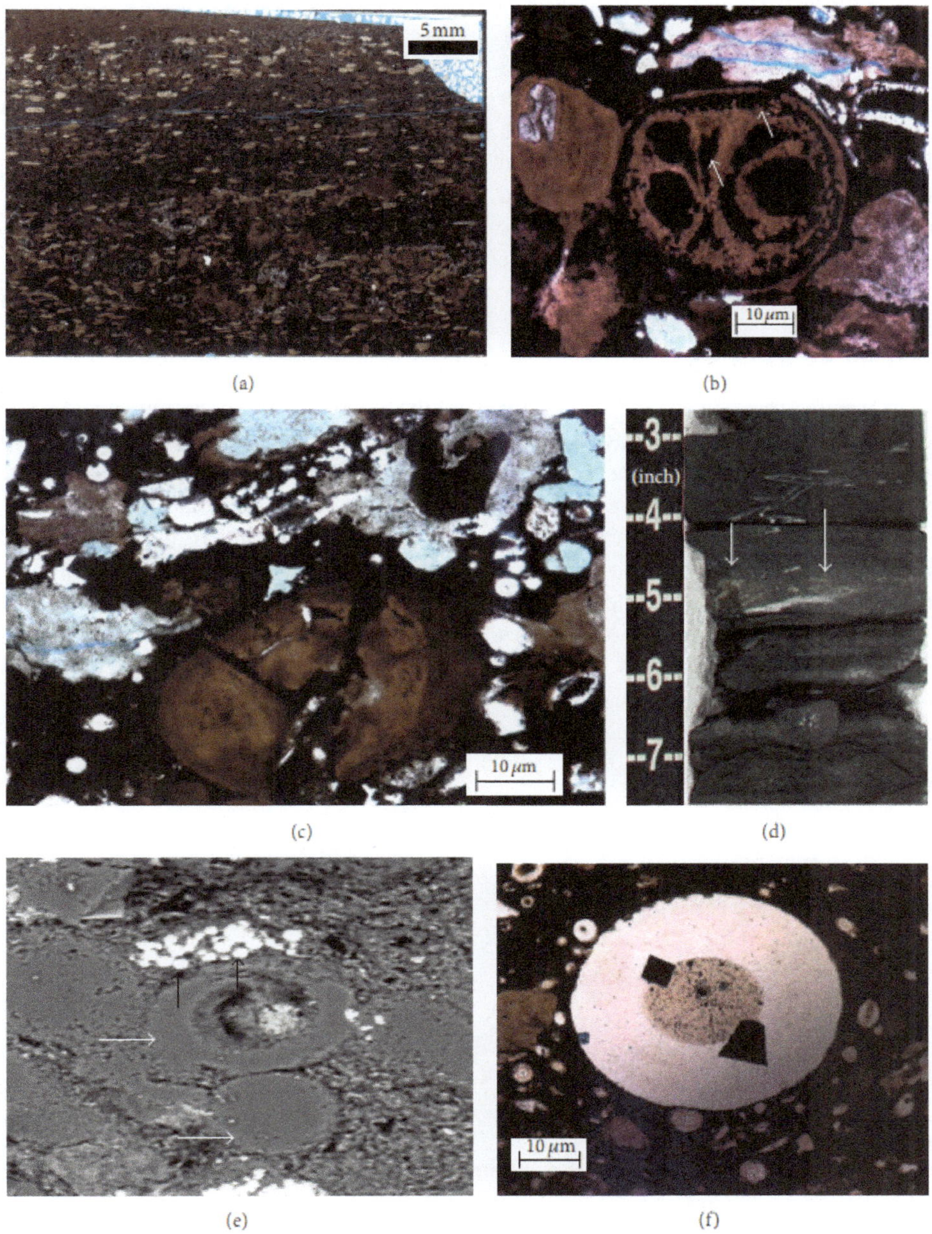

Figure 6: (a) Thin section photo shows concentration of phosphatic pellets associated with erosion surface. (b) Thin section photo shows rounded and subrounded phosphatic pellets; note the pyritic rim inside the phosphatic pellet (white arrows). (c) Thin section photo shows reworked broken phosphatic pellet associated with erosion surface. (d) Core photo shows the extensive pyrite cement of the erosion surface (white arrows). (e) Electron microprobe photo shows framboidal pyrite (black arrows) developed around sponge specula (white arrow) (photo dimensions is 527 × 527 mm). (f) Thin section photo shows fine euhedral pyrite crystal through the core of large sponge spicule associated with erosion surface.

Interpretation

The occurrence of phosphatic pellets is widely documented within the Barnett Shale [9, 28, 29], so it is one of the most important constituents associated with the erosion surfaces. Phosphatic pellets are important indicators of extensive reworking of the underlying substrate. The nucleus of these pellets shows multiple coatings through extensive reworking and transporting processes. The size, roundness, and broken edges of the phosphatic pellets, which are associated with erosion surfaces, imply that the energy conditions were sufficiently high to erode materials from the underlying sediments and agitate individual particles forming pellets.

Pyrite

Description

Pyrite occurs in erosion surfaces dominantly (Figure 6(d)) as a conspicuous accessory component disseminated in most of the phosphatic pellets (Figure 6(b)), shale rip-up clasts (Figure 4(f)), and reworked concretions (Figure 5(e)). Pyrite is present mainly in the form of framboidal pyritic pellets that may aggregate to larger pellets (Figure 6(e)) and disseminated fine euhedral crystals (Figure 6(f)).

Interpretation

The presence of pyrite is an indication of geochemical conditions during and/or after development of erosion surface [18, 30–

32]. Furthermore, pyrite indicates that the erosion surfaces resulting from major sea-level drop [7]. The abundance of fine crystalline pyrite in phosphatic pellets and shale intraclasts indicate deposition from an euxinic (anoxic and sulfidic) water column [30] and hence reducing depositional environment. Pyritization of a thick zone that has taken place beneath an erosion surface may represent a significant hiatus associated with the erosion event [33]. Pyrite framboids are characteristic of early diagenesis and crystallize in the water column, settle to the bottom, and cease to grow [18, 34].

ORIGIN OF EROSION SURFACES

Origin of erosion surfaces was the focus of several works; for example, Baird et al. [35] suggested internal waves during transgression. Allersma [36] and Rine and Ginsburg [37] proposed the migration of mud waves (mud banks) as an origin for erosion surfaces whereas wave scouring in front of an advancing shoreline is also an accepted origin for the regressive surface of marine erosion and ravinement surface [38, 39]. However, Schieber [7] refers to the origin of erosion surfaces in the Chattanooga Shale, storm-induced currents. However, strong bottom-current activity could be the reason for gravel lag formation as stated by Howe et al. [40]. Additionally, storm wave action can cause erosion of the sea floor as proposed by Caron et al. [24].

The erosion events in the Barnett Shale exhibit variable relief (5–60 mm), different geometry for upper and lower bounding contacts, and a wide range of erosional features with variable grain size above and below the surface (Table 1). Clearly, several factors control such variations, including the energy conditions, rate of relative sea-level fluctuation, rate of sedimentation, sediment influx, and the lithofacies type of the underlying bed [41–45].

Although, in most cases, sediment starvation condition is the favored interpretation for development of erosion surfaces in organic-rich shale [46–49], the detrital components associated with these surfaces were likely input to distal basin from the proximal shelf as a result of wave processes eroding and reworking the underlying deposits (Figure 7(a)) [50–52]. Furthermore, the lower sharp to irregular contacts associated with rounded to sub rounded intraclasts indicate that the sea bed was being reworked (Figure 7(b)), and consequently higher-energy conditions are suggested for the development of these surfaces than previously thought [22].

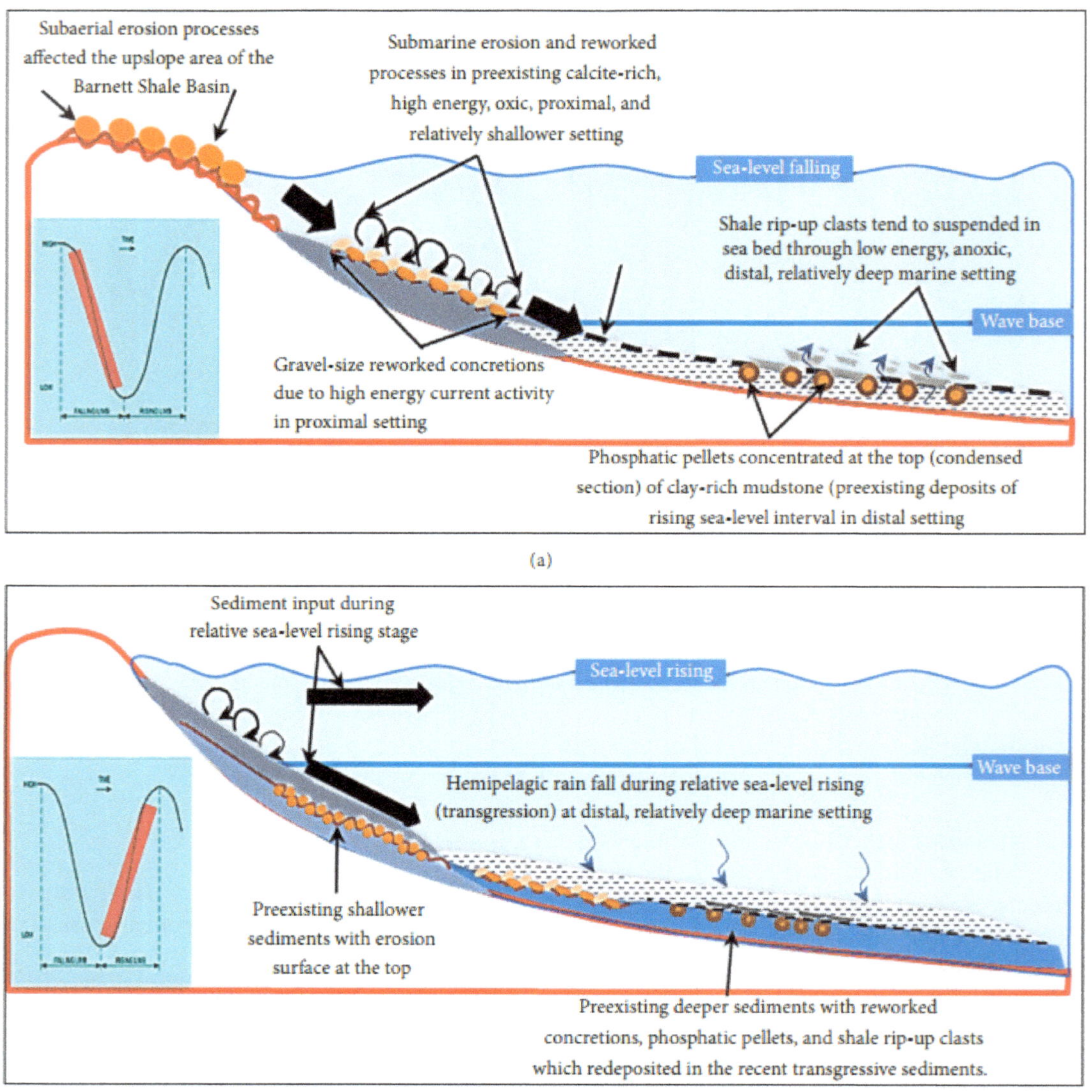

Figure 7: (a) Schematic diagram illustrates the development of erosion surface and their related features during relative sea-level fall in both proximal and distal muddy depositional settings. (b) Schematic diagram shows the transportation of the eroded and reworked materials down-dip during the relative sea-level rise to form erosion surface under the transgressive sediments.

During periods of relative sea-level fall, the action of storm currents and waves possibly caused intensive winnowing and reworking of the underlying deep marine mudstones (Figure 7(a)) [8]. The resulting erosion surfaces which may remove the underlying regressive surface of erosion have commonly been interpreted as transgressive surface of erosion [53, 54].

In some cases, the absence of erosion materials around erosion surfaces suggests either that these materials were transported to somewhere else in the basin, possibly by an unidirectional current strong enough to erode firm mud [1], or that the Barnett Shale was below the fairweather base level during a period of rapid rise of sea level which is insufficient time for sediment-water interaction [45]. However, the sharp to irregular erosion surface at the top of clay-rich shale facies (condensed section) (Figure 8(a)) may indicate that the erosion processes may have removed the expected, underlying, highstand lithofacies at this particular locality.

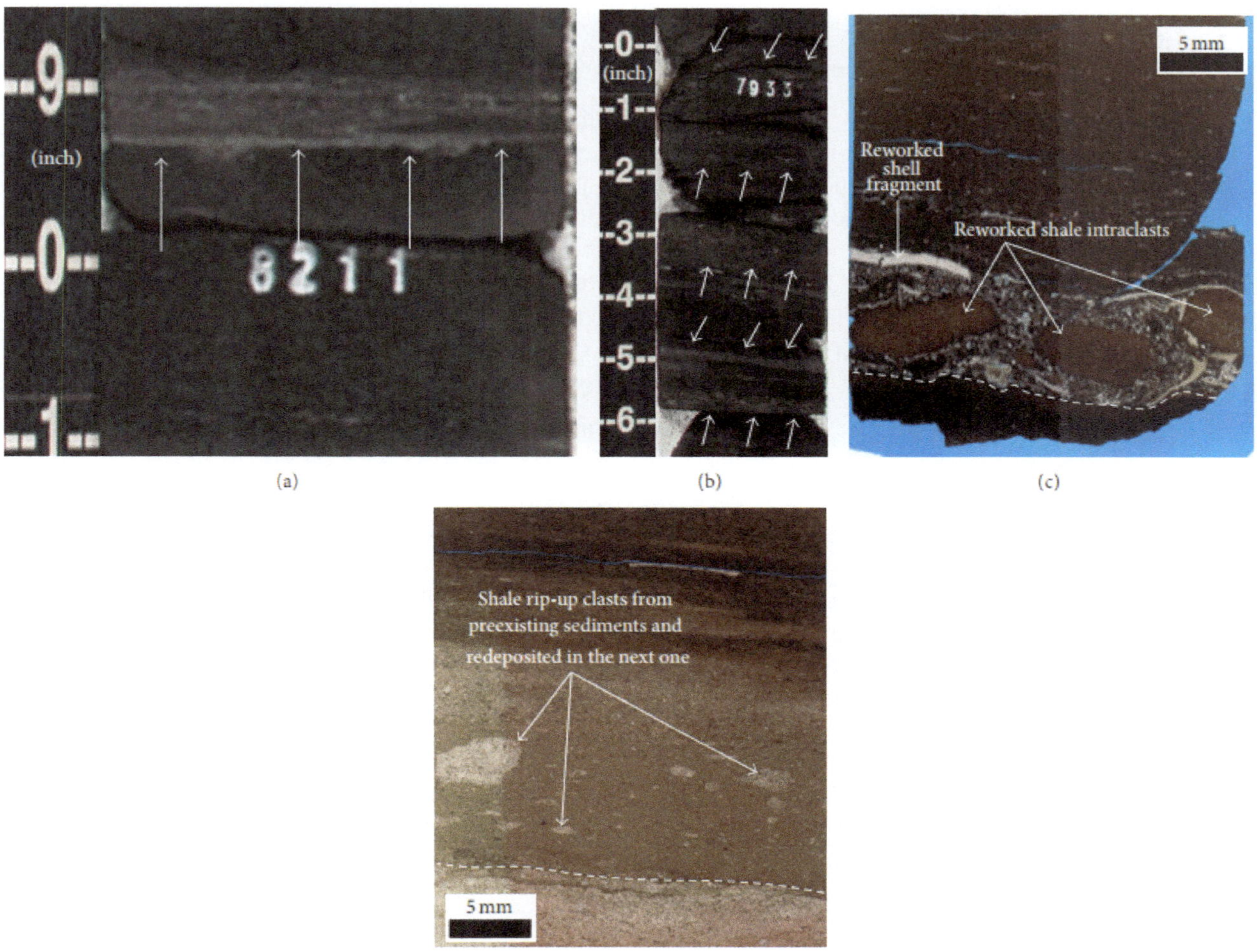

Figure 8: (a) Core photo shows sharp erosional contact at the topmost part of clay-rich mudstone facies. (b) Core photo shows multiple occurrence of erosional surfaces within less than one foot. (c) Thin section photo shows erosional surface and its related features indicated relatively proximal, shallow, and high energy muddy depositional setting (note the lower irregular contact (white dashed line), the pebbly-size reworked shale intraclasts, and reworked shell fragment). (d) Thin section photo shows erosional surface and its related features indicated relatively distal, deep, and low energy muddy depositional settings (note the lower sharp contact (white dashed line) and shale rip-up clasts).

Because the Barnett Shale is mainly a clay-rich mudstone, thus the grain size variation of the substrate has little influence on the properties of the shale as in more proximal settings. However, the high energy conditions suggested for deposition of the Barnett Shale [2, 9, 18, 21] may generate erosion surfaces with high relief [45]. Noteworthy, the frequent occurrence of erosion surfaces throughout the Barnett Shale (Figure 8(b)) suggests that the episodic storm currents and/or hyperpycnal flows are the major carriers for eroded and transported the erosional products from proximal to distal areas [2].

The development of erosion surfaces might have been influenced by water depth, sediment consistency, and slope where these conditions are expressed geologically as lateral facies changes associated with sedimentation breaks. The lateral variation from proximal (high energy preferred) to distal (low energy preferred) marine depositional settings can be indicated from the nature of erosion surfaces (Table 2). For example, the irregular nature, pebbly to sand size reworked intraclasts, reworked shelly laminae (Figure 8(c)), and the abundance of bioturbation indicate relatively shallower settings (Figures 7(a) and 7(b)), whereas sharp nature, shale rip-up clasts (Figure 8(d)), highly compacted shelly laminae, phosphatic-rich laminae, dolomitic and pyritic diagenetic cement, and the absence of bioturbation represent the relatively distal and deeper marine settings (Figures 7(a)and 7(b)).

Table 2: Characteristic features of erosion surfaces in both the proximal and the distal muddy settings.

Characters	Proximal muddy settings	Distal muddy settings
Boundary nature	Irregular, scoured, cutting down, gradational to irregular	Sharp to irregular
Lithofacies	Calcareous-rich mudstone facies with dolomitic cementation	Clay-rich mudstone facies with pyritic cementation
Reworked concretions	Rounded to subrounded calcareous shale intraclasts	Rounded to subrounded phosphatic pellets
Shelly laminae	Reworked shell fragments	Compacted shelly laminae
Shale ripup clasts	Rarely occurred	Elongated, flaky with sharp edges
Energy levels	High energy enough to include reworked shale intraclasts and other components	Low energy where the shale rip-up clasts are suspended in the sea bed layer

The significance of erosion surfaces within the organic-rich Barnett Shale is not only that it aids regional sequence stratigraphic analysis [9, 29] but it also represents erosion of organic-rich muddy sediments so prediction of where these sediments have been re deposited should be considered as a future development target.

EROSION SURFACES AS MULTISCALE STRATIGRAPHIC SURFACES

Since the accommodation space is likely available in the distal marine environments during the time intervals of relative sea-level fall and rise, then the identification of sequential events of deposition, nondepositions and/or erosion especially within fine-grained, organic-rich shale poses a unique challenge. However, the relative sea-level fluctuations are often expressed as subtle changes in bottom water oxygen levels, influx of terrigenous materials, biogenic productivity, and deposition of highly organic-looserich condensed sections [55]. The above discussed erosional features and their associated surfaces are candidates at least in part as boundaries between different genetic types of deposits [56] but on different scales depending on changes in base-level and/or sediment supply. Subsequently, the studied erosion surfaces of the Barnett Shale reveal different strengths and/or duration which can be grouped into three different scales of stratigraphic surfaces including sequence-scale surfaces, parasequence-scale surfaces, and event-scale surfaces.

Sequence-Scale Erosion Surfaces

The sequence-scale erosional surfaces that result from a major erosional event [7] are the largest scale of the identified unconformable surfaces. This type of erosional surface is used to establish sequence stratigraphic framework for the Barnett Shale [9, 29]. Such surfaces have relatively high relief (commonly >25 mm and up to 60 mm) and have an irregular sharp boundary associated with a sandy to silty pyritic lag (Figure 9(a)). The most distinctive features of the sequence-scale erosional surfaces are the rounded to subrounded, reworked, intraclasts and highly compacted, horizontally oriented, and shelly laminae (Figure 9(a)) and (Table 3).

Table 3: Characteristic features of multi-scale erosional surfaces.

Characters	Sequence-scale erosion surface	Parasequence-scale erosion surface	Event-scale erosion surface
Relief	More than 25 mm	Range between 8 and 20 mm	<5 mm
Surface nature	Irregular, scoured, cutting down	Sharp, irregular to gradational	Sharp
Lithofacies	siliceous non calcareous mudstone and Siliceous calcareous mudstone	Siliceous non calcareous mudstone and Siliceous calcareous mudstone	Clay-rich mudstone facies
Reworked concretions	Rounded to subrounded shale intraclasts	Subrounded phosphatic pellets	Calcareous concretions
Shelly laminae	Highly compacted and horizontally oriented shell fragments	Traces of thin disconnected shelly laminae	Rarely occurred
Shale rip-up clasts	Rarely occurred	Common	Rarely occurred
Spectral gamma ray pattern	Abrupt change (decreasing and increasing)	Marking the end of upward increasing or decreasing spectral Gamma Ray pattern	Does not Apply
Energy levels	High energy enough to reworked shale intraclasts and other components	Low energy where the shale rip-up clasts suspended in the sea bed layer	Low bottom energy levels

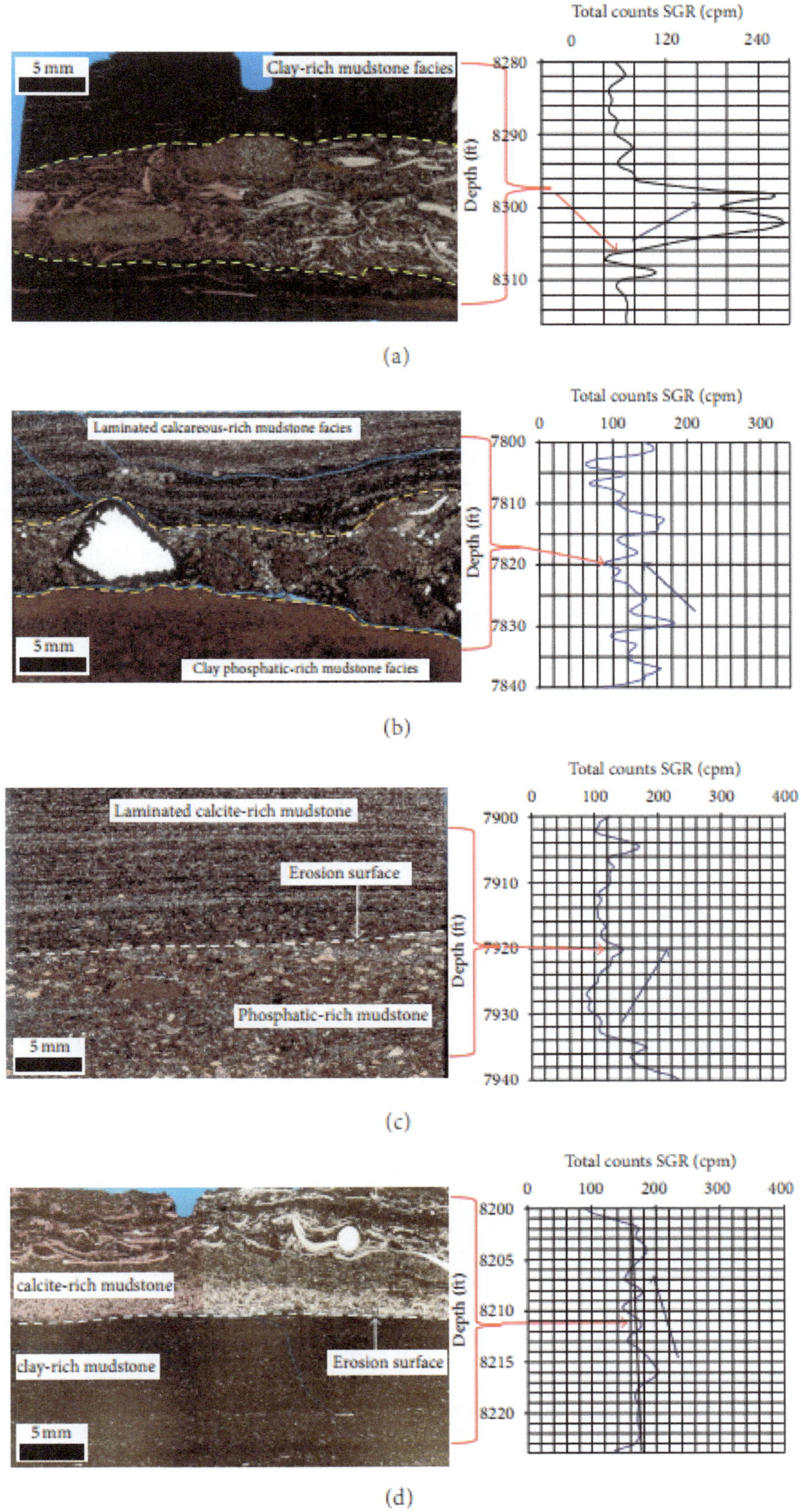

Figure 9: (a) Thin section photo shows sequence-scale erosion surface with irregular nature, pyritic reworked intraclasts, and reworked shell fragments with the opposite spectral gamma ray response that shows the abrupt upward increasing above the erosion surface. (b) Thin section photo shows irregular sequence-scale erosion surfaces; note the underlying clay phosphatic-rich facies and the overlying laminated calcareous facies. The opposite log is the spectral gamma ray value of this interval showing the coincide upward-decreasing pattern. (c) Thin section photo shows parasequence-scale erosion surface with underlying phosphatic-rich mudstone facies with the opposite Spectral Gamma Ray response that shows the upward decreasing above the erosion surface. (d) Thin section photo shows irregular parasequence-scale erosion surfaces; note the underlying clay phosphatic-rich facies and the overlying laminated calcareous facies. The opposite log is the spectral gamma ray value of this interval showing coinciding upward-decreasing pattern.

The lithology underlying these surfaces are commonly siliceous noncalcareous mudstone (Figure 9(b)) or siliceous calcareous mudstone, whereas the overlying lithofacies show an increase in calcite content and an increasing grain size of the detrital components (Figure 9(b)). This lithological variation bounding the sequence-scale erosional surfaces reflect a lateral facies changes from clay rich to calcite rich and the vertical change from relatively deeper to relatively shallower facies which is interpreted as a major change in depositional conditions.

The sequence-scale erosional surfaces are characterized by an abrupt change on gamma ray log (Figures 9(a)–9(d)). In such a case, when the erosion surface coms at the base of upward increasing spectral gamma ray (SGR) interval (Figure 9(a)), then it is interpreted as transgressive surface of erosion (TSE) [9]. Alternatively, the erosion surface associated with abrupt decreasing in SGR values (Figure 9(b)) is a sequence boundary [21].

Parasequence-Scale Erosion Surfaces

The erosion surfaces at this scale represent a major shift in depositional settings accompanied by corresponding changes in environmental energy and sediment supply during either sea-level rise or fall [57]. Examples of changes in depositional trends include the change from sedimentation to erosion and/or starvation and vice versa and also the change from a regression to a transgression and vice versa [58].

The parasequences-scale erosion surfaces are characterized by medium relief (ranging between 8 and 20 mm) with sharp to irregular contacts underlying with phosphatic-rich to clay-rich non calcareous mudstone facies (Figure 9(c)) of relatively deeper depositional environment and overlain with laminated calcareous-rich mudstone facies of relatively shallower depositional settings [9, 18, 21]. Additionally, this type of erosion surfaces represents the turnaround point on SGR curve as a result of the lithofacies change from clay, phosphatic-rich mudstone below to siliceous calcareous mudstone above these surfaces (Figure 9(c)) (Table 3). The parasequence-scale erosion surface may consistent with a marine flooding surface and/or maximum flooding surface which results from an abrupt increase in water depth [59, 60] when it is associated with shale rip-up clasts, traces of compacted and horizontally oriented shell fragments (Figure 9(d)) as well as diagenetic dolomite and pyrite.

Event-Scale Erosion Surfaces

These are the smallest scale types of erosional surfaces that mark changes in sedimentation regimes and may reflect minor or short-duration erosional events [7]. Such events are commonly distinguished by flat, low relief (<5 mm), sharp contacts, and shale on shale boundary with minimal evidence of erosion (Figure 10(a)) (Table3). The sharp nature of these erosional surfaces is attributed to abrupt lithological changes and the presence of calcareous concretions at surfaces as a result of sediment starvation or a low sedimentation rate reflecting low depositional energy [61]. The lithofacies change is markedly around this type of erosional surface with subsequent deposition of siliceous non calcareous mudstone (Figure 10(b)) to a siliceous calcareous mudstone (Figure 10(c)). Generally, the high frequency and diversity of erosion surfaces through the Barnett Shale give a unique view into the short-duration stratigraphic intervals that were previously difficult to detect in fine grained rock.

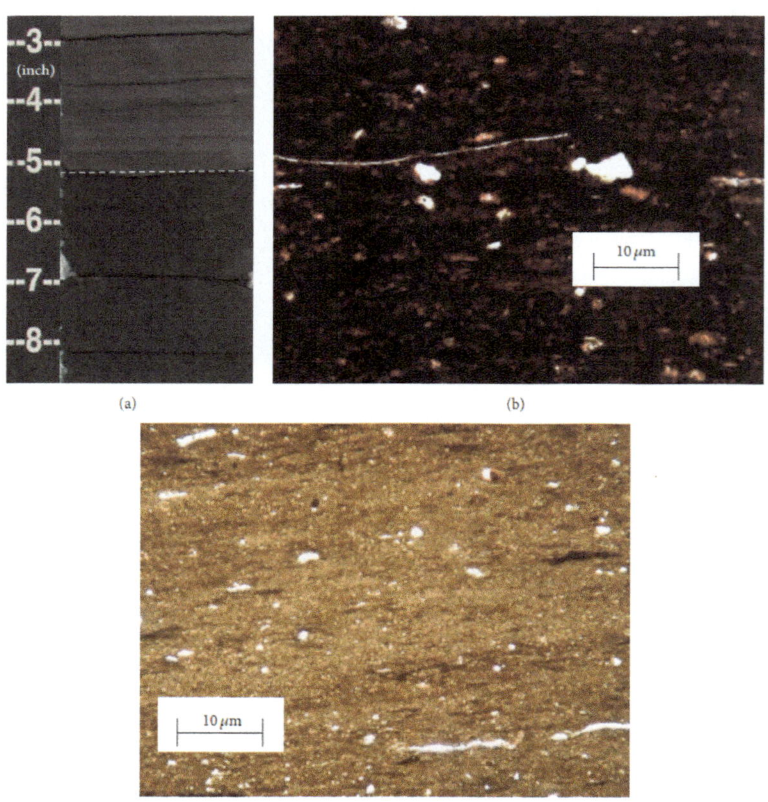

Figure 10: (a) Core photo shows an event-scale erosion surface with sharp nature (dashed line) between laminated black shale and an overlying dark laminated gray shale bed. (b) Thin section photo shows siliceous noncalcareous mudstone facies below the event-scale erosion surface in (a). (c) Thin section photo shows siliceous calcareous mudstone facies above the event-scale erosion surface in (a).

Summary and Conclusion

The erosion surfaces are direct indications of major environmental changes that may include nondeposition and/or erosional events. The erosion surfaces are relatively abundant throughout the Barnett Shale which punctuate the stacking pattern and erode previously deposited strata to an unknown extent.

The erosion surfaces in the organic-rich Barnett Shale exhibit variable relief (5–60 mm) with different geometry of upper and lower bounding contacts and a wide range of variable features including shelly laminae, shale rip-up clasts, reworked intraclasts, phosphatic pellets, and diagenetic minerals (dolomite and pyrite). Several factors control such variations, including the energy conditions at deposition, rate of relative sea-level fluctuation, rate of sedimentation, sediment influx, and the lithofacies type of the underlying and the overlying beds. The alternative occurrence of erosion surfaces throughout the Barnett Shale suggests episodic storm currents which transport eroded particles and intraclasts from proximal to distal areas.

The erosional features and associated surfaces are in part boundaries between different genetic types of deposits but on different scales reflecting the dependence on either base-level and/or sediment supply. Accordingly, the studied erosional surfaces of the Barnett Shale can be grouped into three different scales of sequence stratigraphic surfaces: sequence-scale surfaces, parasequence-scale surfaces, and event-scale surfaces.

ACKNOWLEDGMENTS

The author is thankful to the ConocoPhillips School of Geology and Geophysics, University of Oklahoma, where the labwork of this research has been conducted. The author is also thankful to the Faculty of Petroleum and Mining Engineering, Suez University, for providing the sabbatical leave. The constructive reviews and editorial comments of Roger M. Slatt, Gungoll Family Chair Professor of Petroleum Geology and Geophysics and Director of Reservoir Characterization Institute, University of Oklahoma, are gratefully acknowledged.

REFERENCES

1. J. Schieber, "Evidence for high-energy events and shallow-water deposition in the Chattanooga Shale, Devonian, central Tennessee, USA," Sedimentary Geology, vol. 93, no. 3-4, pp. 193–208, 1994.
2. M. O. Abouelresh and R. M. Slatt, "Shale depositional processes: example from the paleozoic Barnett Shale, Fort Worth Basin, Texas, USA," Central European Journal of Geoscience, vol. 3, no. 4, pp. 398–409, 2011.
3. W. J. Kennedy and H. C. Klinger, "Hiatus-concretions and hardground horizons in the Cretaceous of Zululand," Paleontology, pp. 539–549, 1972.
4. W. J. Kennedy and R. E. Garrison, "Morphology and genesis of nodular chalks and hardgrounds in the Upper Cretaceous of Southern England," Sedimentology, vol. 22, pp. 311–386, 1975.
5. W. J. Kennedy and R. E. Garrison, "Morphology and genesis of nodular phosphates in the Cenomanian Glauconitic Marl of southeast England," Lethaia, vol. 8, pp. 339–360, 1975.
6. G. C. Baird, "Pebbly phosphorites in shale, a key to recognition of a widespread submarine discontinuity in the Middle Devonian of New York," Journal of Sedimentary Petrology, vol. 2, pp. 545–555, 1978.
7. J. Schieber, "Sedimentary features indicating erosion, condensation, and hiatuses in the Chattanooga Shale of Central Tennessee: relevance for sedimentary and stratigraphic evolution," in Shales and Mudstones, J. Schieber, W. Zimmerle, and P. Sethi, Eds., vol. 1, pp. 187–215, 1998.
8. D. Nummedal and D. J. P. Swift, "Transgressive stratigraphy at sequence-bounding unconformities: some principles derived from Holocene and Cretaceous examples," in Sea-Level Fluctuation and Coastal Evolution, D. Nummedal, O. H. Pilkey, and J. D. Howard, Eds., vol. 42, pp. 358–370, SEPM Special Publication, 1987.
9. M. O. Abouelresh and R. M. Slatt, "Lithofacies and sequence stratigraphy of the Barnett Shale in east-central Fort Worth Basin, Texas," AAPG Bulletin, vol. 96, no. 1, pp. 1–22, 2012.
10. S. L. Montgomery, D. M. Jarvie, K. A. Bowker, and R. M. Pollastro, "Mississippian Barnett Shale, Fort Worth basin, north-central Texas: Gas-shale play with multi-trillion cubic foot potential," AAPG Bulletin, vol. 89, no. 2, pp. 155–175, 2005. ·
11. P. T. Flawn, A. Goldstein Jr., P. B. King, and C. E. Weaver, "The Ouachita system: University of Texas, Bureau of Economic Geology," Report 6120, 1961.
12. J. L. Walper, "Plate tectonic evolution of the Fort Worth Basin," in Petroleum Geology of the Fort Worth Basin and Bend Arch Area, C. A. Martin, Ed., pp. 237–251, Dallas Geological Society, Dallas, Tex, USA, 1982.
13. J. K. Arbenz, "Structure framework of the ouachita mountains," in Stratigraphic and Structural Evolution of the Ouachita Mountains and Arkoma Basin, Southeastern Oklahoma and West-Central Arkansas: Application To Petroleum Exploration, N. H. Suneson, Ed., pp. 4–40, Oklahoma Geological Survey, Circular, 2009.
14. J. D. Henry, "Stratigraphy of the Barnett Shale (Mississippian) and associated reefs in the northern Fort Worth Basin," in Petroleum Geology of the Fort Worth Basin and Bend Arch Area, C. A. Martin, Ed., pp. 157–178, Dallas Geological Society, Dallas, Tex, USA, 1982.
15. J. W. Flippin, "The stratigraphy, structure, and economic aspects of the Paleozoic strata in Erath County, north central Texas," in Petroleum Geology of the Fort Worth Basin and Bend Arch Area, C. A. Martin, Ed., pp. 129–155, Dallas Geological Society, Dallas, Tex, USA, 1982.
16. R. S. Kier, L. F. Brown, and E. F. McBride, The Mississippian and Pennsylvanian (Carboniferous) Systems in the United States: Texas, vol. 14, Bureau of Economic Geology, University of Texas at Austin, Geological Circular, Austin, Texas, 1980.
17. W. J. Mapel, R. B. Johnson, G. O. Bachman, and K. L. Varnes, "Southern midcontinent and southern Rocky Mountains region," in Paleotectonic Investigations of

the Mississippian System in the United States, L. C. Craig and C. W. Connor, Eds., vol. 1010, pp. 161–187, Geological Survey Professional Paper, 1979.

18. R. G. Loucks and S. C. Ruppel, "Mississippian Barnett Shale: Lithofacies and depositional setting of a deep-water shale-gas succession in the Fort Worth Basin, Texas," AAPG Bulletin, vol. 91, no. 4, pp. 579–601, 2007. ·

19. H. D. Rowe, R. G. Loucks, S. C. Ruppel, and S. M. Rimmer, "Mississippian Barnett Formation, Fort Worth Basin, Texas: Bulk geochemical inferences and Mo-TOC constraints on the severity of hydrographic restriction," Chemical Geology, vol. 257, no. 1-2, pp. 16–25, 2008. ·

20. B. U. Haq and S. R. Schutter, "A chronology of paleozoic sea-level changes," Science, vol. 322, no. 5898, pp. 64–68, 2008. ·

21. P. Singh, Lithofacies and sequence-stratigraphic framework of the Barnett Shale, northeast Texas [Ph.D. thesis], University of Oklahoma, Norman, Oklahoma, 2008.

22. J. Macquaker, "Micro-Textural Analyses of Fine-Grained Sediments and the Roles that Advective Sediment Transport and Suspension Settling Processes Play in the Deposition of Fine-Grained Organic Carbon-Rich Sediments. OR Just How Shaky are the Current Depositional Models that Seek to Explain the Origin of Source Rocks / shale Gas Reservoirs?" in Proceedings of the Critical assessment of shale resource plays (ex. abs.), American Association Of Petroleum Geologists/ Society of Economic Geologists/Society of Petroleum Geologists/ Society of Petrophysicists and Well-Log Analysis Hedberg Conference, Austin, Texas, USA, December 2010.

23. P. E. Potter, J. B. Maynard, and P. J. Depetris, Mud and Mudstones: Introduction and Overview, Springer, Berlin, Germany, 2005.

24. V. Caron, C. S. Nelson, and P. J. J. Kamp, "Transgressive surfaces of erosion as sequence boundary markers in cool-water shelf carbonates," Sedimentary Geology, vol. 164, no. 3-4, pp. 179–189, 2004. ·

25. A. P. Heward, "A review of wave-dominated clastic shoreline deposits," Earth Science Reviews, vol. 17, no. 3, pp. 223–276, 1981.

26. S. M. Kidwell, "The stratigraphy of shell concentrations," in Taphonomy, Releasing the Data Locked in the Fossil Record, P. A. Allison and D. E. G. Briggs, Eds., pp. 211–290, Plenum, New York, NY, USA, 1991.

27. F. T. Fursich, W. Oschmann, I. B. Singh, and A. K. Jaitly, "Hardgrounds, reworked concretion levels and condensed horizons in the Jurassic of western India: their significance for basin analysis," Journal of Geological Society, vol. 149, no. 3, pp. 313–331, 1992.

28. J. J. Hickey and B. Henk, "Lithofacies summary of the Mississippian Barnett Shale, Mitchell 2 T.P. Sims well, Wise County, Texas," AAPG Bulletin, vol. 91, no. 4, pp. 437–443, 2007. ·

29. P. Singh, R. M. Slatt, G. Borges et al., "Reservoir characterization of unconventional gas shale reservoirs: example from the Barnett Shale, Texas, U.S.A," Oklahoma City Geological Society Shale Shaker, vol. 60, no. 1, pp. 15–31, 2009.

30. R. T. Wilkin, H. L. Barnes, and S. L. Brantley, "The size distribution of framboidal pyrite in modern sediments: an indicator of redox conditions," Geochimica et Cosmochimica Acta, vol. 60, no. 20, pp. 3897–3912, 1996. ·

31. R. T. Wilkin, M. A. Arthur, and W. E. Dean, "History of water-column anoxia in the Black Sea indicated by pyrite framboid size distributions," Earth and Planetary Science Letters, vol. 148, no. 3-4, pp. 517–525, 1997.

32. D. Bond, P. B. Wignall, and G. Racki, "Extent and duration of marine anoxia during the Frasnian-Famennian (Late Devonian) mass extinction in Poland, Germany, Austria and France," Geological Magazine, vol. 141, no. 2, pp. 173–193, 2004. ·

33. K. L. Bann, C. R. Fielding, J. A. MacEachern, and S. C. Tye, "Differentiation of estuarine and offshore marine deposits using integrated ichnology and sedimentology: Permian Pebbly Beach Formation, Sydney Basin, Australia," in The Application of Ichnology To Palaeoenvironmental and Stratigraphic Analysis, D. McIlroy, Ed., vol. 228, pp. 179–211, Geological Society, London, UK, 2004.

34. J. P. Herbin, C. Muller, J. R. Geyssant, F. Melieres, I. E. Penn, and Y. Group, "Variation of the distribution of organic matter within a transgressive system tract: Kimmeridge Clay (Jurassic), England," in Source Rocks in a Sequence Stratigraphic Framework, B. J. Katz and L. M. Pratt, Eds., vol. 37, pp. 67–100, The American Association of Petroleum Geologists, 1993.

35. G. C. Baird, C. E. Brett, and W. T. Kirchgasser, "Genesis of black shale-roofed discontinuities in the Devonian Genesee Formation, Western New York State," in Devonian of the World, N. J. McMillan, A. F. Embry, and D. J. Glass, Eds., vol. 2, pp. 357–375, Canadian Society of Petroleum Geologists, Calgary, Canada, 1988.

36. E. Allersma, "Mud on the oceanic shelf of Guiana," in Proceedings of the Symposium on Investigation and Resources of the Caribbean Sea and Adjacent Regions, pp. 193–203, UNESCO, Paris, France.

37. J. M. Rine and R. N. Ginsburg, "Depositional facies of a mud shoreface in Suriname, South America: a mud analogue to sandy, shallow-marine deposits," Journal of Sedimentary Petrology, vol. 55, no. 5, pp. 633–652, 1985.

38. A. G. Plint, "Sharp-based shoreface sequences and "offshore bars" in the Cardium Formation of Alberta, their relationship to relative changes in sea level," in Sea Level Changes: An Integrated Approach, C. K. Wilgus, B. S. Hastings, C. G. St. C. Kendall, H. W. Posamentier, C. A. Ross, and J. C. Van Wagoner, Eds., vol. 42, pp. 357–370, SEPM Special Publication, 1988.

39. D. Nummedal, G. W. Riley, and P. L. Templet, "High-resolution sequence architecture: a chronostratigraphic model based on equilibrium profile studies," in Sequence Stratigraphy and Facies Association, H. Posamentier, C.

P. Summerhayes, B. U. Haq, and C. P. Allen, Eds., vol. 18, pp. 55–68, SEPM Special Publication, 1993.

40. J. A. Howe, M. S. Stoker, and K. J. Woolfe, "Deep-marine seabed erosion and gravel lags in the northwestern Rockall Trough, North Atlantic Ocean," Journal of the Geological Society, vol. 158, no. 3, pp. 427–438, 2001.

41. J. A. MacEachern, B. A. Zaitlin, and S. G. Pemberton, "High-resolution sequence stratigraphy of early transgressive deposits, Viking Formation, Joffre Field, Alberta, Canada," AAPG Bulletin, vol. 82, no. 5 A, pp. 729–756, 1998.

42. S. A. J. Pattison, "Sequence stratigraphic significance of sharp-based lowstand shoreface deposits, Kenilworth Member, Book Cliffs, Utah," AAPG Bulletin, vol. 79, no. 3, pp. 444–462, 1995.

43. S. M. Kidwell and P. J. Brenchley, "Patterns in bioclastic accumulation through the Phanerozoic: changes in input or in destruction?" Geology, vol. 22, no. 12, pp. 1139–1143, 1994.

44. R. G. Walker, "Sedimentary and tectonic origin of a transgressive surface of erosion: Viking Formation, Alberta, Canada," Journal of Sedimentary Research B, vol. 65, pp. 209–221, 1995. ·

45. I. G. Hwang and P. L. Heller, "Anatomy of a transgressive lag: Panther Tongue Sandstone, Star Point Formation, central Utah," Sedimentology, vol. 49, no. 5, pp. 977–999, 2002.

46. A. Hallam, "Eustatic cycles in the Jurassic," Palaeogeography, Palaeoclimatology, Palaeoecology, vol. 23, pp. 1–32, 1978.

47. M. Elrick and J. F. Read, "Cyclic ramp-to-basin carbonate deposits, Lower Mississippian, Wyoming and Montana: a combined field and computer modeling study," Journal of Sedimentary Petrology, vol. 61, no. 7, pp. 1194–1224, 1991.

48. P. B. Wignall and J. R. . Maynard, "The sequence stratigraphy of transgressive black shales," in Source Rocks in a Sequence stratigraphic framework, B. J. Katz and L. Pratt, Eds., vol. 37, pp. 35–47, The American Association of Petroleum Geologists, Studies in Geology, 1993.

49. P. B. Wignall, Black Shales, Clarendon Press Oxford, 1994.

50. D. J. P. Swift and J. A. Thorne, "Sedimentation on continental margins I: a general model for shelf sedimentation," in Shelf Sand and Sandstone Bodies, D. J. P. Swift, G. F. Oertel, R. W. Tillman, and J. A. Thorne, Eds., vol. 14, pp. 3–31, International Association of Sedimentologists, 1991.

51. D. J. P. Swift, S. Phillips, and J. A. Thorne, "Sedimentation on continental margins I: a general model for shelf sedimentation," in Shelf Sand and Sandstone Bodies, D. J. P. Swift, G. F. Oertel, R. W. Tillman, and J. A. Thorne, Eds., vol. 14, pp. 153–187, International Association of Sedimentologists, 1991.

52. R. G. Walker and A. G. Plint, "Wave- and storm-dominated shallow marine systems," in Facies Models: Response To Sea Level Change, R. G. Walker and N. P. James, Eds., pp. 219–238, Geological Association of Canada, 1992.

53. J. P. Bhattarcharya, "The expression and interpretation of marine flooding surfaces and erosional surfaces in core, examples from the upper Cretaceous Dunvegan Formation, Alberta Foreland basin, Canada," inSequence Stratigraphy and Facies Associations, H. W. Posamentier, C. P. Summerhayes, B. U. Haq, and G. P. Allen, Eds., vol. 18 of International Association of Sedimentologists, Special Publication, pp. 125–160, Blackwell Publishing, 1993.

54. A. F. Embry, "Sequence boundaries and sequence hierarchies: problems and proposals," in Sequence Stratigraphy: Advances and Applications for Exploration and Production in Northwest Europe. Stavanger, R. J. Steel, V. L. Felt, E. P. Johannesen, and C. Mathieu, Eds., pp. 1–11, Elsevier, Amsterdam, The Netherlands, 1995.

55. K. M. Husley, Lithofacies characterization and sequence stratigraphic framework for some gas-bearing shales within the Horn River Basin, Northeastern British Colombia [M.S. thesis], University of Oklahoma, Norman, Oklahoma, 2011.

56. O. Catuneanu, V. Abreu, J. P. Bhattacharya, et al., "Toward the standardization of the sequence stratigraphy," Earth-Science Reviews, vol. 92, no. 1-2, pp. 1–33, 2009.

57. O. Catuneanu, Principles of Sequence Stratigraphy, Elsevier, Amsterdam, The Netherlands, 2006.

58. A. Embry, E. Johannessen, D. Owen, B. Beauchamp, and P. Gianolla, "Sequence stratigraphy as a "Concrete" stratigraphic discipline," Report of the ISSC Task Group on Sequence Stratigraphy, 2007.

59. K. M. Bohacs and J. R. Schwalbach, "Sequence stratigraphy of fine-grained rocks with special reference to the Monterey Formation," in Sequence Stratigraphy in Fine-Grained Rocks: Examples From the Monterey Formation, J. R. Schwalbach and K. M. Bohacs, Eds., vol. 70, Pacific Section SEPM Guidebook, 1992.

60. M. R. Slatt, Stratigraphic Reservoir Characterization for Petroleum Geologists, Geophysicists, and Engineers, Elsevier, 2006.

61. R. Raiswell and D. E. Canfield, "Sources of iron for pyrite formation in marine sediments," The American Journal of Science, vol. 298, no. 3, pp. 219–245, 1998.

INDEX

A

air pressure 89, 90, 91, 92, 93, 94, 95, 96, 97, 98
amplitude variation with offset (AVO) 68
analogical material 141, 142, 144
Ando gabbro (AG) 81
artificial neural network (ANN) 28

B

barites-iron-sand cementation analogical (BISA) 140
basic structural units (BSU) 155, 158
bioturbation 68, 175
boxcounting method 129
bulk density (BD) 28
bulk modulus (BM) 28

C

capillary bundle model 89, 96
coal structure 155, 156, 159, 160, 161, 164
computerized numerical control (CNC) 143

D

data acquisition system (DAS) 144
decarbonation 2, 14
diagenesis 68
dynamic mechanical analysis (DMA) 120

E

Eastern Quxi granite (EQG) 81
electron paramagnetic resonance (EPR) 155, 157
Elementary Representative Volume 131
erosion zone 170

G

gamma ray (GR) 28
geomechanical model 140, 141
geomechanical modeling 28
Geomechanical well logs 27, 28
global economy 139
Grating ruler multipoints displacement measuring system (GRDS) 144

H

half cycle 111
Hercynian orogenesis 128, 130, 131
high-resolution transmission electron microscopy (HRTEM) 155, 157
hydraufracturing stress measurement 80
hydraulic aperture 100, 103, 104
hydraulic system 143
Hydrocarbon 99

I

industrial waste 119
instantaneous input energy (IIE) 111

J

Joint Roughness Coefficients (JRC) 133

K

karstification 53, 54

L

latent heat 39
Lhasa granite (LG) 81
linear displacement indicators 100

M

Main Transcurent Shear Zone (MTZ) 128
mean absolute percentage error (MAPE) 29
microdiscontinuities 130, 133
micropores 155, 159, 162, 164, 166
multipoint displacement measuring System 144

N

National Laboratory for Civil Engineering 141
nondeposition 167, 179
nondrainage process 93
Northeastern Yangbajain granite (NYG) 81
Northwestern Quxi granite (NQG) 81
nuclear magnetic resonance (NMR) 155, 157

P

Pacaya Volcano 1, 2, 15, 16, 17
peak particle velocity (PPV) 107
phenomenon 139, 140, 149, 150, 151, 152, 153, 154
physical hardening 119
point load test (PLT) 23
Poisson's ratio (PR) 28

R

Restoring force 109
Rock Quality Designation index (RQD) 22
Rolling Thin Film Oven Test 120

S

Sanisotropy 68
sedimentological 169
Senegal-Mali fault (SMF) 128
shear modulus (SM) 28
Shelly laminae 170, 171
signal translating system (STS) 144
single degree of freedom (SDOF) 107, 108
source rock reservoirs (SRR) 27
South Ando granite (SAG) 81, 83
Strategic Highway Research Program 124

T

tectonically deformed coal (TDC) 155
Tectonically deformed coal (TDC 155
Temperature 158, 164
thermal energy 37, 38, 40, 42, 44, 47, 48, 50
Total Core Recovery (TCR) 22
transgressive surface of erosion (TSE) 178

U

ultramicropores 159, 162, 164
unconfined compressive strength (UCS) 62
unconstrained compressive strength (UCS) 142
uniaxial compressive strength tests (UCS) 2
uniaxial compressive strength (UCS) 19, 22, 25
Uniaxial compressive strength (UCS) 5

V

vertical transverse isotropy (VTI) 70
vitrinite reflectance anisotropy (VRA) 158

W

Wangxia Cliff 53, 54, 60, 61
Wudaoliang limestone (WL) 81

X

Xidatan granite (XTG) 81
X-ray diffraction (XRD) 155, 157

Y

Young's modulus (YM) 28